Identification and Functional Characterization of Novel Venom Components

Identification and Functional Characterization of Novel Venom Components

Special Issue Editor

Steven D. Aird

MDPI • Basel • Beijing • Wuhan • Barcelona • Belgrade • Manchester • Tokyo • Cluj • Tianjin

Special Issue Editor
Steven D. Aird
Technical Editor
Japan

Editorial Office
MDPI
St. Alban-Anlage 66
4052 Basel, Switzerland

This is a reprint of articles from the Special Issue published online in the open access journal *Toxins* (ISSN 2072-6651) (available at: https://www.mdpi.com/journal/toxins/special_issues/novel_venom_components).

For citation purposes, cite each article independently as indicated on the article page online and as indicated below:

LastName, A.A.; LastName, B.B.; LastName, C.C. Article Title. *Journal Name* **Year**, *Article Number*, Page Range.

ISBN 978-3-03936-467-1 (Pbk)
ISBN 978-3-03936-468-8 (PDF)

Cover image courtesy of Steven D. Aird.

Contents

About the Special Issue Editor

Steven D. Aird received his B.Sc. in Zoology from Montana State University (1974) and an M.Sc. in Biology from Northern Arizona University (1977). He completed his Ph.D. in zoology at Colorado State University (1984) and then conducted post-doctoral research on snake venom chemistry with the Department of Biochemistry and Molecular Biology at the University of Wyoming from late 1983 until early 1988. After his post-doc, Dr. Aird spent four years researching spider venom chemistry for a start-up pharmaceutical firm. In 1997, he became a Visiting Researcher at the Centro de Ciências Biológicas at Universidade Federal de Pernambuco before transferring to the Programa de Pós-Graduação em Ciências Fisiológicas at Universidade Estadual do Ceará in 1999. After returning to the U.S. in 2001, he served as an Associate Professor of Chemistry and Biology at Norfolk State University from 2002 to 2008 before taking a position with the University of Maryland University College in Okinawa, Japan. In 2012, he was hired as the Technical Editor at the Okinawa Institute of Science and Technology Graduate University, where he continued venom research in addition to his editing responsibilities until 2019. Dr. Aird now works as a freelance technical editor (www.sda-technical-editor.org). Dr. Aird has authored more than 60 scholarly articles and book chapters, including five review articles. ORCID ID: 0000-0003-0645-6677.

Editorial

Introduction to the Toxins Special Issue on Identification and Functional Characterization of Novel Venom Components

Steven D. Aird

Technical Editor, Nakama, Onna-son, Kunigami-gun, Okinawa-ken 904-0401, Japan; sda.tech.ed@icloud.com

Received: 4 May 2020; Accepted: 13 May 2020; Published: 20 May 2020

Throughout most of the 20th century, the toxinological literature consisted largely of pharmacological and functional characterizations of crude venoms and venom constituents, often constituents that could not be identified unambiguously. The advent of amino acid composition analysis in the 1950s enabled the first forays into physical characterizations of purified toxins, although these remained few in number until the 1970s. Then, tryptic and chymotryptic cleavage of venom proteins coupled with manual Edman degradation began to provide the first complete sequences, particularly of three-finger toxins. Polyacrylamide gel electrophoresis and superior resins for liquid chromatography permitted improved purification and better gross structural characterization of venom components. The early 1980s saw the advent of automated Edman degradation, and entire sequences of longer proteins began to appear in the literature. Then, the molecular biology revolution made it possible to generate cDNA sequences of more and larger proteins, followed by mass spectrometry-based proteomics and quantitative high-throughput DNA sequencing and genomics. Today, we face a hitherto unprecedented situation in which our capacity to generate sequence/structural data has completely overwhelmed our capacity to characterize venom constituents functionally. This Special Issue of *Toxins* comprises 11 publications addressing the discovery and functional characterization of novel venom constituents of vertebrate and invertebrate venoms.

Tadokoro et al. [1] review the functional diversity of snake toxins belonging to the Cysteine-Rich Secretory Protein (CRISP) superfamily and find that despite their broad taxonomic distribution, very few of these proteins have been functionally characterized; however, those that have, exhibit diverse activities, inhibiting ion channels and angiogenesis, increasing vascular permeability, and promoting inflammatory responses (leukocyte and neutrophil infiltration).

Pérez-Peinado et al. [2] review snake toxins that have been exploited or are being explored for diagnosis and treatment of various cardiovascular disorders and blood abnormalities. They focus on development of antimicrobial and anticancer drugs, reviewing their principal activities in vitro and in vivo, their structures, mechanisms of action, and their potential as drug leads.

Santibanez-Lopez et al. [3] use new and previously published transcriptomic resources to assess evolutionary relationships of closely related scorpions from the family Hadruridae and their toxins. In particular, they survey potassium channel toxins known as scorpine-like peptides, showing that these peptides exhibit gene duplications. The authors find that more toxin sites are evolving under negative selection than under positive selection.

Delgado-Prudencio et al. [4] examine the means by which many scorpion venom peptides are amidated at their C-termini, a post-translational modification that is essential for correct biological functioning of these peptides. They report the existence of a dual enzymatic α-amidation system, and identify the enzymes in scorpion venom glands responsible.

Ullah et al. [5] examine the structure of snake venom phosphodiesterases, little studied enzymes responsible for unleashing purine nucleosides in the prey to promote hypotension, inhibition of platelet aggregation, edema, and paralysis. They examine the enzyme from *Crotalus adamanteus* venom and report that this PDE comprises a somatomedin B domain, a somatomedin B-like domain, an ectonucleotide pyrophosphatase domain, and a DNA/RNA non-specific domain. They find that snake venom PDEs exhibit high sequence identity, but comparatively low identity with mammalian and bacterial PDEs. Nonetheless, they suggest that intraspecific sequence differences may account for different substrate specificities.

Inamaru et al. [6] follow up on earlier work to report a novel, sixth small serum protein from *Protobothrops flavoviridis*. Using comprehensive genomic analysis, the authors conclude that some SSPs were present in all snake genomes before the divergence of venomous snakes, but that others arose specifically in venomous snakes. The authors report that the evolutionary emergence of SSP genes is probably related to physiological functions of SSPs and that venom SSP composition may reflect snake habitat, prey, and venom differences.

Kuo et al. [7] examine mechanisms and structure–activity relationships of disintegrins TFV-1 and TFV-3, from *Protobothrops flavoviridis* venom, which have very different pro-hemorrhagic tendencies. The find that TFV-1 selectively inhibits Gα13-mediated platelet aggregation without affecting physiological hemostatic processes. It causes neither severe thrombocytopenia nor bleeding in mice and does not induce hypocoagulation in human whole blood. In contrast, TFV-3 and eptifibatide exhibit all of these hemostatic effects.

Heep et al. [8] explore the peptidome of a predatory ant, *Manica rubida*, and observe severe fitness costs in the pea aphid (*Acyrthosiphon pisum*), a common agricultural pest. They identify a novel decapeptide, which, while inactive against bacteria and fungi, reduces aphid survival and reproduction. Both crude venom and the peptide reversibly paralyze injected aphids, inducing loss of body fluids. They suggest that this venom may be a promising source of additional bio-insecticide leads.

Huancahuire-Vega et al. [9] report the purification and biochemical and functional characterization of ACP-TX-I and ACP-TX-II, two phospholipases A_2 (PLA_2) from *Agkistrodon contortrix pictigaster* venom. ACP-TX-I is a monomeric, catalytically inactive K49 PLA_2, exhibiting markedly diminished myotoxic and inflammatory activity, compared to dimeric K49 toxins. ACP-TX-II is an enzymatically active D49 PLA2, exhibiting in vivo local myotoxicity, edema-forming activity, and in vitro cytotoxicity. ACP-TX-I PLA2 is likewise cytotoxic, indicating that cytotoxicity does not require enzymatic activity.

Tourki et al. [10] describe Lebetin 2 (L2), a natriuretic peptide from *Macrovipera lebetinus* venom and report that following myocardial infarction, L2 reduces leukocyte and proinflammatory M1 macrophage infiltration into the infarcted area. It also increased anti-inflammatory M2-like macrophages, inducing a higher density of endothelial cells and cardiomyocytes than BNP. The authors find that L2 has strong, acute, prolonged cardioprotective effects that reduce ischemic reperfusion-induced necrotic and fibrotic effects.

Wu et al. [11] describe the isolation of a novel bradykinin-related peptide from the skin secretion of a ranid frog, *Ordorrana hejiangensis*. They report that when Arginine-4 is substituted with Leucine, this bradykinin agonist is transformed into a bradykinin antagonist.

Conflicts of Interest: The author declares no conflict of interest.

References

1. Tadokoro, T.; Modahl, C.M.; Maenaka, K.; Aoki-Shioi, N. Cysteine-Rich Secretory Proteins (CRISPs) from Venomous Snakes: An Overview of the Functional Diversity in a Large and Underappreciated Superfamily. *Toxins* **2020**, *12*, 175. [CrossRef] [PubMed]
2. Pérez-Peinado, C.; Defaus, S.; Andreu, D. Hitchhiking with Nature: Snake Venom Peptides to Fight Cancer and Superbugs. *Toxins* **2020**, *12*, 255. [CrossRef] [PubMed]
3. Santibáñez-López, C.E.; Graham, M.R.; Sharma, P.P.; Ortiz, E.; Possani, L.D. Hadrurid Scorpion Toxins: Evolutionary Conservation and Selective Pressures. *Toxins* **2019**, *11*, 637.
4. Delgado-Prudencio, G.; Possani, L.D.; Becerril, B.; Ortiz, E. The Dual α-Amidation System in Scorpion Venom Glands. *Toxins* **2019**, *11*, 425. [CrossRef] [PubMed]

5. Ullah, A.; Ullah, K.; Ali, H.; Betzel, C.; ur Rehman, S. The Sequence and a Three-Dimensional Structural Analysis Reveal Substrate Specificity among Snake Venom Phosphodiesterases. *Toxins* **2019**, *11*, 625. [CrossRef] [PubMed]

6. Inamaru, K.; Takeuchi, A.; Maeda, M.; Shibata, H.; Fukumaki, Y.; Oda-Ueda, N.; Hattori, S.; Ohno, M.; Chijiwa, T. Discovery of the Gene Encoding a Novel Small Serum Protein (SSP) of *Protobothrops flavoviridis* and the Evolution of SSPs. *Toxins* **2020**, *12*, 177. [CrossRef] [PubMed]

7. Kuo, Y.-J.; Chung, C.-H.; Pan, T.-Y.; Chuang, W.-J.; Huang, T.-F. A Novel $\alpha_{IIb}\beta_3$ Antagonist from Snake Venom Prevents Thrombosis without Causing Bleeding. *Toxins* **2020**, *12*, 11. [CrossRef] [PubMed]

8. Heep, J.; Skaljac, M.; Grotmann, J.; Kessel, T.; Seip, M.; Schmidtberg, H.; Vilcinskas, A. Identification and Functional Characterization of a Novel Insecticidal Decapeptide from the Myrmicine Ant *Manica rubida*. *Toxins* **2019**, *11*, 562. [CrossRef] [PubMed]

9. Huancahuire-Vega, S.; Hollanda, L.M.; Gomes-Heleno, M.; Newball-Noriega, E.E.; Marangoni, S. ACP-TX-I and ACP-TX-II, Two Novel Phospholipases A_2 Isolated from Trans-Pecos Copperhead *Agkistrodon contortrix pictigaster* Venom: Biochemical and Functional Characterization. *Toxins* **2019**, *11*, 661. [CrossRef] [PubMed]

10. Tourki, B.; Dumesnil, A.; Belaidi, E.; Ghrir, S.; Godin-Ribuot, D.; Marrakchi, N.; Richard, V.; Mulder, P.; Messadi, E. Lebetin 2, a Snake Venom-Derived B-Type Natriuretic Peptide, Provides Immediate and Prolonged Protection against Myocardial Ischemia-Reperfusion Injury via Modulation of Post-Ischemic Inflammatory Response. *Toxins* **2019**, *11*, 524. [CrossRef] [PubMed]

11. Wu, Y.; Shi, D.; Chen, X.; Wang, L.; Ying, Y.; Ma, C.; Xi, X.; Zhou, M.; Chen, T.; Shaw, C. A Novel Bradykinin-Related Peptide, RVA-Thr6-BK, from the Skin Secretion of the Hejiang Frog; Ordorrana hejiangensis: Effects of Mammalian Isolated Smooth Muscle. *Toxins* **2019**, *11*, 376. [CrossRef] [PubMed]

Review

Cysteine-Rich Secretory Proteins (CRISPs) from Venomous Snakes: An Overview of the Functional Diversity in a Large and Underappreciated Superfamily

Takashi Tadokoro [1], Cassandra M. Modahl [2], Katsumi Maenaka [1] and Narumi Aoki-Shioi [2,3,*]

[1] Faculty of Pharmaceutical Sciences, Hokkaido University, Faculty of Pharmaceutical Sciences, Hokkaido University, Kita-12, Nishi-6, Kita-ku, Sapporo 060-0812, Japan; tadokorot@pharm.hokudai.ac.jp (T.T.); maenaka@pharm.hokudai.ac.jp (K.M.)

[2] Department of Biological Sciences, National University of Singapore, Singapore 117543, Singapore; dbscmm@nus.edu.sg

[3] Department of Chemistry, Faculty of Science, Fukuoka University, 19-1, 8-chomeNanakuma, Jonan-ku, Fukuoka 814-0180, Japan

* Correspondence: anarumi@fukuoka-u.ac.jp

Received: 17 February 2020; Accepted: 10 March 2020; Published: 12 March 2020

Abstract: The CAP protein superfamily (Cysteine-rich secretory proteins (CRISPs), Antigen 5 (Ag5), and Pathogenesis-related 1 (PR-1) proteins) is widely distributed, but for toxinologists, snake venom CRISPs are the most familiar members. Although CRISPs are found in the majority of venoms, very few of these proteins have been functionally characterized, but those that have been exhibit diverse activities. Snake venom CRISPs (svCRISPs) inhibit ion channels and the growth of new blood vessels (angiogenesis). They also increase vascular permeability and promote inflammatory responses (leukocyte and neutrophil infiltration). Interestingly, CRISPs in lamprey buccal gland secretions also manifest some of these activities, suggesting an evolutionarily conserved function. As we strive to better understand the functions that CRISPs serve in venoms, it is worth considering the broad range of CRISP physiological activities throughout the animal kingdom. In this review, we summarize those activities, known crystal structures and sequence alignments, and we discuss predicted functional sites. CRISPs may not be lethal or major components of venoms, but given their almost ubiquitous occurrence in venoms and the accelerated evolution of svCRISP genes, these venom proteins are likely to have functions worth investigating.

Keywords: CAP superfamily; ion channel blockage; salivary component; co-factors

Key Contribution: Characterized cysteine-rich secretory proteins (CRISPs), including those that have been isolated from snake venoms and other sources, bind various target molecules, but all of them alter cellular signaling. This is a key feature of CRISPs and other proteins of the large CAP superfamily. Crystal structures and amino acid sequence alignments can be used to infer CRISP functional residues involved in binding.

---

1. Introduction

The CAP protein superfamily (Cysteine-rich secretory proteins (CRISPs), Antigen 5 (Ag5), and Pathogenesis-related 1 (PR-1) proteins), occasionally called the sperm coating protein (SCP) or Tpx-1/Ag5/PR-1/Sc7 (TAPS) family, occurs in a wide range of organisms. This superfamily is defined by a common structural feature, the CAP/PR-1 domain, with a unique α-β-α fold. The CAP/PR-1 domain comprises approximately 150–160 amino acids and includes four signature sequences, as defined in the PROSITE Database (http://www.expasy.ch/prosite/):

CAP1, [GDER][HR][FYWH][TVS][QA][LIVM][LIVMA]Wxx[STN];

CAP2, [LIVMFYH][LIVMFY]xC[NQRHS]Yx[PARH]x[GL]N[LIVMFYWDN];

CAP3 (HNxxR); and

CAP4 (G[EQ]N[ILV]).

Most proteins in this superfamily have only one CAP/PR-1 domain; however, a few species of parasitic helminths have proteins with more than one [1]. The CAP superfamily is extensive. The Pfam database (v32.0) (http://pfam.xfam.org/) contains 20,748 sequences (Pfam ID: PF00188) from 5356 species, ranging from bacteria to eukaryotes, and 39 structures, including a number of identical molecules with different IDs (Protein Data Bank, http://www.rcsb.org/; accessed on 2 January 2020). Sequence information for members of this superfamily continues to grow with the advancement of high-throughput technologies, such as next-generation cDNA sequencing.

Secreted PR-1 proteins were the first known members of the CAP superfamily, described in 1970 from *Nicotiana tabacum* plants infected with tobacco mosaic virus [2]. The abundance of PR-1 proteins increases in tobacco leaves infected with various pathogens [3]. These early results indicated that PR-1 proteins are involved in plant systemic responses to disease. Overexpression of the *PR-1* gene results in increased plant resistance to fungi [4], oomycetes [3,5], and bacteria [6], but not to viruses [7]. Subsequently, PR-1 proteins were found ubiquitously distributed among plants. *PR-1* genes are also associated with abiotic stress responses [8–12], though their expression may also be independent of stress responses [13]. The broad-ranging functions of PR-1 proteins require further investigation, especially after the discovery of PR-1 receptor-like kinases, which may be involved in initiation of signaling cascades [14]. The current hypothesis is that PR-1 proteins possess antimicrobial activity, amplifying defense signals via sterols or effector binding.

Ag5 proteins are abundant in insect venoms and saliva, including venoms of vespids and fire ants [15], and in the saliva of blood-feeding ticks [16], flies [17], and mosquitoes [18]. As one of the major allergens in insect venoms, immunoglobulins from human victims cross-react with Ag5s in venoms of yellow jackets, hornets, and paper wasps [15,19,20]. The function of Ag5 in saliva proteomes of hematophagous arthropods may be to regulate the host immune system and to inhibit coagulation during feeding [21,22]. For example, Ag5s from blood-feeding insects, *Dipetalogaster maxima* and *Triatoma infestans*, strongly inhibit collagen-induced platelet aggregation by interaction with Cu^{2+}, providing redox potential for catalytic removal of O_2, and decreasing inflammation [23].

CRISPs are highly expressed in rodent male reproductive tracts [24,25], with lower levels of expression in neutrophils, plasma, salivary gland, pancreas, ovary, thymus, and colon [26,27]. There has been a lack of consistent nomenclature regarding CRISPs. For example, CRISP-3 localized in seminal plasma is also known as specific granule protein 28 (SGP28), horse seminal plasma protein-3, and Aeg2 (NCBI Gene ID:10321). Three predominantly mammalian CRISPs (CRISP-1 to -3) have been referenced by different names in various studies, and a list of all published nomenclature has been assembled in a review by Adam et al. [28]. Mammalian CRISPs are associated with reproduction, cancer, and immune responses [28,29]. In addition to these activities, CRISPs have been identified as toxins in venom glands of snakes, lizards, spiders, scorpions, and cone snails [30–34]. Interestingly, a CRISP similar to those found in snake venom was also described as a main salivary component of the parasitic Japanese river lamprey (*Lethenteron japonicum*) [35].

CRISPs first appeared in reptile venoms approximately 170 million years ago in the clade Toxicofera [36,37]. Many CRISP orthologs have been found in lizard and snake venoms [38]. A review of venom proteomes confirmed the presence of CRISPs in viperid, elapid, and colubrid venoms, and their absence in atractaspidid venoms and those of some elapids, such as coral snakes [39]. The abundance of snake venom CRISPs (svCRISPs) in crude venom varies from 0.05% to 10%. The svCRISPs ablomin (*Gloydius blomhoffii*), triflin (*Protobothrops flavoviridis*), latisemin (*Laticauda semifasciata*), and tigrin (*Rhabdophis tigrinus*) were some of the first characterized, and were also classified as helothermine-like venom proteins (helveprins). Helothermine is a CRISP isolated from the venom of the Mexican beaded lizard (*Heloderma horridum*) [40]. The most common svCRISP activity has been non-enzymatic inhibition of various membrane channels, but many other activities have also been observed [31,41]. Sensitive "-omic" analyses, predominately transcriptomics and proteomics, have identified a large number of svCRISPs. However, in most cases these proteins have not been isolated or characterized experimentally, and their targets and biological roles remain unknown.

Target binding to alter cellular signaling cascades is a common function of CRISPs and other proteins of the CAP superfamily. In this review, we detail interactions between CAP superfamily proteins and their targets. It is important to view svCRISPs in the larger context of the entire CAP superfamily in order to identify their potential functions in venoms. We review svCRISPs that have been characterized during the past 10 years, examining their molecular surfaces and identifying regions and residues that contribute to their diverse biological activities.

2. Structural Features of Cysteine-Rich Secretory Proteins (CRISPs)

The Protein Data Bank (PDB) contains various CAP superfamily tertiary structures. High-resolution crystal structures of svCRISPs reveal a common secondary structure that includes 16 conserved cysteine residues (Figure 1). CRISPs have two main domains, a CAP/PR-1 domain at the N-terminus and a cysteine-rich (CRD)/ion channel regulatory (ICR) domain at the C-terminus, connected by a hinge region. For structure descriptions in this section, we have used residue numbering from triflin and natrin, well-characterized svCRISPs with published structures (Figure 1).

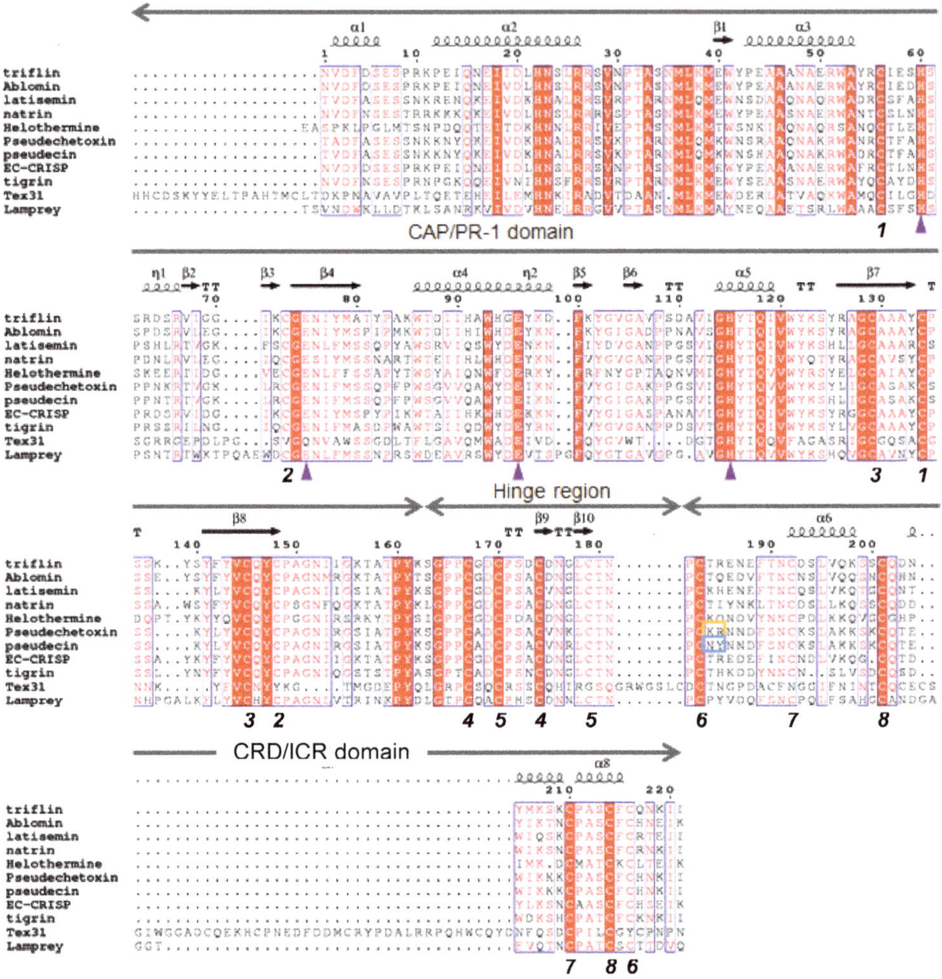

Figure 1. Amino-acid sequence alignments of Cysteine-rich secretory proteins (CRISPs). Highly conserved residues are highlighted in red, and other conserved residues are shown in a red font. Disulfide bridges are indicated below the alignment with black numbers. Identical numbers identify bonded residues. The secondary structure of triflin (PDB ID: 1WVR) is shown above the alignment. Purple triangles indicate conserved residues involved in binding of divalent cations (His60, Glu75, Glu96, and His115 for triflin). Basic residues in pseudetoxin are indicated with a yellow box, whereas the corresponding residues are neutral in pseudecin and are indicated with a blue box.

2.1. CAP/PR-1 Domain

Most CRISP structures contain an N-terminal α-β-α sandwich composed of five α-helixes and eight β-sheets, with five conserved cysteine residues among the 162 residues. The crystal structures of CRISP CAP/PR-1 domains show high similarity with CAP/PR-1 domains in P14a, a plant pathogenesis group-1 protein [42], and Ves v5, from yellow hornet venom [19]. A metal-ion-binding site in the CAP/PR-1 domain of CRISPs is also well conserved (His60, Glu75, Glu96, and His115; triflin numbering). Human CRISPs contain a glycosylation site (Asn-X(except Pro)-Ser/Thr) and a glycosylated form exists [43], but very few svCRISPs share this feature [41].

2.2. Hinge Region

About 20 amino acids (positions 163–182) form the hinge region between the CAP/PR-1 and CRD/ICR domains. The hinge region includes two disulfide bonds.

2.3. Cysteine-Rich Domain (CRD)/Ion Channel Regulatory (ICR) Domain

Three disulfide bonds and a few short α-helixes are well conserved in the CRD/ICR domain (positions 183–221). The CRD/ICR domain of svCRISPs may be important for recognizing ion channels. This has been suggested because the ion-binding motif in kaliotoxin (KTX) and margatoxin (an α-KTX) from buthid scorpion venoms [44], as well as in ShTx and BgK from sea anemone venoms [45], have the same tertiary structure. These peptide toxins, comprising approximately 40 amino acid residues, display high affinity for voltage-gated potassium channels and calcium-activated potassium channels, and possible interaction sites with these target channels have been proposed.

3. CRISP Co-Factors

CAP family members perform various physiological functions by binding to small compounds and proteins in the characteristic concavity of the CAP/PR-1 domain. Plant PR-1 proteins and the yeast CAP proteins, Pry1 and Pry2, bind sterols and lipids to inhibit pathogen proliferation. Sterols are essential for eukaryotes and bacteria, and removing them from membrane surfaces of pathogens inhibits their growth and can even kill them [46]. Lipid-related functions of the CAP superfamily have been summarized by Schneiter et al. in two reviews [47,48].

CRISPs bind divalent cations (Zn^{2+}, Ca^{2+}, and Cd^{2+}), heparin, small peptides (substrates for Tex31, a cone snail CRISP [32]), and proteins (receptors). Five crystal structures of svCRISPs (natrin, triflin, pseudetoxin, pseudecin, and stecrisp) have revealed the presence of divalent ions in the CAP/PR-1 domain (Table 1). In crystal structures of the elapid CRISPs, pseudechetoxin and pseudecin, the CRD/ICR domains and the N-terminal domains form a groove that narrows upon Zn^{2+} binding, consistent with the finding that Zn^{2+} likely influences target molecule recognition [49]. Two Zn^{2+}-binding sites in natrin are responsible for slight conformational differences with and without Zn^{2+}, detected in 3D structure comparisons [50].

Binding of divalent cations alters CRISP activity. Zn^{2+} enhances the binding of natrin to heparin, resulting in increased expression of adhesion molecules on endothelial cells (ECs). It has been proposed that the heparin-binding site is located opposite the Zn^{2+}-binding site in natrin [50]. Ca^{2+} (1 mM) increases cleavage activity of Tex31 (from *Conus textile*) 5-fold, but this increase was not observed with Zn^{2+} or Mg^{2+} [32]. It is interesting that members of the salivary antigen-5/CAP family from hematophagous insects are Cu^{2+}-dependent antioxidant enzymes in competition assays, although these proteins also bind other divalent metals (Mn^{2+}, Ni^{2+}, Co^{2+}, and Zn^{2+}) depending on their presence in the buffers used [23]. Thus, there is clear evidence that divalent cations affect bioactivity not only of svCRISPs, but also other members of the CAP superfamily. However, whether conformational changes induced by cation-binding are correlated with their activity requires corroboration.

Table 1. Targets and biological effects of cysteine-rich secretory proteins from buccal glands or venom.

Animals	Name	Species	Target (Interaction Molecules)	Biological Effect (or Related Investigation)	Accession Numbers	Ref.
Lamprey	Lamprey CRISP Buccal gland secretion protein-2 (BGSP-2) Cysteine-rich buccal gland protein (CRBGP)	*Lethenteron japonicum* (*Lampetra japonica*)	• Voltage-dependent Na$^+$ channels • Integrin β2 (CD11s/CD18)	Ca^{2+} channel blocker-like properties Anti-angiogenic activities Permeability Inhibition of adhesion, proliferation, migration, and invasion of cells (HUVEC; IC50 = 4.0 μM and Hela cell; IC50 = 6.7 μM) Non-fibrinogenolytic activity Activity of immunosuppressant (neutrophil inhibitory factor) Inhibition of Na$^+$ channels in hippocampal neurons (12 μM) Inhibition K$^+$ channels in hippocampal neurons (120 μM)	A4PIZ5	[35–51,51–56]
Cone snail	Tex31	*Conus textile*	N.D.	Proteinase	Q7YT83	[32]
Lizard	Helothermine	*Heloderma horridum* salivary secretion	• Ryanodine receptors • Ca^{2+} channels • K$^+$ channels	Lethargy, partial paralysis of rear limbs and lowering of body temperature Blockage of receptors (Cerebellar Granule Cells) Inhibition of K$^+$ channels (IC50 = 0.52 μM) Inhibition of Ca^{2+} channels (IC50 = 0.25 μM) Inhibition of skeletal ryanodine receptors (about 1.0 μM)	Q91055	[40,57–59]
Snake						
(Vipers)	Ablomin	*Gloydius blomhoffi.*	N.D.	Ca^{2+} channel blocker-like properties	Q8JI40	[60]
	Piscivorinc	*Agkistrodon piscivorus*	N.D.	Ca^{2+} channel blocker-like properties	AY181982	[61]
	Catrin	*Crotalus atrox*	N.D.	Ca^{2+} channel blocker-like properties	AY181983	[61]
	Triflin	*Protobothrops flavoviridis*	N.D.	Ca^{2+} channel blocker-like properties	Q8JI39 1WVR, 6IMF (with inhibitor)	[60]
	Stecrisp	*Trimeresurus stejnegeri* (*Viridovipera stejnegeri*)	N.D.	No proteolysis activity (unlikeTex31)	P60623 1RC9	[62]
	Bj-CRP	*Bothrops jararaca*	Component C3 and C4	Bind and cleaved to component C3 and C4 Lack of effect of K$^+$ channel blockage activity (1.0 μM) (Kv1.1 to Kv1.6, Shaker-IR, Kv3.1, Kv7.1, Kv7.2, Kv7.4 and Kv10.1)	N.D. (partial sequence)	[63]
	Hellerin	*Crotalus oreganus helleri*	N.D.	To increase trans-epithelial permeability Cytotoxicity against HUVEC (cytotoxic concentration CC50 = 2.3 μM)	G9DCH4	[64]
	EC-CRISP	*Echis Carinatus Sochureki*	N.D.	Binding to HUVEC cell Anti-angiogenic Activities (10-20 μg/ml, <1 μM)	P0DMT4	[65]
	Crovirin	*Crotalus viridis viridis*	N.D.	Anti-protozoan activity against *Trypanosoma cruzi* and *Leishmania amazonensis*	N.D (partial sequence)	[66]

Table 1. *Cont.*

Animals	Name	Species	Target (Interaction Molecules)	Biological Effect (or Related Investigation)	Accession Numbers	Ref.
(Elapid)	Pseudechetoxin (PsTx)	*Pseudechis australis*	Cyclic nucleotide-gated ion channels	Inhibition of CNGA1 subunit (apparent Ki = 70 nM) Inhibition of CNGA2 subunits (apparent Ki = 15 nM)	Q8AVA4 2DDA	[49,67]
	Pseudecin	*Pseudechis porphyriacus*	Cyclic nucleotide-gated ion channels	Ca^{2+} channel blocker-like properties	Q8AVA3 2DDB, 2EPF(Zn2+)	[49,60]
	Latisemin	*Laticauda semifasciata*	N.D.	Ca^{2+} channel blocker-like properties	Q8JI38	[60]
	Ophanin	*Ophiophagus hannah*	N.D.	Ca^{2+} channel blocker-like properties	AY181984	[61]
	Natrin	*Naja Naja atra*	Bkca Kv1.3 Ryanodine receptors Heparin	High-conductance calcium-activated potassium (BKCa) channel (34.4 nM) Inhibition of Kv1.3 (10–200 nM) Inhibition of ryanodine receptors (1 μM, Kd = 1.5–2.3 nM) Inflammatory Modulator (<1.0 μM) Non-proteolytic activity (BSA, neurotensin, Tex31 substrate, kenetensin)	Q7T1K6 1XX5, 1XTA, 2GIZ 3MZ8 (Zn2+)	[68–70]
(Colubrid)	Tigrin	*Rhabdophis tigrinus*	N.D.	Non-Ca^{2+} channel blocker-like properties	Q8JGT9	[60]
	Patagonin	*Philodryas patagoniensis*	N.D.	Non-Ca^{2+} channel blocker-like properties Non-fibrinogenolytic activity Skeletal myotoxic activity (43 and 87 μM)	N.D. (only N-terminal sequence)	[71]

Tex31 is located in Pathogenesis-related 1 (PR-1) members. N.D.; not determined or did not investigate. Ca^{2+} channel blocker-like properties; inhibited depolarization-induced contraction of rat-tail arterial smooth muscle. All accession numbers are from UniProt (https://www.uniprot.org/) and Protein Data Bank (PDB, https://www.rcsb.org/pdb/home/sitemap.do).

4. Proteins That Bind to CRISPs

Human CRISP-3 occurs in seminal plasma at high concentrations (14.8 g/mL) [72], but its function remains unknown. CRISP-3 is a promising bio-marker candidate for prostate cancer because the concentration of this protein increases >50-fold in pre-malignant prostate lesions and in primary tumors compared to normal prostatic epithelium [73]. To understand the physiological activity of CRISP-3, molecules with which it interacts have been identified, namely a prostate secretory protein of 94 amino acids (PSP94) (also known as a β-microseminoprotein; MSP) [74] and α-1B glycoprotein in human plasma (A1BG) [75]. Both bind to CRISP-3 with high affinity (K_D = 6.28 × 10^{-11} M with PSP94 and K_D = 2.8 × 10^{-9} M with A1BG). Identification of interaction surfaces between CRISP-3 and these binding proteins from mammals has been a focus of attention due to the medical relevance of CRISP-3 [76,77]. However, experimental evidence has been limited to NMR titration and mutagenesis analysis [78,79]. Recently, we determined the structure of a complex between PSP94 and CRISP family proteins that provided insight into CRISP-3 binding [80].

Small serum protein-2 (SSP-2) was identified in the serum of *Protobothrops flavoviridis* as an endogenous inhibitor against triflin (svCRISP) [81]. We built a binding model by superimposing SSP-2 onto PSP94, because PSP94 and SSP-2 are structurally similar and interact strongly with triflin across species [82]. The previously published PSP94–CRISP-3 model based on NMR titration showed that the N-terminal Greek key motif and the C-terminal β8 strand of PSP94 interact with the N-terminal CAP/PR-1 domain and hinge region of CRISP-3, respectively, in a parallel manner [78]. Our structure is upside-down compared to the other model, but the same surface of PSP94 interacts with the concave CAP/PR-1 domain of triflin (Figure 2A). In addition to the β5 and β8 strands, other key structural elements of PSP94 involved in complex formation are likely to be conserved. In PSP94, the β1 and β8 strands at the N- and C-termini are aligned in a linear manner and form an edged binding surface, whereas the β1 and β5 strands of SSP-2 form the binding surface. SSP-2 has a shorter C-terminal region compared with PSP94, so the N- and C-termini of SSP-2 are located on opposite sides. Consequently, this is in contrast to the N- and C- termini of PSP94, which are located on the same side. We hypothesize that formation of a parallel β-sheet between the SSP-2 β5 strand and the triflin β4 strand may allow the SSP-2 β1 strand to fit into the cavity between the CAP/PR-1 and CRD/ICR domains of triflin, thereby blocking the Zn^{2+} binding site and stabilizing the interaction. These findings indicate that our model provides significant structural insight into the human PSP94–CRISP-3 complex, which has been debated for many years.

The SSP-2–CRISP-3 complex model reveals that the N-terminal alanine of SSP-2 penetrates the metal-binding site of triflin, and that the CRD/ICR domain is shifted compared to the position of the CRD/ICR domain in triflin. This observation agrees with the conformational change of the CRD/ICR domain in the presence or absence of Zn^{2+}, which has been documented for another svCRISP, pseudecine (Figure 2B). We have evidence that the binding of SSP-2 dramatically suppresses the channel inhibition activity of triflin (unpublished data). The structure of our complex also indicates that the concave region of the triflin CAP/PR-1 domain was fully occupied by the entire SSP-2 molecule, whereas direct interaction at the CRD/ICR domain was limited. Thus, SSP-2 may inhibit activity of several svCRISPs, because the concave region of this family shows great conservation. SSP-2 binding prevents Zn^{2+} binding to the concave region of the PR-1/CAP domain (Figure 2B). The structure-based alignment of venom CRISPs and human CRISP-3 show that the contact residues identified in our complex are relatively well conserved among CRISPs, suggesting the relevance of the binding ability of PSP94 to a wide range of CRISPs, including svCRISPs. As described above, divalent cations affect some CRISP activities (Section 3). Therefore, binding of the side face of the β-sheets of SSP-2 to both CAP/PR-1 and CRD/ICR domains of triflin might be important for suppression of triflin functions.

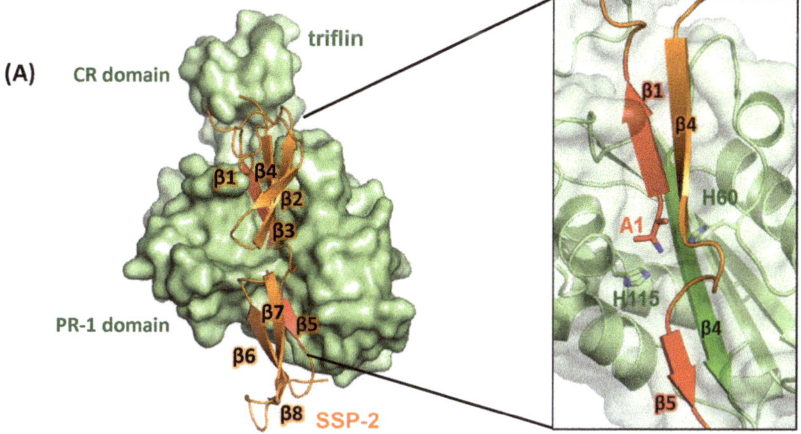

triflin in complex with SSP-2

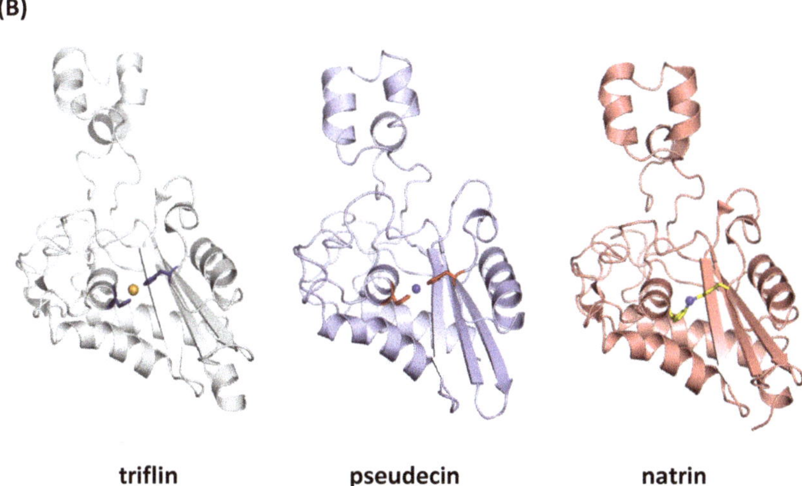

Figure 2. Inhibition of the divalent cation binding site by the serum inhibitor Small serum protein-2 (SSP-2). (**A**) Our complex structure of SSP-2-triflin (PDB ID: 6IMF) clearly indicates that the inhibitor occupies and blocks the conserved divalent cation binding site, which is functionally important. The inset is a focused view of the β1 and β5 strands of SSP-2. Ala1 of SSP-2 and His60 and His 115 of triflin are shown as stick models. (**B**) The same view of the apo-triflin structure (PDB ID: 1WVR, left), Pseudecin (PDB ID: 2FPF, middle) and natrin (PDB ID: 3MZ8, right) are shown. Divalent cations are bound at the conserved location via histidines, indicated with stick models. Structures were prepared using PyMOL (https://pymol.org/).

5. Isolation and Characterization of Snake Venom CRISPs

Snake venom CRISPs have proven difficult to express recombinantly and to fold properly in bacteria and yeast systems due to their eight disulfide bonds. Therefore, the first step in characterizing svCRISPs is usually to isolate from crude venom. Normal-phase and reversed-phase high-performance liquid chromatography (HPLC), have been used to purify venom CRISPs. The svCRISP natrin from

Naja, triflin from *Protobothrops flavoviridis*, ablomin from *G. blomhoffii*, latisemin from *L. semifasciata*, tigrin from *R. tigrinus*, kaouthin-1 and kaouthin-2 from *Naja kaouthia* [83], and patagonin from *Philodryas patagoniensis* were purified by size exclusion chromatography, ion exchange chromatography, or heparin affinity chromatography. Pseudechetoxin from *Pseudechis australis*, TJ-CRVP from *Trimeresurus jerdonii*, and NA-CRVPs from *Naja atra* [84], as well as helothermine from the saliva of *Heloderma horridum* [40] and Tex31, from a homogenized extract of *Conus textile* [32], were purified by reversed-phase HPLC (RP-HPLC) with an acetonitrile buffer containing 0.1% (*v/v*) trifluoroacetic acid for the final purification step. It is possible that RP-HPLC purification in acidic buffers may affect CRISP activity and tertiary structure, although Tex31 which was purified by this method retained its proteolytic activity. In our own lab, we compared CD spectra of triflin purified by normal-phase and RP-HPLC and found no differences (unpublished). Therefore, we concluded that svCRISPs are very stable, even in strong acids (<pH 2.0).

In our work, two serum CRISPs were discovered in *Protobothrops flavoviridis* and *G. blomhoffii* venoms, which we designated as serotriflin and seroablomin, respectively [85]. We also reported complexes between serum CRISPs and other proteins in the blood of snakes. Formation of these complexes is pH-dependent, so they might not be found if acidic RP-HPLC buffers are used, but neutral or basic volatile buffer systems can probably be used.

The greatest challenge with CRISP isolation is that it is difficult to conclude if the biological activity is retained given that, for most, their activities are unknown. So far, no svCRISPs have proven lethal to mammals. A key feature of this family of proteins is that, although they bind various target molecules, all affect cellular signaling.

5.1. Ancestral CRISP Activity

Ito et al. and Xiao et al. characterized CRISPs from buccal glands of lampreys. These proteins included a CRISP from *L. japonicum* and buccal gland secretion protein-2 (CRBGP-2) from *Lampetra japonica* [35,51]. Lamprey CRISPs exhibited similar pharmacological effects under almost the same conditions and concentrations as svCRISPs, such as blocked depolarization-induced contraction of rat-tail arterial smooth muscle at 1 μM and suppression of angiogenesis related to EC apoptosis via microfilament disorganization [51–54], although their selectivity differed from that of svCRISPs [51]. Recombinant PR-1/CAP retained both cytotoxic activity against human umbilical vein endothelial cells (HUVECs) and anti-angiogenic activity. In 2011, a lamprey CRISP was demonstrated as a neutrophil inhibitory factor, and its inhibitory effect was caused by binding to α β 2 integrin receptors [55]. These observations suggested that some physiological activities of svCRISPs have been conserved from ancestral vertebrates.

5.2. Myotoxicity

Patagonin, from *Philodryas patagoniensis* venom, caused skeletal myotoxicity in murine gastrocnemius muscle, including muscle necrosis, edema, and inflammatory infiltration of polymorphonuclear leukocytes without smooth muscle contraction, as well as proteolytic activity, hemorrhage, and inhibition of platelet aggregation [70]. The authors hypothesized that the molecular mechanism by which patagonin induced muscle necrosis may be associated with binding to ion channels and speculated that this might be a general property of svCRISPs. They also suggested that tigrin (*R. tigrinus*) may cause skeletal myotoxicity because patagonin and tigrin are both from rear-fanged snakes and have high sequence similarity.

5.3. Ion Channel Inhibition

Several venom CRISPs from viperids and elapids target ion channels [31]. One of the best characterized svCRISPs is natrin, isolated from *N. atra*, the crystal structure and receptor targets of which are known. Natrin has an inhibitory effect on high-conductance calcium-activated potassium (BKca) Kv1.3 channels, as well as calcium release channel/ryanodine receptors (RyR) [67]. In 2010,

Wang et al. demonstrated that <1 M natrin activated ECs to promote monocytic cell adhesion in a heparin sulfate- and Zn^{2+}-dependent manner via increased expression of adhesion molecules (VCAM-1 and ICAM-1) and E-selectin as an inflammatory modulator [50]. They proposed that the mechanism involved binding of natrin to heparin in the presence of Zn^{2+}. A cryo-EM study showed that the CRD/ICR domain of natrin is crucial for binding to ryanodine receptor 1 (RyR1, a Ca^{2+} release channel) [68]. However, sequence comparisons among svCRISPs suggest that the amino acid residues 42–44, 57–59, and 63–65 in the CAP/PR-1 domain may also be important for target channel recognition (Figure 1) [49,83]. These regions are putative interaction sites that target ion-channels, and are variable svCRISP residues. Indeed, the CRD/ICR domain of pseudechetoxin, a cyclic nucleotide-gated (CNG) channel blocker, did not inhibit CNG channels alone [49].

5.4. Anti-Protozoal Activity

Crovirin from *Crotalus viridis* has anti-protozoan activity against *Trypanosoma cruzi* and *Leishmania amazonensis* with low IC_{50} and LD_{50} values (1.10–2.38 g/mL), but was non-toxic to mice in an ex vivo assay measuring creatine kinase activity after an injection of 10 µg/mL of crovirin [71]. A considerably higher concentration of crovirin (20 g/mL) displayed limited toxicity to mammalian cells. The mechanism responsible for this activity was not investigated.

5.5. Anti-Angiogenic Activity

A CRISP from *Echis carinatus sochureki* venom, EC-CRISP, is a negative regulator of angiogenesis. The HPLC fraction did not interact with cancer cells, such as the glioma cell line, LN18, but showed pro-adhesive properties for normal ECs, such as HUVECs. At concentrations of 10–20 g/mL (<1 M), EC-CRISP interacted with ECs without affecting fibronectin, vitronectin, collagen type I, or laminin, and was transported into the cytoplasm. This toxin inhibited the MAPK Erk1/2 signaling pathway induced by vascular endothelial growth factor (VEGF), but had no effect on two other MAP kinases, p38 and SAPK/JNK [65]. However, the target receptor of ES-CRISP is still unknown.

5.6. Vascular Permeability Regulator

The effect of hellerin, from *Crotalus oreganus helleri* venom, on vascular permeability was demonstrated in vivo and in vitro. Trans-capillary leakage was observed 30 min after mice were subcutaneously injected with hellerin (70 nM), but leakage was approximately half that of the vascular permeability produced by vascular endothelial growth factor A (VEGF-A, 50 nM) [64]. Hellerin (2 M) reduced the viability of HUVECs by 50%, and hellerin-treated HUVEC cells had rounded cell shapes and detached from the substrate. In human dermal lymphatic endothelial cells (HDLECs) and human dermal blood endothelial cells (HDBECs), hellerin (675 nM) increased trans-epithelial permeability and decreased the level of F-actin.

5.7. Inflammation Regulator and Protease Activity

The *Bothrops jararaca* svCRISP, Bj-CRP, did not show proteolytic, hemorrhagic, coagulant, or potassium channel inhibition [63]. However, Bj-CRP increased leukocyte and neutrophil infiltration in mice 1 and 4 hr after i.p. injection, but this activity did not increase over 24 h following the injection. This inflammatory response might be related to the observed increase in IL-6 expression 1 h after injection, but no increase in TNF-, IL-10, or NO was observed. Bj-CRP cleaved C3 and C4 weakly and bound to components C3 and C4, which resulted in increased levels of C3a, C4a, and C5a. These results indicate that Bj-CRP modulates hemolytic activity associated with the complement pathways because >50 g of Bj-CRP reduced hemolytic activity.

6. Functional Sites Identified in CRISPs

CRISPs are highly conserved proteins in snake venoms; however, few svCRISPs have known biological activities (Table 1). Amino acid alignments between CRISPs with known structures and activities would be useful for understanding other CRISPs. CRISP putative functional sites have been proposed based on amino acid sequence conservation and variability. However, it is still necessary to characterize the biological activities of novel CRISPs because structure-function relationships for these toxins are yet to be completely understood.

6.1. Potential Functional Sites Responsible for Protease Activity

Cone snail Tex31, a PR-1 protein, showed proteolytic activity against synthesized peptides such as Ac-KLEKR-*p*NA. Tex31 is thought to cleave the propeptide region of conotoxin TxVIA as a native substrate [32]. Stecrisp, a svCRISP isolated from *Trimeresurus stejnegeri* venom, had no proteolytic activity under almost the same conditions. Based on tertiary structures, the authors proposed that Ser80 in Tex31, which is in a loop region, may be in close proximity to the highly conserved divalent cation site (His130 and Glu115 in Tex31), and this forms a catalytic triad responsible for cleavage [32]. Unlike Tex31, the residue at position 80 in stecrisp is proline (Pro80 in the reference [62]).

Patagonin and Lamprey CRISP lacked cleavage activity against fibrinogen [70]. Natrin lacked proteolytic activity against bovine serum albumin (BSA), neurotensin, a Tex31 substrate, or kenetensin [50]. Although Bj-CRP did not cleave azocasein, fibrinogen, or fibrin, it showed low catalytic activity against C3 and C4 [63]. This result might be similar to the general activation of component C3 by C3-convertase, a serine protease. C3-convertase cleaves component C3 to C3a and C3b, at a cleavage site between Arg726 and Ser727 (–LAR^{726}S^{727}NLD–) of component C3 [86]. This cleavage site is similar to the leucine at position P4 of the preferred substrate sequences of Tex31, counting from the C-terminal end. A sequence comparison between Bj-CRP and Tex31 could not be performed because the Bj-CRP sequence is not yet available. There are still many unknown sequences, such as the non-proteolytic patagonin. Thus, functional sites responsible for CRISP proteolytic activity are unclear because of the limited number of reports, and because their proteolytic activity against different substrates has not been comprehensively evaluated. Likewise, protease activity of mammalian, fungal, and plant CAP superfamily proteins is still lacking.

6.2. Potential Domains and Functional Sites Responsible for Ion Channel Inhibition

Seven CRISPs inhibit high-potassium-induced contraction of smooth muscle, with three of these (piscivorin, ophanin, and catrin) demonstrating a weaker effect at 1 µM [61]. Three CRISPs (tigrin, patagonin, and natrin) did not show inhibition when tested (Table 1). The current hypothesis is that suppression of muscle contraction is a result of the restriction of the Ca^{2+} incurrent by CRISP channel blockade; however, there is no direct evidence to support this. As previously discussed, the CRD/ICR domain is one of the CRISP regions potentially responsible for targeting ion-channels. This has been suggested because this domain is very similar to peptides that block ion channels, including peptide toxins KTX, α-KTX, and ShTx (43). CRD/ICR domain sequences between CRISPs that cause muscle contraction and those that inhibit contractions were compared (Figure 3A). Interestingly, there is a deletion of two residues in tigrin and an insertion of five residues (GAGGT) in lamprey CRISP in this region. These differences are also present only in the surface-exposed ICR motif of the CRD/ICR domain. (Figure 3B). However, there are no significant feature differences among active ion channel inhibitors and those that are inactive. There are likely multiple residues responsible for this activity in the CAP/PR-1 domain and/or hinge region.

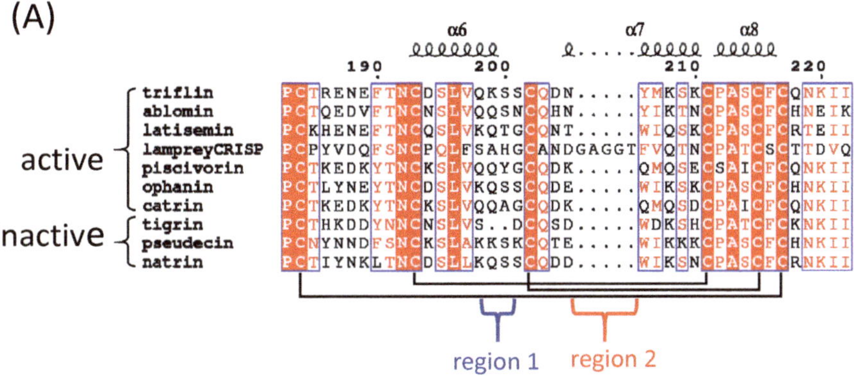

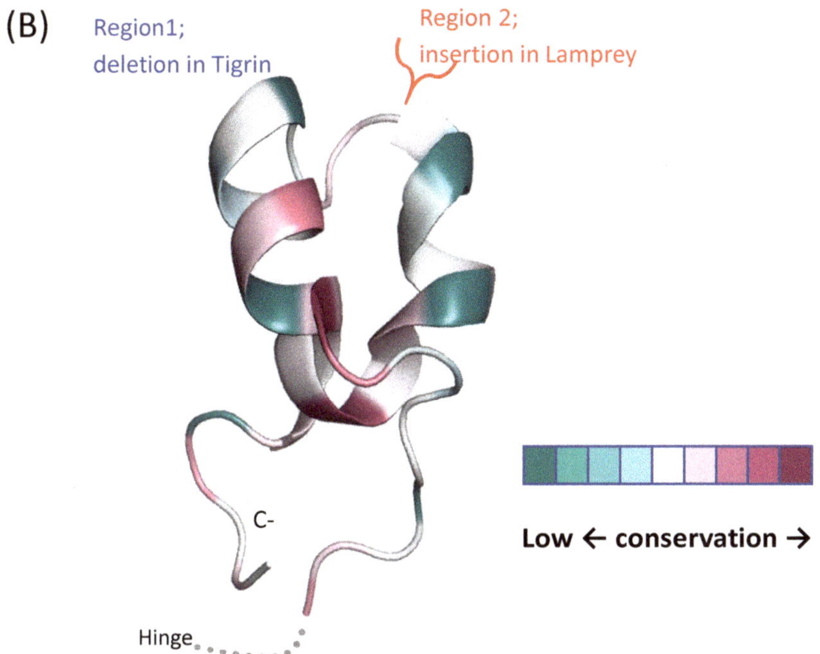

Figure 3. Comparisons between CRD/ICR domain regions of CRISPs that inhibit ion channels and those that do not. (**A**) Amino-acid sequence alignment of CRD/ICR regions of CRISP family proteins, showing highly conserved residues highlighted in red, and other conserved residues in a red font. Cysteine residues forming disulfide bridges are indicated by black brackets. The secondary structure of triflin (PDB ID: 1WVR) is shown above the alignment. (**B**) Structural conservation in the CRD/ICR region of snake venom CRISPS (svCRISPs0 that inhibit high-potassium-induced contraction of smooth muscle is shown on a triflin scaffold (PDB ID: 1WVR). Conservation scores were calculated with the Consurf server using default settings. Conservation scores are graded on a nine-point scale, from the most variable positions (turquoise) to the most conserved positions (maroon). The structure was prepared using PyMOL (https://pymol.org/).

Helothermine and natrin bind to ion channel targets with high affinity (<1 μM). These targets include K$^+$ channels, ryanodine receptors [58,67,69], and voltage-dependent Ca^{2+} channels [57]. Not all CRISPs have been identified as having ion channel targets. Bj-CRISP had no effect on 13 voltage-gated potassium channels tested using two-electrode voltage-clamping on *Xenopus* oocytes. CNG channels were identified as pseudechetoxin and pseudecin targets. Interestingly, these toxins had different affinities, even though they differ by only seven residues. Two basic residues in pseudechetoxin (Lys184 and Arg185) seem to contribute to higher CNG binding affinity; the affinity of pseudechetoxin is 10-fold greater than that of pseudecin, with neutral residues (Asn184 and Tyr185) in this region (Figure 1) [60]. Matsunaga et al. discussed sequence variation in the concave region between the N- and C-terminal domains of svCRISPs. They suggested that variation in svCRISP activity toward different ion channels can be explained by charge distribution differences on the surfaces of svCRISPs [83]; however, there is no experimental evidence identifying the binding sites of these toxins.

7. svCRISP Evolution

CRISPs of toxicoferan reptiles have experienced positive selection, and more in snakes than in lizards [87]. In contrast, mammalian CRISPs appear constrained by negative selection. Episodes of rapid gene divergence are seen for svCRISPs in elapids and rear-fanged colubrids at all codon positions, in comparison to weaker gene divergence in viperid and boid snakes, and anguid, helodermatid, and iguanid lizards [88]. Evolution of toxicoferan CRISPs may be linked to snake predatory behavior. Elapid CRISPs manifest lower levels of positive selection (ω values), potentially due to the presence of highly toxic neurotoxins in these venoms for prey incapacitation. Viperids and rear-fanged colubrids have comparatively fewer toxic venom components, and have CRISPs manifesting higher levels of positive selection. At the protein-level, positive selection was identified at sites on the molecular surface, primarily in the CRD/ICR domain [87]. Future characterizations of CRISP activities with their corresponding functional sites may provide better insight into their role in prey capture and how this impacts their evolution.

Although 1-6 gene copies are observed for CRISPs in snake genomes [89,90], CRISP gene positions in snake genomes appear to be conserved. For genomes of both *Naja naja* and *Crotalus viridis*, svCRISP genes are located on chromosome 1 [90,91], potentially close to the centromere. This genome location may promote sequence conservation and may explain why expression of some toxins varies for different snake families, such as is observed for phospholipase A$_2$s and three-finger toxins, but svCRISPs are present at consistent levels in venoms of most venomous snakes.

Expression levels of toxin and non-toxin homologs of svCRISPs have been evaluated for one opisthoglyph and for proteroglyphs and solenoglyphs [89–92]. At least two CRISP genes are expressed in snake venom glands, but differences in expression levels of each are apparent. Molecular evolution and functionality may not be equal for these two genes. Therefore, interpretations based on genome sequences alone require caution. Expression levels should also be considered to avoid misrepresenting the significance of venom CRISPs.

8. Conclusions and Future Directions

Two decades have passed since the discovery of venom CRISPs [58]. Many are now known and have been investigated in various ways, including protein and nucleotide sequencing, and characterization of molecular weights and isoelectric points. A few have been functionally characterized. However, new technologies generating large amounts of sequence data have completely overwhelmed our capacity to functionally characterize the many venom constituents being reported. This is especially difficult when it has yet to become apparent what biological roles certain toxins provide in venoms, as we can only document the activities for which we assay.

It is possible for toxins to have conserved amino acid sequences, functions, and mechanisms associated with their structures. This information can provide insights into their surface features and even into specific residues involved in binding. Conserved features may preserve selectivity and/or specificity. Structures and functions of CRISPs have been reviewed here, including those from lampreys. Ion channel-blocking activity and/or disruption of EC conditions may be an activity of svCRISPs that has been conserved evolutionarily. Further, it is important to consider co-factors and interacting proteins that regulate these activities in order to comprehensively understand biological effects of svCRISPs. Venom proteins are remarkably stable, with high selectivity and affinity for targets. Exploring toxins, even non-lethal ones like CRISPs, can improve our understanding of how proteins target specific channels, receptors, or substrates. This can be useful for the development of therapeutics, not only to treat snake envenoming, but also other maladies.

svCRISP characterization faces several challenges. The first is that these proteins occur at low levels in crude venoms. Moreover, eight disulfide bonds make them difficult to biosynthesize in large quantities for biological assays or structure determinations. An additional struggle lies in designing proper assays to characterize svCRISP activity, as a wide diversity of activities have been documented. We propose that the range of activities known from the CAP superfamily as a whole should dictate assays to be used in characterizing novel svCRISPs. Functional characterization of CRISPs lags far behind the number of genomic, transcriptomic, and proteomic CRISP sequences. An integrated approach to study snake venoms is required. Researchers involved in '-omics' need to collaborate with labs that specialize in structure-function relationships to execute more comprehensive studies. In the future, we encourage toxinologists to characterize svCRISPs functionally.

Author Contributions: N.A.-S. and K.M. conceived and designed the content of this article; C.M.M. analyzed and summarized sequences of the CRISP proteins; T.T. contributed analysis tools and making protein tertiary structures; N.A.-S., C.M.M., T.T. and K.M. wrote this review article. All authors have read and agreed to the published version of the manuscript.

Funding: This research was funded by Japan society of the promotion of science grant number 17KK0179 and JBC31025 and N.S was funded by Fukuoka University Foundation (181042 and 197104).

Acknowledgments: We thank R. Manjunatha Kini for help with literature search and helpful comments on the manuscript sequence alignments. We thank Margaret Biswas from Edanz Group (www.edanzediting.com/ac) for editing a draft of this manuscript.

Conflicts of Interest: The authors declare no conflicts of interest.

References

1. Cantacessi, C.; Gasser, R.B. SCP/TAPS proteins in helminths–Where to from now? *Mol. Cell. Probes* **2012**, *26*, 54–59. [CrossRef] [PubMed]
2. Van Loon, L.C.; Van Kammen, A. Polyacrylamide disc electrophoresis of the soluble leaf proteins from Nicotiana tabacum var. 'Samsun' and 'Samsun NN'. *Virology* **1970**, *40*, 199–211. [CrossRef]
3. Alexander, D.; Goodman, R.M.; Gut-Rella, M.; Glascock, C.; Weymann, K.; Friedrich, L.; Maddox, D.; Ahl-Goy, P.; Luntz, T.; Ward, E. Increased tolerance to two oomycete pathogens in transgenic tobacco expressing pathogenesis-Related protein 1a. *Proc. Natl. Acad. Sci. USA* **1993**, *90*, 7327–7331. [CrossRef] [PubMed]
4. Kiba, A.; Nishihara, M.; Nakatsuka, T.; Yamamura, S. Pathogenesis-Related protein 1 homologue is an antifungal protein in Wasabia japonica leaves and confers resistance to Botrytis cinerea in transgenic tobacco. *Plant Biotechnol.* **2007**, *24*, 247–253. [CrossRef]
5. Sarowar, S.; Kim, Y.J.; Kim, E.N.; Kim, K.D.; Hwang, B.K.; Islam, R.; Shin, J.S. Overexpression of a pepper basic pathogenesis-Related protein 1 gene in tobacco plants enhances resistance to heavy metal and pathogen stresses. *Plant Cell Rep.* **2005**, *24*, 216–224. [CrossRef]
6. Shin, S.H.; Park, J.-H.; Kim, M.-J.; Kim, H.-J.; Oh, J.S.; Choi, H.K.; Jung, H.W.; Chung, Y.S. An Acidic PATHOGENESIS-RELATED1 Gene of Oryza grandiglumis is Involved in Disease Resistance Response Against Bacterial Infection. *Plant Pathol. J.* **2014**, *30*, 208–214. [CrossRef]
7. Carr, J.P.; Beachy, R.N.; Klessig, D.F. Are the PR1 proteins of tobacco involved in genetically engineered resistance to TMV? *Virology* **1989**, *169*, 470–473. [CrossRef]
8. Hon, W.C.; Griffith, M.; Mlynarz, A.; Kwok, Y.C.; Yang, D. Antifreeze Proteins in Winter Rye Are Similar to Pathogenesis-Related Proteins. *Plant Physiol.* **1995**, *109*, 879–889. [CrossRef]

9. Zeier, J.; Pink, B.; Mueller, M.; Berger, S. Light conditions influence specific defence responses in incompatible plant-pathogen interactions: Uncoupling systemic resistance from salicylic acid and PR-1 accumulation. *Planta* **2004**, *219*. [CrossRef]

10. Seo, P.J.; Lee, A.-K.; Xiang, F.; Park, C.-M. Molecular and Functional Profiling of Arabidopsis Pathogenesis-Related Genes: Insights into Their Roles in Salt Response of Seed Germination. *Plant Cell Physiol.* **2008**, *49*, 334–344. [CrossRef]

11. Liu, W.-X.; Zhang, F.-C.; Zhang, W.-Z.; Song, L.-F.; Wu, W.-H.; Chen, Y.-F. Arabidopsis Di19 Functions as a Transcription Factor and Modulates PR1, PR2, and PR5 Expression in Response to Drought Stress. *Mol. Plant* **2013**, *6*, 1487–1502. [CrossRef] [PubMed]

12. Kothari, K.S.; Dansana, P.K.; Giri, J.; Tyagi, A.K. Rice Stress Associated Protein 1 (OsSAP1) Interacts with Aminotransferase (OsAMTR1) and Pathogenesis-Related 1a Protein (OsSCP) and Regulates Abiotic Stress Responses. *Front. Plant Sci.* **2016**, *7*. [CrossRef] [PubMed]

13. Memelink, J.; Linthorst, H.J.M.; Schilperoort, R.A.; Hoge, J.H.C. Tobacco genes encoding acidic and basic isoforms of pathogenesis-Related proteins display different expression patterns. *Plant Mol. Biol.* **1990**, *14*, 119–126. [CrossRef] [PubMed]

14. Breen, S.; Williams, S.J.; Outram, M.; Kobe, B.; Solomon, P.S. Emerging Insights into the Functions of Pathogenesis-Related Protein 1. *Trends Plant Sci.* **2017**, *22*, 871–879. [CrossRef] [PubMed]

15. King, T.P.; Spangfort, M.D. Structure and Biology of Stinging Insect Venom Allergens. *Int. Arch. Allergy Immunol.* **2000**, *123*, 99–106. [CrossRef] [PubMed]

16. Mans, B.J.; Andersen, J.F.; Francischetti, I.M.B.; Valenzuela, J.G.; Schwan, T.G.; Pham, V.M.; Garfield, M.K.; Hammer, C.H.; Ribeiro, J.M.C. Comparative sialomics between hard and soft ticks: Implications for the evolution of blood-Feeding behavior. *Insect Biochem. Mol. Biol.* **2008**, *38*, 42–58. [CrossRef] [PubMed]

17. Charlab, R.; Valenzuela, J.G.; Rowton, E.D.; Ribeiro, J.M.C. Toward an understanding of the biochemical and pharmacological complexity of the saliva of a hematophagous sand fly Lutzomyia longipalpis. *Proc. Natl. Acad. Sci. USA* **1999**, *96*, 15155–15160. [CrossRef]

18. Calvo, E.; Dao, A.; Pham, V.M.; Ribeiro, J.M.C. An insight into the sialome of Anopheles funestus reveals an emerging pattern in anopheline salivary protein families. *Insect Biochem. Mol. Biol.* **2007**, *37*, 164–175. [CrossRef]

19. Henriksen, A.; King, T.P.; Mirza, O.; Monsalve, R.I.; Meno, K.; Ipsen, H.; Larsen, J.N.; Gajhede, M.; Spangfort, M.D. Major venom allergen of yellow jackets, Ves v 5: Structural characterization of a pathogenesis-Related protein superfamily. *Proteins Struct. Funct. Genet.* **2001**, *45*, 438–448. [CrossRef]

20. Müller, U.R.; Johansen, N.; Petersen, A.B.; Fromberg-Nielsen, J.; Haeberli, G. Hymenoptera venom allergy: Analysis of double positivity to honey bee and *Vespula* venom by estimation of IgE antibodies to species-Specific major allergens Api m1 and Ves v5. *Allergy* **2009**, *64*, 543–548. [CrossRef]

21. Ribeiro, J.M.C.; Francischetti, I.M.B. Role of arthropod saliva in blood feeding: Sialome and Post-Sialome Perspectives. *Annu. Rev. Entomol.* **2003**, *48*, 73–88. [CrossRef] [PubMed]

22. dos Santos-Pinto, J.R.A.; dos Santos, L.D.; Andrade Arcuri, H.; Castro, F.M.; Kalil, J.E.; Palma, M.S. Using Proteomic Strategies for Sequencing and Post-Translational Modifications Assignment of Antigen-5, a Major Allergen from the Venom of the Social Wasp Polybia paulista. *J. Proteome Res.* **2014**, *13*, 855–865. [CrossRef] [PubMed]

23. Assumpção, T.C.F.; Ma, D.; Schwarz, A.; Reiter, K.; Santana, J.M.; Andersen, J.F.; Ribeiro, J.M.C.; Nardone, G.; Yu, L.L.; Francischetti, I.M.B. Salivary Antigen-5/CAP Family Members Are Cu^{2+}-Dependent Antioxidant Enzymes That Scavenge O_{2-} and Inhibit Collagen-induced Platelet Aggregation and Neutrophil Oxidative Burst. *J. Biol. Chem.* **2013**, *288*, 14341–14361. [CrossRef] [PubMed]

24. Cameo, M.S.; Blaquier, J.A. Androgen-Controlled specific proteins in rat epididymis. *J. Endocrinol.* **1976**, *69*, 47–55. [CrossRef]

25. Kierszenbaum, A.L.; Lea, O.; Petrusz, P.; French, F.S.; Tres, L.L. Isolation, culture, and immunocytochemical characterization of epididymal epithelial cells from pubertal and adult rats. *Proc. Natl. Acad. Sci. USA* **1981**, *78*, 1675–1679. [CrossRef]

26. Kratzschmar, J.; Haendler, B.; Eberspaecher, U.; Roosterman, D.; Donner, P.; Schleuning, W.-D. The Human Cysteine-Rich Secretory Protein (CRISP) Family. Primary Structure and Tissue Distribution of CRISP-1, CRISP-2 and CRISP-3. *Eur. J. Biochem.* **1996**, *236*, 827–836. [CrossRef]

27. Udby, L.; Cowland, J.B.; Johnsen, A.H.; Sørensen, O.E.; Borregaard, N.; Kjeldsen, L. An ELISA for SGP28/CRISP-3, a cysteine-Rich secretory protein in human neutrophils, plasma, and exocrine secretions. *J. Immunol. Methods* **2002**, *263*, 43–55. [CrossRef]

28. Koppers, A.J.; Reddy, T.; O'Bryan, M.K. The role of cysteine-Rich secretory proteins in male fertility. *Asian J. Androl.* **2011**, *13*, 111–117. [CrossRef]

29. Gibbs, G.M.; Roelants, K.; O'Bryan, M.K. The CAP Superfamily: Cysteine-Rich Secretory Proteins, Antigen 5, and Pathogenesis-Related 1 Proteins—Roles in Reproduction, Cancer, and Immune Defense. *Endocr. Rev.* **2008**, *29*, 865–897. [CrossRef]

30. Gibbs, G.M.; O'Bryan, M.K. Cysteine rich secretory proteins in reproduction and venom. *Soc. Reprod. Fertil. Suppl.* **2007**, *65*, 261–267.

31. Yamazaki, Y.; Morita, T. Structure and function of snake venom cysteine-Rich secretory proteins. *Toxicon* **2004**, *44*, 227–231. [CrossRef] [PubMed]

32. Milne, T.J.; Abbenante, G.; Tyndall, J.D.A.; Halliday, J.; Lewis, R.J. Isolation and Characterization of a Cone Snail Protease with Homology to CRISP Proteins of the Pathogenesis-Related Protein Superfamily. *J. Biol. Chem.* **2003**, *278*, 31105–31110. [CrossRef] [PubMed]

33. Undheim, E.A.; Sunagar, K.; Herzig, V.; Kely, L.; Low, D.H.; Jackson, T.N.; Jones, A.; Kurniawan, N.; King, G.F.; Ali, S.A.; et al. A proteomics and transcriptomics investigation of the venom from the barychelid spider Trittame loki (brush-Foot trapdoor). *Toxins* **2013**, *5*, 2488–2503. [CrossRef] [PubMed]

34. Romero-Gutierrez, T.; Peguero-Sanchez, E.; Cevallos, M.A.; Batista, C.V.F.; Ortiz, E.; Possani, L.D. A Deeper Examination of Thorellius atrox Scorpion Venom Components with Omic Techonologies. *Toxins* **2017**, *9*, 399. [CrossRef] [PubMed]

35. Ito, N.; Mita, M.; Takahashi, Y.; Matsushima, A.; Watanabe, Y.G.; Hirano, S.; Odani, S. Novel cysteine-Rich secretory protein in the buccal gland secretion of the parasitic lamprey, Lethenteron japonicum. *Biochem. Biophys. Res. Commun.* **2007**, *358*, 35–40. [CrossRef] [PubMed]

36. Fry, B.G.; Casewell, N.R.; Wuster, W.; Vidal, N.; Young, B.; Jackson, T.N. The structural and functional diversification of the Toxicofera reptile venom system. *Toxicon* **2012**, *60*, 434–448. [CrossRef]

37. Fry, B.G.; Vidal, N.; Norman, J.A.; Vonk, F.J.; Scheib, H.; Ramjan, S.F.R.; Kuruppu, S.; Fung, K.; Blair Hedges, S.; Richardson, M.K.; et al. Early evolution of the venom system in lizards and snakes. *Nature* **2006**, *439*, 584–588. [CrossRef]

38. Barua, A.; Mikheyev, A.S. Many Options, Few Solutions: Over 60 My Snakes Converged on a Few Optimal Venom Formulations. *Mol. Biol. Evol.* **2019**, *36*, 1964–1974. [CrossRef]

39. Tasoulis, T.; Isbister, G. A Review and Database of Snake Venom Proteomes. *Toxins* **2017**, *9*, 290. [CrossRef]

40. Morrissette, J.; Krätzschmar, J.; Haendler, B.; el-Hayek, R.; Mochca-Morales, J.; Martin, B.M.; Patel, J.R.; Moss, R.L.; Schleuning, W.D.; Coronado, R. Primary structure and properties of helothermine, a peptide toxin that blocks ryanodine receptors. *Biophys. J.* **1995**, *68*, 2280–2288. [CrossRef]

41. Mackessy, S.P.; Heyborne, W.H. Cysteine-Rich secretory proteins in reptile venoms. In *Handbook of Venoms and Toxins of Reptiles*; Mackessy, S.P., Ed.; CRC Press/Taylor & Francis Group: Boca Raton, FL, USA, 2010; pp. 325–334.

42. Niderman, T.; Genetet, I.; Bruyere, T.; Gees, R.; Stintzi, A.; Legrand, M.; Fritig, B.; Mosinger, E. Pathogenesis-Related PR-1 Proteins Are Antifungal (Isolation and Characterization of Three 14-Kilodalton Proteins of Tomato and of a Basic PR-1 of Tobacco with Inhibitory Activity against Phytophthora infestans). *Plant Physiol.* **1995**, *108*, 17–27. [CrossRef] [PubMed]

43. Anklesaria, J.H.; Pandya, R.R.; Pathak, B.R.; Mahale, S.D. Purification and characterization of CRISP-3 from human seminal plasma and its real-Time binding kinetics with PSP94. *J. Chromatogr. B* **2016**, *1039*, 59–65. [CrossRef] [PubMed]

44. Crest, M.; Jacquet, G.; Gola, M.; Zerrouk, H.; Benslimane, A.; Rochat, H.; Mansuelle, P.; Martin-Eauclaire, M.F. Kaliotoxin, a novel peptidyl inhibitor of neuronal BK-type Ca(2+)-activated K+ channels characterized from Androctonus mauretanicus mauretanicus venom. *J. Biol. Chem.* **1992**, *267*, 1640–1647. [PubMed]

45. Madio, B.; King, G.F.; Undheim, E.A.B. Sea Anemone Toxins: A Structural Overview. *Mar. Drugs* **2019**, *17*, 325. [CrossRef] [PubMed]

46. Gupte, M.; Kulkarni, P.; Ganguli, B. Antifungal antibiotics. *Appl. Microbiol. Biotechnol.* **2002**, *58*, 46–57. [CrossRef]

47. Schneiter, R.; Di Pietro, A. The CAP protein superfamily: Function in sterol export and fungal virulence. *Biomol. Concepts* **2013**, *4*. [CrossRef]

48. Darwiche, R.; El Atab, O.; Cottier, S.; Schneiter, R. The function of yeast CAP family proteins in lipid export, mating, and pathogen defense. *FEBS Lett.* **2018**, *592*, 1304–1311. [CrossRef]

49. Suzuki, N.; Yamazaki, Y.; Brown, R.L.; Fujimoto, Z.; Morita, T.; Mizuno, H. Structures of pseudechetoxin and pseudecin, two snake-Venom cysteine-Rich secretory proteins that target cyclic nucleotide-Gated ion channels: Implications for movement of the C-terminal cysteine-Rich domain. *Acta Crystallogr. Sect. D Biol. Crystallogr.* **2008**, *64*, 1034–1042. [CrossRef]

50. Wang, Y.-L.; Kuo, J.-H.; Lee, S.-C.; Liu, J.-S.; Hsieh, Y.-C.; Shih, Y.-T.; Chen, C.-J.; Chiu, J.-J.; Wu, W.-G. Cobra CRISP Functions as an Inflammatory Modulator via a Novel Zn^{2+}- and Heparan Sulfate-Dependent Transcriptional Regulation of Endothelial Cell Adhesion Molecules. *J. Biol. Chem.* **2010**, *285*, 37872–37883. [CrossRef]

51. Xiao, R.; Li, Q.-W.; Perrett, S.; He, R.-Q. Characterisation of the fibrinogenolytic properties of the buccal gland secretion from Lampetra japonica. *Biochimie* **2007**, *89*, 383–392. [CrossRef]

52. Jiang, Q.; Liu, Y.; Duan, D.; Gou, M.; Wang, H.; Wang, J.; Li, Q.; Xiao, R. Anti-Angiogenic activities of CRBGP from buccal glands of lampreys (Lampetra japonica). *Biochimie* **2016**, *123*, 7–19. [CrossRef] [PubMed]

53. Chi, S.; Xiao, R.; Li, Q.; Zhou, L.; He, R.; Qi, Z. Suppression of neuronal excitability by the secretion of the lamprey (Lampetra japonica) provides a mechanism for its evolutionary stability. *Pflügers Arch.-Eur. J. Physiol.* **2009**, *458*, 537–545. [CrossRef] [PubMed]

54. Jiang, Q.; Liu, Y.; Gou, M.; Han, J.; Wang, J.; Li, Q.; Xiao, R. Data for the inhibition effects of recombinant lamprey CRBGP on the tube formation of HUVECs and new blood vessel generation in CAM models. *Data Brief* **2016**, *6*, 661–667. [CrossRef] [PubMed]

55. Xue, Z.; Bai, J.; Sun, J.; Wu, Y.; Yu, S.Y.; Guo, R.Y.; Liu, X.; Li, Q.W. Novel neutrophil inhibitory factor homologue in the buccal gland secretion of Lampetra japonica. *Biol. Chem.* **2011**, *392*. [CrossRef] [PubMed]

56. Xiao, R.; Pang, Y.; Li, Q.W. The buccal gland of Lampetra japonica is a source of diverse bioactive proteins. *Biochimie* **2012**, *94*, 1075–1079. [CrossRef] [PubMed]

57. Nobile, M.; Noceti, F.; Prestipino, G.; Possani, L. Helothermine, a lizard venom toxin, inhibits calcium current in cerebellar granules. *Exp. Brain Res.* **1996**, *110*. [CrossRef]

58. Nobile, M.; Magnelli, V.; Lagostena, L.; Mochca-Morales, J.; Possani, L.D.; Prestipino, G. The toxin helothermine affects potassium currents in newborn rat cerebellar granule cells. *J. Membr. Biol.* **1994**, *139*. [CrossRef]

59. Mochca-Morales, J.; Martin, B.M.; Possani, L.D. Isolation and characterization of Helothermine, a novel toxin from Heloderma horridum horridum (Mexican beaded lizard) venom. *Toxicon* **1990**, *28*, 299–309. [CrossRef]

60. Yamazaki, Y.; Brown, R.L.; Morita, T. Purification and Cloning of Toxins from Elapid Venoms that Target Cyclic Nucleotide-Gated Ion Channels. *Biochemistry* **2002**, *41*, 11331–11337. [CrossRef]

61. Yamazaki, Y.; Hyodo, F.; Morita, T. Wide distribution of cysteine-rich secretory proteins in snake venoms: Isolation and cloning of novel snake venom cysteine-rich secretory proteins. *Arch. Biochem. Biophys.* **2003**, *412*, 133–141. [CrossRef]

62. Guo, M.; Teng, M.; Niu, L.; Liu, Q.; Huang, Q.; Hao, Q. Crystal Structure of the Cysteine-Rich Secretory Protein Stecrisp Reveals That the Cysteine-Rich Domain Has a K^+ Channel Inhibitor-Like Fold. *J. Biol. Chem.* **2005**, *280*, 12405–12412. [CrossRef] [PubMed]

63. Lodovicho, M.E.; Costa, T.R.; Bernardes, C.P.; Menaldo, D.L.; Zoccal, K.F.; Carone, S.E.; Rosa, J.C.; Pucca, M.B.; Cerni, F.A.; Arantes, E.C.; et al. Investigating possible biological targets of Bj-CRP, the first cysteine-Rich secretory protein (CRISP) isolated from Bothrops jararaca snake venom. *Toxicol. Lett.* **2017**, *265*, 156–169. [CrossRef] [PubMed]

64. Suntravat, M.; Cromer, W.E.; Marquez, J.; Galan, J.A.; Zawieja, D.C.; Davies, P.; Salazar, E.; Sánchez, E.E. The isolation and characterization of a new snake venom cysteine-Rich secretory protein (svCRiSP) from the venom of the Southern Pacific rattlesnake and its effect on vascular permeability. *Toxicon* **2019**, *165*, 22–30. [CrossRef] [PubMed]

65. Lecht, S.; Chiaverelli, R.A.; Gerstenhaber, J.; Calvete, J.J.; Lazarovici, P.; Casewell, N.R.; Harrison, R.; Lelkes, P.I.; Marcinkiewicz, C. Anti-Angiogenic activities of snake venom CRISP isolated from Echis carinatus sochureki. *Biochim. Biophys. Acta BBA Gen. Subj.* **2015**, *1850*, 1169–1179. [CrossRef]

66. Adade, C.M.; Carvalho, A.L.O.; Tomaz, M.A.; Costa, T.F.R.; Godinho, J.L.; Melo, P.A.; Lima, A.P.C.A.; Rodrigues, J.C.F.; Zingali, R.B.; Souto-Padrón, T. Crovirin, a Snake Venom Cysteine-Rich Secretory Protein (CRISP) with Promising Activity against Trypanosomes and Leishmania. *PLoS Negl. Trop. Dis.* **2014**, *8*, e3252. [CrossRef]

67. Brown, R.L.; Haley, T.L.; West, K.A.; Crabb, J.W. Pseudechetoxin: A peptide blocker of cyclic nucleotide-Gated ion channels. *Proc. Natl. Acad. Sci. USA* **1999**, *96*, 754–759. [CrossRef]

68. Wang, J.; Shen, B.; Guo, M.; Lou, X.; Duan, Y.; Cheng, X.P.; Teng, M.; Niu, L.; Liu, Q.; Huang, Q.; et al. Blocking Effect and Crystal Structure of Natrin Toxin, a Cysteine-Rich Secretory Protein from *Naja atra* Venom that Targets the BK $_{Ca}$ Channel. *Biochemistry* **2005**, *44*, 10145–10152. [CrossRef]

69. Zhou, Q.; Wang, Q.-L.; Meng, X.; Shu, Y.; Jiang, T.; Wagenknecht, T.; Yin, C.-C.; Sui, S.-F.; Liu, Z. Structural and Functional Characterization of Ryanodine Receptor-Natrin Toxin Interaction. *Biophys. J.* **2008**, *95*, 4289–4299. [CrossRef]

70. Wang, F.; Li, H.; Liu, M.-N.; Song, H.; Han, H.-M.; Wang, Q.-L.; Yin, C.-C.; Zhou, Y.-C.; Qi, Z.; Shu, Y.-Y.; et al. Structural and functional analysis of natrin, a venom protein that targets various ion channels. *Biochem. Biophys. Res. Commun.* **2006**, *351*, 443–448. [CrossRef]

71. Peichoto, M.E.; Mackessy, S.P.; Teibler, P.; Tavares, F.L.; Burckhardt, P.L.; Breno, M.C.; Acosta, O.; Santoro, M.L. Purification and characterization of a cysteine-Rich secretory protein from *Philodryas patagoniensis* snake venom. *Comp. Biochem. Physiol. Part C Toxicol. Pharmacol.* **2009**, *150*, 79–84. [CrossRef]

72. Magdaleno, L.; Gasset, M.A.; Varea, J.; Schambony, A.M.; Urbanke, C.; Raida, M.; Töpfer-Petersen, E.; Calvete, J.J. Biochemical and conformational characterisation of HSP-3, a stallion seminal plasma protein of the cysteine-Rich secretory protein (CRISP) family. *FEBS Lett.* **1997**, *420*, 179–185. [CrossRef]

73. Kosari, F.; Asmann, Y.W.; Cheville, J.C.; Vasmatzis, G. Cysteine-Rich secretory protein-3: A potential biomarker for prostate cancer. *Cancer Epidemiol. Prev. Biomark.* **2002**, *11*, 1419–1426.

74. Udby, L.; Lundwall, Å.; Johnsen, A.H.; Fernlund, P.; Valtonen-André, C.; Blom, A.M.; Lilja, H.; Borregaard, N.; Kjeldsen, L.; Bjartell, A. β-Microseminoprotein binds CRISP-3 in human seminal plasma. *Biochem. Biophys. Res. Commun.* **2005**, *333*, 555–561. [CrossRef] [PubMed]

75. Udby, L.; Sørensen, O.E.; Pass, J.; Johnsen, A.H.; Behrendt, N.; Borregaard, N.; Kjeldsen, L. Cysteine-Rich Secretory Protein 3 Is a Ligand of α_1B-Glycoprotein in Human Plasma [†]. *Biochemistry* **2004**, *43*, 12877–12886. [CrossRef]

76. Reeves, J.R.; Xuan, J.W.; Arfanis, K.; Morin, C.; Garde, S.V.; Ruiz, M.T.; Wisniewski, J.; Panchal, C.; Tanner, J.E. Identification, purification and characterization of a novel human blood protein with binding affinity for prostate secretory protein of 94 amino acids. *Biochem. J.* **2005**, *385*, 105–114. [CrossRef]

77. Anklesaria, J.H.; Jagtap, D.D.; Pathak, B.R.; Kadam, K.M.; Joseph, S.; Mahale, S.D. Prostate Secretory Protein of 94 Amino Acids (PSP94) Binds to Prostatic Acid Phosphatase (PAP) in Human Seminal Plasma. *PLoS ONE* **2013**, *8*, e58631. [CrossRef]

78. Ghasriani, H.; Fernlund, P.; Udby, L.; Drakenberg, T. A model of the complex between human β-Microseminoprotein and CRISP-3 based on NMR data. *Biochem. Biophys. Res. Commun.* **2009**, *378*, 235–239. [CrossRef]

79. Breed, A.A.; Gomes, A.; Roy, B.S.; Mahale, S.D.; Pathak, B.R. Mapping of the binding sites involved in PSP94–CRISP-3 interaction by molecular dissection of the complex. *Biochim. Biophys. Acta BBA Gen. Subj.* **2013**, *1830*, 3019–3029. [CrossRef]

80. Shioi, N.; Tadokoro, T.; Shioi, S.; Okabe, Y.; Matsubara, H.; Kita, S.; Ose, T.; Kuroki, K.; Terada, S.; Maenaka, K. Crystal structure of the complex between venom toxin and serum inhibitor from Viperidae snake. *J. Biol. Chem.* **2019**, *294*, 1250–1256. [CrossRef]

81. Aoki, N.; Sakiyama, A.; Deshimaru, M.; Terada, S. Identification of novel serum proteins in a Japanese viper: Homologs of mammalian PSP94. *Biochem. Biophys. Res. Commun.* **2007**, *359*, 330–334. [CrossRef]

82. Hansson, K.; Kjellberg, M.; Fernlund, P. Cysteine-Rich secretory proteins in snake venoms form high affinity complexes with human and porcine β-Microseminoproteins. *Toxicon* **2009**, *54*, 128–137. [CrossRef] [PubMed]

83. Matsunaga, Y.; Yamazaki, Y.; Hyodo, F.; Sugiyama, Y.; Nozaki, M.; Morita, T. Structural Divergence of Cysteine-Rich Secretory Proteins in Snake Venoms. *J. Biochem.* **2009**, *145*, 365–375. [CrossRef] [PubMed]

84. Jin, Y.; Lu, Q.; Zhou, X.; Zhu, S.; Li, R.; Wang, W.; Xiong, Y. Purification and cloning of cysteine-Rich proteins from Trimeresurus jerdonii and Naja atra venoms. *Toxicon* **2003**, *42*, 539–547. [CrossRef]

85. Aoki, N.; Sakiyama, A.; Kuroki, K.; Maenaka, K.; Kohda, D.; Deshimaru, M.; Terada, S. Serotriflin, a CRISP family protein with binding affinity for small serum protein-2 in snake serum. *Biochim. Biophys. Acta BBA Proteins Proteom.* **2008**, *1784*, 621–628. [CrossRef] [PubMed]

86. Del Tordello, E.; Vacca, I.; Ram, S.; Rappuoli, R.; Serruto, D. Neisseria meningitidis NalP cleaves human complement C3, facilitating degradation of C3b and survival in human serum. *Proc. Natl. Acad. Sci. USA* **2014**, *111*, 427–432. [CrossRef] [PubMed]

87. Sunagar, K.; Johnson, W.E.; O'Brien, S.J.; Vasconcelos, V.; Antunes, A. Evolution of CRISPs Associated with Toxicoferan-Reptilian Venom and Mammalian Reproduction. *Mol. Biol. Evol.* **2012**, *29*, 1807–1822. [CrossRef] [PubMed]

88. Manceau, M.; Marin, J.; Morlon, H.; Lambert, A. Model-Based Inference of Punctuated Molecular Evolution. *BioRxiv* **2019**. [CrossRef]

89. Perry, B.W.; Card, D.C.; McGlothlin, J.W.; Pasquesi, G.I.M.; Adams, R.H.; Schield, D.R.; Hales, N.R.; Corbin, A.B.; Demuth, J.P.; Hoffmann, F.G.; et al. Molecular Adaptations for Sensing and Securing Prey and Insight into Amniote Genome Diversity from the Garter Snake Genome. *Genome Biol. Evol.* **2018**, *10*, 2110–2129. [CrossRef]

90. Suryamohan, K.; Krishnankutty, S.P.; Guillory, J.; Jevit, M.; Schröder, M.S.; Wu, M.; Kuriakose, B.; Mathew, O.K.; Perumal, R.C.; Koludarov, I.; et al. The Indian cobra reference genome and transcriptome enables comprehensive identification of venom toxins. *Nat. Genet.* **2020**, *52*, 106–117. [CrossRef]

91. Schield, D.R.; Card, D.C.; Hales, N.R.; Perry, B.W.; Pasquesi, G.M.; Blackmon, H.; Adams, R.H.; Corbin, A.B.; Smith, C.F.; Ramesh, B.; et al. The origins and evolution of chromosomes, dosage compensation, and mechanisms underlying venom regulation in snakes. *Genome Res.* **2019**, *29*, 590–601. [CrossRef]

92. Shibata, H.; Chijiwa, T.; Oda-Ueda, N.; Nakamura, H.; Yamaguchi, K.; Hattori, S.; Matsubara, K.; Matsuda, Y.; Yamashita, A.; Isomoto, A.; et al. The habu genome reveals accelerated evolution of venom protein genes. *Sci. Rep.* **2018**, *8*, 11300. [CrossRef] [PubMed]

 *toxins*

Review

Hitchhiking with Nature: Snake Venom Peptides to Fight Cancer and Superbugs

Clara Pérez-Peinado, Sira Defaus and David Andreu *

Department of Experimental and Health Science, Universitat Pompeu Fabra, Barcelona Biomedical Research Park, 08003 Barcelona, Spain; clara.perez@upf.edu (C.P.-P.); sira.defaus@upf.edu (S.D.)
* Correspondence: david.andreu@upf.edu; Tel.: +34-9-3316-0868

Received: 28 February 2020; Accepted: 9 April 2020; Published: 15 April 2020

Abstract: For decades, natural products in general and snake venoms (SV) in particular have been a rich source of bioactive compounds for drug discovery, and they remain a promising substrate for therapeutic development. Currently, a handful of SV-based drugs for diagnosis and treatment of various cardiovascular disorders and blood abnormalities are on the market. Likewise, far more SV compounds and their mimetics are under investigation today for diverse therapeutic applications, including antibiotic-resistant bacteria and cancer. In this review, we analyze the state of the art regarding SV-derived compounds with therapeutic potential, focusing on the development of antimicrobial and anticancer drugs. Specifically, information about SV peptides experimentally validated or predicted to act as antimicrobial and anticancer peptides (AMPs and ACPs, respectively) has been collected and analyzed. Their principal activities both in vitro and in vivo, structures, mechanisms of action, and attempts at sequence optimization are discussed in order to highlight their potential as drug leads.

Keywords: snake venoms; antimicrobial peptides; anticancer peptides; cathelicidin; defensin; crotamine; snake venom proteins; snake venom peptides

---

1. Introduction

Snakes are arguably among if not the most despised creatures in the entire animal kingdom. With some exceptions (serpents protected and revered, even worshipped, in ancient and some contemporary societies), for the vast majority of people, snakes epitomize harm, evil, and treachery. Examples are found in scriptures and ancient tradition (the serpent of Eden, the popular belief of Cleopatra's asp-assisted suicide), classic literature (e.g., Poe's *The cask of amontillado*, Doyle's *The Speckled Band*) or colloquial language ("he's a snake in the grass"). Beside this general perception of snakes incarnating terror and deceit, the ancient view of the serpent as a dual archetype of good and evil (e.g., Moses' brass serpent, the rod of Asclepius, Hindu mythology) has not only survived, but paradoxically, become increasingly supported by evidence that, when properly used, snake venom (SV) compounds can actually help save lives rather than ending them [1].

SVs are cocktails of toxins refined over millions of years of evolution, sometimes encoding several powerful biological effects in a single component. SVs should, therefore, be considered as rich libraries of bioactive molecules with potential to treat human disorders. A pioneering example was the development of Captopril® from a peptide discovered in the venom of the Brazilian pit viper *Bothrops jararaca* [2,3], and subsequent development of angiotensin-converting enzyme (ACE) inhibitors that have had a significant impact in hypertension control since the 1980s. Similarly, leads for other SV-derived drugs are mostly peptides and enzymes (Table 1). To date, most SV-based medicines have been approved for cardiovascular maladies or as diagnostic tools for blood-related abnormalities (Table 1). On the other hand, new uses currently being investigated in clinical and preclinical trials focus on other conditions, such as sclerosis, chronic pain, infections, or cancer (Table 1), thus revealing the potential of snake toxins to serve as molecular scaffolds in diverse therapeutic fields.

Here we focus on the potential relevance of SV compounds to two major therapeutic challenges of the 21st century: multi-drug-resistant bacteria and cancer. First, we are frightfully close to running out of effective antibiotics in the war against pathogenic bacteria, due to factors such as over-prescription, non-compliance, veterinary use, etc., fueling the emergence of superbugs [4,5] Recent forecasts of human costs associated with antimicrobial resistance are sobering: by 2050, the number of deaths attributed to antimicrobial resistance will increase from the current 700,000 a year to 10 million, becoming the main cause of death [6,7]. Second, new cancer therapies are urgently needed to complement and/or replace current chemotherapeutic drugs with their numerous contraindications. Cancer is now the second leading cause of death worldwide, accounting for one out of six deaths [8]. Annual statistics from the Global Cancer Observatory reported 18 million new cases and 9.6 million deaths in 2018, with the annual incidence expected to rise to 29.5 million cases and 16.4 million deaths globally by 2040 [9–11]. In light of the foregoing, the present work reviews SV-derived antimicrobial and anticancer peptides (SV-AMPs, and SV-ACPs, respectively), examining their biological activities, selectivity, structures, mechanisms of action, and overall therapeutic potential.

Table 1. Snake venom-derived products (drugs and diagnostic tools) currently on the market or in clinical and pre-clinical trials. Information was extracted from literature and updated on the corresponding company website and/or the U.S. National Library of Medicine (https://clinicaltrials.gov/ct2/home). *Withdrawn.

Commercial Drugs [12–17]			
Drug (Commercial Name)	**Source Organism**	**Target**	**Therapeutic Use**
Captopril (Capoten®)	*Bothrops jararaca*	Angiotensin-converting enzyme (ACE)	Hypertension
Enalapril (Vasotec®)	*Bothrops jararaca*	ACE	Hypertension
Tirofiban (Aggrastat®)	*Echis carinatus*	Glycoprotein IIb/IIIa	Acute coronary syndromes
Eptifibatide (Integrilin®)	*Sistrurus miliarius barbouri*	Glycoprotein IIb/IIIa	Acute coronary syndromes
Batroxobin (Defibrase®)	*Bothrops* sp.	Fibrinogen	Infarction / Ischemia / Microcirculation dysfunctions
Platelet gel (Plateltex-Act®)	*Bothrops atrox*	Fibrinogen	Platelet-induced tissue-healing
Fibrin sealant (Vivostat®)	*Bothrops moojeni*	Fibrinogen	Autologous fibrin sealant in surgery
Haemocoagulase (Reptilase®)	*Bothrops atrox*	Fibrinogen Factor X / Prothrombin	Hemorrhage
Ximelagatran (Exanta®)*	Cobra venom	Thrombin	Atrial fibrillation / Blood clotting
Ancrod (Viprinex®)*	*Agkistrodon rhodostoma*	Fibrinogen	Heparin-induced thrombocytopenia

Drugs in Clinical and Pre-Clinical Trials [12–17]				
Drug Name	**Phase**	**Source Organism**	**Target**	**Therapeutic Use**
Fibrolase (Alfimeprase)*	III	*Agkistrodon contortrix*	Fibrinogen	Stroke and catheter occlusion
Crotoxin	I	*Crotalus durissus terrificus*	Unknown	Cancer
Cenderitide	II	*Dendroaspis angusticeps*	Natriuretic peptide receptor	Heart failure
RPI-MN	I	*Naja atra*	Nicotinic acetylcholine receptor	HIV / Amyotrophic lateral sclerosis / Herpes simplex keratitis
RPI-78M	I/II	*Naja atra*	Nicotinic acetylcholine receptor	Multiple sclerosis / Herpes simplex infections / Adrenomyeloneuropathy
RPI-78	preclinical	*Naja atra*	Nicotinic acetylcholine receptor	Pain / Rheumatoid arthritis
Prohanin	preclinical	*Ophiophagus hannah*	Nitric oxide synthase	Chronic pain
Oxynor	preclinical	*Oxyuranus scutellatus*	Unknown	Wound healing
Natriuretic peptides	preclinical	*Oxyuranus microlepidotus*	Natriuretic peptide receptor	Heart failure
Textilinin-1™	preclinical	*Pseudonaja textilis*	Plasmin	Preoperative bleeding
Vicrostatin	preclinical	Chimeric; *Echis carinatus / Agkistrodon contortrix*	Integrin receptor	Cancer
Haempatch™	preclinical	*Pseudonaja textilis*	Prothrombin	Blood loss during vascular trauma
CoVase™	preclinical	*Pseudonaja textilis*	Factor Xa	Hemorrhage
Contortrostatin	preclinical	*Agkistrodon contortrix*	Integrin	Breast cancer

Table 1. *Cont.*

Diagnostic Tools [12–17]			
Product	**Source Organism**	**Target**	**Clinical Assessment of:**
Protac®	*Agkistrodon contortix*	Protein C activation	Protein C
Reptilase®	*Bothrops jararaca*	Fibrinogen	Fibrinogen
Ecarin clotting time	*Echis carinatus*	Prothrombin	Meizothrombin
Textarin®/Ecarin ratio	*Pseudonaja textilis*	Prothrombin	Lupus anticoagulant
Russell's viper venom-factor X	*Daboia russelii*	Factor X	Factor X
Dilute Russell's Viper Venom Time	*Daboia russelii*	Factor X, Factor V	Lupus anticoagulant
Taipan Venom Time	*Oxyuranus scutellatus*	Prothrombin	Lupus anticoagulant
Pefakit® APCR Factor V Leiden	*Daboia russelii / Notechis scutatus scutatus*	Factor V / Protein C / Prothrombin	Resistance to activated protein C
Botrocetin®	*Bothrops sp*	Factor VIIIa	von Willebrand Factor

2. Antibacterial and Antitumoral Activity of Snake Venoms (SVs)

Envenomation is associated with a remarkably low incidence of microbial infections [18]. This is paradoxical if one considers that snakebites are puncture wounds. Thus, it is reasonable to suspect that SVs contain antibacterial agents [19]. Indeed, whole crotalid venoms possess antimicrobial activity against bacteria commonly found in snake oral cavities, such as *Pseudomonas aeruginosa* (minimal inhibitory concentration (MIC) = 80–160 µg/mL) and *Alcaligenes faecalis* (MIC = 5–20 µg/mL), and also toward human pathogens such as *Staphylococcus aureus* (MIC = 5–40 µg/mL) and *Escherichia coli* (MIC = 80–160 µg/mL) [20]. Additional studies demonstrated similar effects of SVs from a large variety of snakes, including viperids (*Agkistrodon rhodostoma, Bothrops atrox, B. jararaca, Bothrops alternatus* and *Daboia russellii russellii*, MICs <20 µg/mL) and elapids (*Pseudechis australis*, MIC = 40 µg/mL) [2,21–23].

In the late 1980s, elapid, crotalid, and viperid venoms were found to act against melanoma and chondrosarcoma cell lines [24]. Venoms of *B. jararaca* and *Crotalus durissus terrificus* also exhibited antitumoral properties, presumably by direct action on tumor cells and by modulating inflammatory responses [25,26]. *Ophiophagus hannah* venom demonstrated strong anti-cancer properties in vitro against pancreatic tumor cells, as well as anti-angiogenic activity in vivo [27].

Due to the therapeutic potential of SVs, efforts were focused on isolating and characterizing active compounds responsible for antibacterial and antitumoral activities. Examples included enzymes and proteins, the principal functions of which are not directly related to microbial or tumor clearance, but which nevertheless demonstrate antimicrobial, antifungal, antitumoral, and immunomodulatory properties. This is the case of L-amino acid oxidase (LAAO) and phospholipases A$_2$s (PLA$_2$s) [28,29]. Other SV components displaying various activities include metalloproteinases [30–32], cardio-, neuro- or myotoxins [33–36], disintegrins [37] and lectins [38–40], among others [41–43]. These compounds exhibit various mechanisms of action, including direct toxic action (PLA$_2$s), free radical generation (LAAO), induction of apoptosis (PLA$_2$s, LAAO and metalloproteinases), and anti-angiogenesis (disintegrins and lectins) [41]. SV-derived cationic AMPs and ACPs complete this arsenal of anti-infective and antitumoral components [44,45].

2.1. Snake Venom-Derived Antimicrobial and Anticancer Peptides (SV-AMPs and ACPs)

Cationic peptides, including AMPs and ACPs, encompass a sequence-diverse family of peptides characterized by a net positive charge and a high content of hydrophobic residues [46]. The initial function proposed for cationic peptides was to act as AMPs against a broad spectrum of Gram-positive and Gram-negative bacteria, fungi, and parasites [47,48]. Subsequent studies have demonstrated antiviral and anti-biofilm properties and anticancer and immune modulatory activities [49,50]. Due to their ability to translocate across lipid membranes, cationic peptides have also been exploited as delivery vectors (Figure 1).

In particular, AMPs and ACPs contribute to microbe/tumor cell clearance by three complementary means: (i) direct membrane disruption; (ii) interference with key intracellular processes such as nucleic acid and protein synthesis; (iii) immune-cell function recruitment or activation via a broad array of functions, with the ultimate goal of clearing pathogens or tumor cells [51–56]. In addition, anti-angiogenesis [57,58] and suppression of metastasis [59] also contribute to tumor control by ACPs. Naturally occurring cationic peptides are present in every kingdom and phylum, including plants [60], animals [61], fungi [62] and bacteria [63], and they have also been identified in SV.

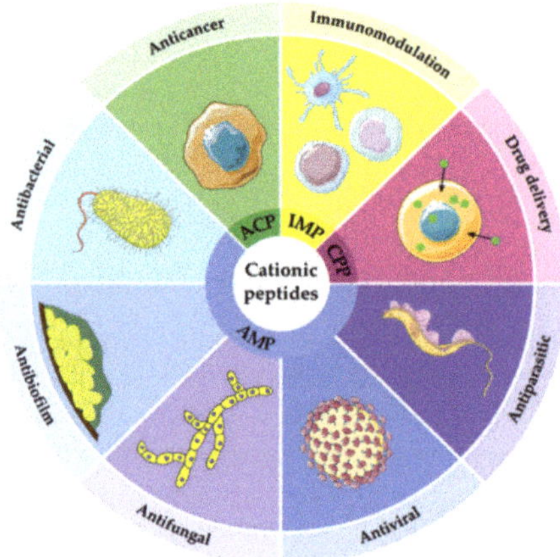

Figure 1. Principal functions of cationic peptides. ACP, anticancer peptides; AMP, antimicrobial peptides; CPP, cell-penetrating peptides; IMP, immunomodulatory peptides. This figure was prepared using the image repository Smart Servier Medical Art (available at: https://smart.servier.com).

2.1.1. SV-Cathelicidins (SV-CATHs)

Possibly the largest family of SV-AMPs and -ACPs described to date are the cathelicidins (CATHs), a group of structurally diverse bioactive peptides with antimicrobial, anticancer, and immunomodulatory functions, acting as effector molecules of the innate immune system [64,65]. Members of the CATH family possess highly homologous pre- and pro-regions comprising the N-terminal signal peptide and the cathelin (cathepsin L inhibitor)-like domain [66,67] (Figure 2a). In contrast, the C-terminal domain, which encodes the mature bioactive peptide, is diverse in amino acid (aa) sequence and higher structure [66,67].

While most mature CATHs are linear, 25–35-residue, amphipathic α-helical peptides, some family members (protegrins) are smaller, 12–18-residue peptides, displaying β-hairpin structures stabilized by disulfide bonds. Others consist of sequences enriched in specific aa, such as Trp-rich indolicidin [68]. Despite this conformational and compositional variability, most CATHs share certain physicochemical properties, including a generally positive charge and amphipathicity.

CATHs have been found in humans (LL-37) and other mammals [69–74], and in fish, birds, or reptiles [75–77]. SV-derived CATHs (SV-CATHs) were first identified by Zhao et al. in 2008 from elapid venom gland cDNA libraries [76]. To date, 25 SV-CATHs have been identified, as well as fragments derived from parental SV-CATHs (see Table 2 for a summary of mature SV-CATHs and their principal activities, and Table S1 for SV-CATH precursors).

Overall, SV-CATH precursors range from 184 to 194 aa, with some exceptions (Table S1). The signal peptide domain comprises the ~22 N-terminal aa residues, followed by 65–66 aa residues forming the cathelin domain. The 30–34 C-terminal aa residues encode the mature peptide, sometimes preceded by a Glu-rich domain of 9 to 29 aa, responsible for inactivation of the mature peptide, by inducing conformational changes [78]. CATHs from *Python bivittatus*, *Thamnophis sirtalis*, and *Protobothrops mucrosquamatus* exhibit exceptions to the aforementioned domain features, presenting different lengths or even being absent in some cases (Table S1).

Snake-derived CATH precursors generally display >50% sequence identity, excluding four of the six CATHs predicted from the boid, *P. bivittatus* (Pb-CATH 2, Pb-CATH 4, Pb-CATH 5 and Pb-CATH 6), and one predicted from the colubrid, *T. sirtalis* (Ts-CATH 4) (Figure S1). Sequence analysis (Figure 2b) reveals conserved residues in the N-terminal part of the signal domain, as well as most of the cathelin domain and in the C-terminal end encoding the mature active peptide. Table 2 summarizes the principal biological activities of mature SV-CATHs and Table 3 compares experimentally validated SV-CATHs in terms of biological activity, hemolysis, and selectivity for microbial/tumor cells, rather than healthy eukaryotic cells.

- **Oh-CATH:** identified in the venom gland of the king cobra (*O. hannah*), this was the first predicted SV-CATH to be synthesized and experimentally validated as an AMP [76] (Table 2). Oh-CATH exerts strong salt-resistant, antibacterial activity against Gram-positive and Gram-negative bacteria (MICs in the 1–20 μg/mL range) with weak hemolysis (~10% hemolysis observed at 200 μg/mL) [76,79]. It is apparently membrane-active and is an inhibitor of ATP-synthase [80,81]. A collection of analogs designed by Zhang et al., to study the structure-function relationships of Oh-CATH, suggested that the four N-terminal aa residues are responsible for cytotoxicity toward eukaryotic cells, while the C-terminal 10 strongly influence antimicrobial activity [79]. Accordingly, OH-CATH30, the most promising analog, lacking the four N-terminal aa, also named OH-CATH(5-34), was further characterized and optimized. Then, OH-CATH30 was tested against a panel of 584 clinical isolates of 14 different species, showing antibacterial activity against 85% of them and overall higher efficacy against Gram-positive strains [82].

Li et al. successfully downsized OH-CATH30 by removing the 10 C-terminal aa and by including single-aa mutations, giving rise to OH-CM6, which displays almost identical activity against a panel of Gram-positive and Gram-negative bacteria [81]. Although OH-CATH30 and OH-CM6 displayed potent antibacterial activity (MICs from 1.56 to 12.5 μg/mL against clinical isolates from *E. coli*, *P. aeruginosa* and methicillin-resistant *S. aureus* (MRSA)), both were inactive against *Candida albicans* strains (MICs >200 μg/mL) [81]. L-forms of both peptides conserve some antimicrobial activity in the presence of 25% serum, probably due to the membrane-targeting mechanism before enzymatic degradation occurs, but pre-incubation in 100% serum results in total activity loss after 4 h. In contrast, activity of their D-amino acid forms remained unchanged even after 12 h pre-incubation in 100% serum [81]. In vivo studies carried out by the same group revealed that intraperitoneal (*i.p.*) LD_{50}s of OH-CATH30 and OH-CM6 in mice were 120 and 100 μg/g, respectively. Both peptides were able to rescue infected mice in a bacteremia model induced by drug-resistant *E. coli* at 10 μg/g as well as decreased TNF-α production in a mouse model of neutropenic thigh infection [81].

Table 2. Mature snake cathelicidins (CATHs) identified to date and their properties. Information was extracted from literature or the National Center for Biotechnology Information (NCBI). A unified name was given to each CATH, according to the binomial initials of the snake. Biological and hemolytic activities of experimentally validated CATHs. Hemolytic activity denotes causing 10% hemolysis (low: HC_{10} >50 µg/mL; medium, 50 > HC_{10} > 10 µg/mL; high, HC_{10} <10 µg/mL) n.d.: non detectable activity.

Source Organism	Unified Name	Common name	Mature Peptide Sequence	Length	Hemolysis	Activity	Ref
Ophiophagus hannah	Oh-CATH	KF-34	KRFKKFFKKLKNSVKKRAKKFFKKPRVIGVSIPF	34	Medium	G+ and G- bacteria.	[76,82–84]
Bungarus fasciatus	Bf-CATH	Cath-BF	KRFKKFFRKLKKSVKKRAKEFFKKPRVIGVSIPF	34	High	G+ and G- bacteria.	[85,86]
	Bf-CATH30	BF-30, cathelicidin-BF, C-BF, cathelicidin-WA, CWA	KFFRKLKKSVKKRAKEFFKKPRVIGVSIPF	30	High	G+ and G- bacteria, fungi and tumor cells. Anti-inflamatory. Activation of innate immunity.	[87–97]
Naja atra	Na-CATH	-	KRFKKFFKKLKNSVKKRAKKFFKKPKVIGVTFPF	34	Low	G+ and G- bacteria.	[98–104]
Hydrophis cyanocinctus	Hc-CATH	-	KFFKRLLKSVRRAVKKFRKKPRLIGLSTLL	30	Low	G+ and G- bacteria and fungi. Anti-inflamatory. Inactive against tumor cells.	[105,106]
Crotalus durissus terrificus	Cdt-CATH	crotalicidin, Ctn	KRFKKFFKKVKKSVKKRLKKIFKKPMVIGVTIPF	34	High	G+ and G- bacteria, fungi, parasites and tumor cells. Overall proinflammatory.	[84,107–110]
Bothrops atrox	Ba-CATH	batroxicidin, BatxC	KRFKKFFKKLKNSVKKRVKKFFRKPRVIGVTFPF	34	High	G+ and G- bacteria and parasites. Overall proinflammatory.	[84,110,111]
Pseudonaja textilis	Pt-CATH1	Pt-CRAMP1	KRFKKFFMKLKKSVKKRVMKFFKKPMVIGVTFPF	34	High	G+ and G- bacteria.	[84]
	Pt-CATH2	Pt-CRAMP2	KRFKKFFRKLKKSVKKRVKKFFKKPRVIGVTIPF	34	n.d.	n.d.	[84]
Lachesis muta rhombeata	Lmr-CATH	lachesicidin	KRFKKFFKKVKKSVKKRLKKIFKKPMVIGVTFPF	34	n.d.	n.d.	[84]
Bothrops lutzi	Bl-CATH	lutzicidin	KRFKKFFKKLKNNVKKRVKKFFRKPRVIGVTIPF	34	n.d.	n.d.	[84]

Table 2. *Cont.*

Source Organism	Unified Name	Common name	Mature Peptide Sequence	Length	Hemolysis	Activity	Ref
Python bivittatus	Pb-CATH1	CATHPb1	KRFKKFFRKIKKGFRKIFKKTKIFIGGTIPI	31	Low	G+ and G- bacteria and fungi. Chemotactic. Anti-inflammatory.	[112]
	ΔPb-CATH1	ΔPb-CATH1	RVKRFKKFFRKIKKGFRKIFKKTKIFIG	28	Medium	G+ and G- bacteria.	[113]
	Pb-CATH2	CATHPb2	KRNGFRKFMRRLKKFFAGGGSSIAHIKLH	29	Low	G+ and G- bacteria and fungi. Chemotactic. Weakly anti-inflammatory.	[112]
	ΔPb-CATH2	Pb-CATH3	HRVKRNGFRKFMRRLKKFFAGG	22	Medium or low	G+ and G- bacteria.	[113]
	Pb-CATH3	CATHPb3	KRFQNFFRELEKKFREFFRVYRITIGATIRF	31	Low	Inactive against G+ and G- bacteria and fungi. Immunomodulatory inactive.	[112]
	Pb-CATH4	CATHPb4	TRSRWRRFIRGAGRFARRYGWRIALGLVG	29	Medium or high	G+ and G- bacteria and fungi. Weakly anti-inflammatory.	[112]
	ΔPb-CATH4	ΔPb-CATH4	TRSRWRRFIRGAGRFARRYGWRIA	24	Medium	G+ and G- bacteria and tumor cells.	[113]
	Pb-CATH5	CATHPb5	SPPQAMGFPPQVNVEHYIPASYSVAALTVTEEE	33	Low	Inactive against G+ and G- bacteria and fungi. Immunomodulatory inactive.	[112]
	Pb-CATH6	CATHPb6	RAAPQRRLRAMARLKKFAEAGGADPDSGGLR ARFPER	37	Low	Inactive against G+ and G- bacteria and fungi. Weakly anti-inflammatory.	[112]
Sinonatrix annularis	Sa-CATH	-	KFFKKLKKSVKKHVKKFFKKPKVIGVSIPF	30	Low	G+ and G- bacteria and fungi. Anti-inflamatory.	[114]
Crotalus durissus cascavella	Cdc-CATH1	Cas-CATH isoform 1	KRFKKFFKKVKKSVKKRLKKIFKKPIFKKVKKSVK KRLKKIFKKPMVIGVTIPF	54	n.d.	n.d.	NCBI
	Cdc-CATH2	Cas-CATH isoform 2	KRFKKFFKKVKKSVKKRLKKIFKKPMVIGVSIPF	34	n.d.	n.d.	NCBI
	Cdc-CATH3	Cas-CATH isoform 3	KRFKKFFKKVKKSVKKRLKKIFKKPMVIGVTIPF	34	n.d.	n.d.	NCBI
Thamnophis sirtalis	Ts-CATH1	-	KRFKKFFKKIKKSVKKRVKKLFKKPRVIPISIPF	34	n.d.	n.d.	NCBI
	Ts-CATH3	-	KKRRRIRVQITVKITFKI	18	n.d.	n.d.	NCBI
	Ts-CATH4	-	KKGLKKLFKRKKVVAGYVTA	20	n.d.	n.d.	NCBI
Protobothrops mucrosquamatus	Pm-CATH	-	KRFAGFFQFVVGVSFRF	17	n.d.	n.d.	NCBI

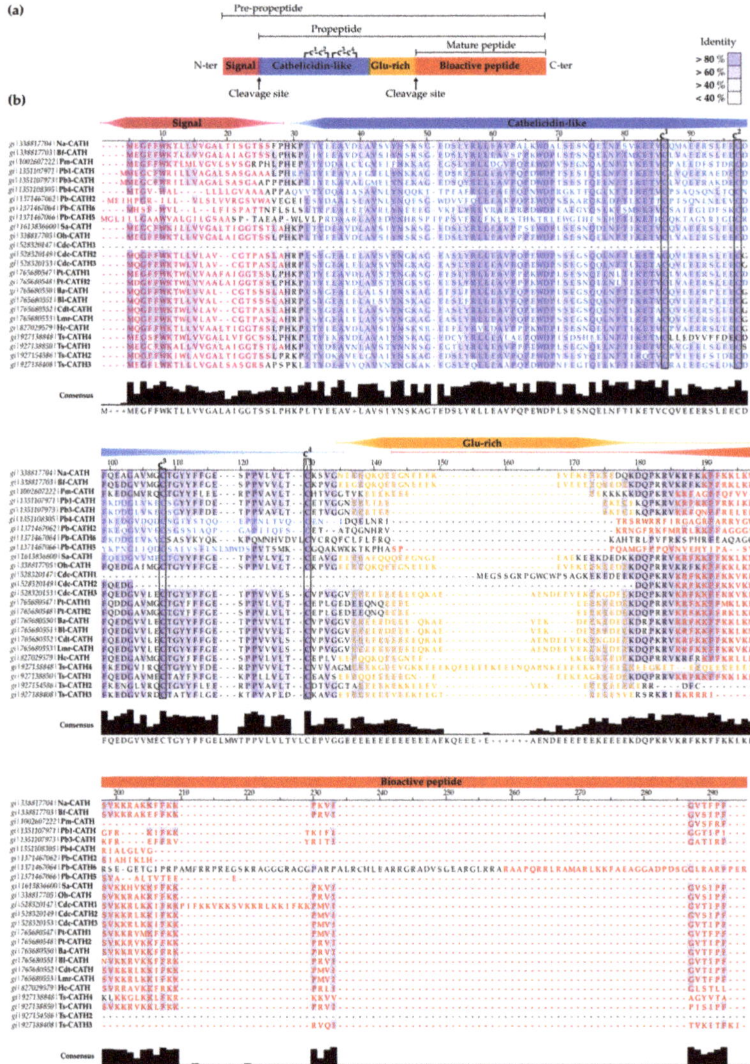

Figure 2. Snake-derived CATHs. (**a**) Schematic representation of CATH precursor structure. Domains are depicted in different colors and cleavage sites are highlighted. The Cys-pairing pattern is also annotated. (**b**). Alignment of CATH precursor sequences. Extended sequence information is available in Table S1. This multiple sequence alignment was performed using Clustal Omega [115] and represented with Jalview v2.11.0 software [116]. Residues were colored to match domains displayed in (**a**). Domain annotation was done using UniProt [117], National Center for Biotechnology Information (NCBI, https://www.ncbi.nlm.nih.gov/protein) or by homology (Table S1). The background is colored according to the percentage of residues in agreement with the consensus sequence appearing at the bottom.

- **Na-CATH:** this CATH from the venom gland of the Chinese cobra *N. atra* [76] also demonstrated powerful, salt-resistant, antimicrobial activity against Gram-positive and Gram-negative bacteria, including *Francisella novicida* (the non-virulent strain in humans related to *Francisella tularensis*, the causative agent of tularemia) [98], *E.coli, Aggregatibacter actinomycetemcomitans,*

Bacillus cereus [99], *P. aeruginosa* [100], and *S. aureus* [101] at low concentrations (EC_{50} < 3 µg/mL). Na-CATH is also active against *Mycobacterium smegmatis* [102], *Burkholderia thailandensis* (closely related to *B. pseudomallei*, the causative agent of melioidosis) [103] and *Bacillus anthracis* (anthrax) [104]. The last two strains are of particular relevance due to their potential use as biological weapons. Indeed, in vivo studies using wax moth larvae demonstrated that Na-CATH was able to rescue 100% of waxworms after *B. anthracis* Sterne infection at low peptide concentrations [104]. In addition, Na-CATH is not only active against planktonic bacteria, but also inhibits *S. aureus* and *B. thailandensis* biofilm formation [101,102], while inducing minimal hemolysis (<2% at 100 µg/mL) [99]. However, Na-CATH did not inhibit *Pseudomonas* biofilm formation [100].

Structurally, Na-CATH folds into a well-defined amphipathic α-helix between residues Phe3 and Lys23 in the presence of trifluoroethanol. The remaining 11-residue tail, consisting mostly of aromatic and hydrophobic residues, does not present a defined structure, but appears to interact with lipid membranes [118]. Na-CATH contains an 11-residue sequence [KR(F/A)KKFFKK(L/P)K] known as an ATRA motif, repeated twice and almost totally shared by other SV-CATHs (Table 2). The ATRA motif, when tested by itself, is also active, but the Pro at position 10 dramatically reduces its antibacterial potency, probably by destabilizing its helical structure, while a Phe to Ala change at position 3 does not impair activity [98,99].

Du et al. postulated that Na-CATH is able to disrupt bacterial membrane-like liposomes via membrane thinning or transient-pore formation [118]. This hypothesis was further confirmed in vitro by Gupta el al. and Juba et al., who described membrane depolarization and transient-pore formation in *Mycobacterium smegmatis* [103], as well as in *E.coli* and *B. cereus* [119] after Na-CATH treatment. However, Samuel et al. suggested a more detailed mechanism changing from membrane disruption to pore-based lysis, depending on liposome lipid composition and phase [120].

- **Bf-CATH:** unlike other 34-aa AMPs predicted as SV-CATHs, the purified peptide from *Bungarus fasciatus* venom gland is a 30-aa peptide lacking the four N-terminal residues (Table 2). This difference in length suggests different enzymatic processing rather than the proposed elastase-like protease cleavage at the conserved site (Val) or post-processing of the 34-residue precursor [97]. Expression of Bf-CATH is widespread, including stomach, trachea, skin, muscle, heart, kidney, lung, brain, intestine, spleen, liver, ovary and venom glands [97].

Like other SV-CATHs, Bf-CATH has a random-coil conformation in aqueous solution, but adopts an α-helical structure in hydrophobic or membrane-like environments, specifically from residues Phe2 to Phe18 [97]. Bf-CATH presents potent antimicrobial activity against a broad range of clinically isolated, drug-resistant Gram-negative and Gram-positive bacteria, as well as saprophytic fungi [97]. In vitro and in vivo analyses demonstrate that Bf-CATH partially retains antibacterial activity in the presence of human serum, but not in gastrointestinal fluids. Its fluorescein-labeled analog can be absorbed into the mouse circulatory system within 30 min after intraperitoneal injection, without tissue accumulation (totally excreted after 24 h) [121]. Interestingly, Bf-CATH was less prone to induce bacterial resistance than classical antibiotics, such as ciprofloxacin or gentamicin, when administered at sublethal concentrations [94]. Bf-CATH demonstrated membrane disruptive antibacterial activity [94], as well as the capacity to inhibit secretion of pro-inflammatory molecules, such as TNF-α, IL-8, IL-1, or MCP-1 (monocyte chemoattractant protein-1) and to inhibit $O_2 \cdot^-$ production induced by *Propionibacterium acnes* (acne vulgaris) [95].

Table 3. Properties of mature snake-derived CATHs. Hemolytic activity (HC_{10}, µg/mL) and biological activity (minimal inhibitory concentration (MIC) or IC_{50}, µg/mL) against representative or clinically isolated (CI) microorganisms and tumor cells were collected from published sources. If more than one value was available in the literature, a range is given. Selectivity ratio was calculated as HC_{10}/MIC or HC_{10}/IC_{50}. MIC values against reference strains were used for selectivity ratio calculation, unless denoted as (*). The most restrictive value (highest MIC or IC_{50} and lowest HC_{10}) of an interval was chosen for selectivity ratio calculation. n.d.: non detectable activity.

Microorganism: / Peptide:	Oh-CATH	Bf-CATH	Bf-CATH30	Hc-CATH	Cdt-CATH	Ba-CATH	Pt-CATH1	ΔPb-CATH1	ΔPb-CATH2	ΔPb-CATH4	Pb-CATH1	Pb-CATH2	Pb-CATH3	Pb-CATH4	Pb-CATH5	Pb-CATH6	Sa-CATH
Hemolysis (HC_{10})																	
	~200	~31	50 - >400	>200	~104	~53	~13	~64	>64	~64	>100	>100	>100	<100	>100	>100	>200
Gram-negative bacteria (MIC)																	
E. coli ATCC 25922	0.25–8	8	2.3–8	2.3	0.25–0.78	0.25–0.78	2	2	3	1	9.4	37.5	n.d.	18.8	n.d.	n.d.	18.8
E. coli (CI)	2–20		0.6–16	2.3–9.4	16	16	16			4.7	9.4	75	n.d.	18.8	n.d.	n.d.	75
Selectivity ratio	25	4	6	>87	133	68	7	32	>21	64	>11	>3	-	<5	-	-	>11
Gram-positive bacteria (MIC)																	
S. aureus ATCC 25923	4–64	16	16 > 400	4.7–25.8	32	32	32	>128	>128	>128	37.5	n.d.	n.d.	18.8	n.d.	n.d.	75
S. aureus (CI)	8–64	32–64	16 > 400	4.7 > 200	32	32	32				4.7–37.5	75	n.d.	18.8	n.d.	n.d.	
Selectivity ratio	3	2	<0.1	>8	3	2	0.4	<0.5	0.5	<0.5	>3	-	-	<5	-	-	>3
Fungi (MIC)																	
C. albicans ATCC 2002			4.7														
C. albicans (CI)				2.3–4.7	10–40						9.4–18.8	18.8–37.5	n.d.	9.4–18.8	n.d.	n.d.	18.8–37.5
Selectivity ratio			11	43*	3*						>5*	>3*	-	<5*	-	-	>5*
Tumor cells (IC_{50})																	
PC-3 (prostate cancer)			70.2	n.d.													
U937 (leukemia)					<4												
MCF-7 (breast cancer)	n.d.		353							~64							
HepG2 (liver cancer)				n.d.													
Selectivity ratio	-		0.7	-	>26					1							
Ref.	[76,84]	[85]	[93,94,96,97]	[105,106]	[84,107,108,122,123]	[84]	[84]	[113]	[113]	[113]	[112]	[112]	[112]	[112]	[112]	[112]	[114]

Color code:

	Low (>50)	Medium (50–10)	High (>10)
Hemolysis:			
Biological activity (MIC/IC_{50}):	Low (>50)	Medium (50–10)	High (<10)
	Non-selective (ratio < 1)		Selective (ratio > 1)

In vivo studies by Wang et al. showed an overall anti-inflammatory effect of Bf-CATH and reduced *P. acnes*-induced granulomatous inflammation, revealing its therapeutic potential to treat acne [95]. In vivo, Bf-CATH reduced bacterial loads and colonization in murine models of *P. aeruginosa* and *Salmonella typhimurium* infections, attenuating symptoms and intestinal alterations [94,121]. Bf-CATH also demonstrated in vivo prevention of intestinal barrier dysfunction in mouse and piglet models of lipopolysaccharide-induced endotoxemia/inflammation, presumably by downregulating TNF-α expression through the NF-κB signaling pathway [89,92]. Parallel downregulation of NF-κB signaling and activation of the signal transducer and activator of transcription 1 (STAT-1) in weanling piglets seems to help suppress intestinal inflammation and to enhance phagocytosis of immune cells, and modulation of intestinal immune responses during stress-related inflammatory processes, such as weaning [91]. A recent study by Liu et al. suggested that pre-treatment with Bf-CATH ameliorates *P. aeruginosa*-induced pneumonia by enhancing NETosis (activation and release of neutrophil extracellular traps), confirming the immunomodulatory activity of Bf-CATH [88].

Anti-tumor activity of Bf-CATH was also investigated, showing potent in vitro activity against mouse melanoma cells (IC_{50} ~7 μM) and in vivo inhibition of mouse melanoma cell proliferation, migration, and angiogenesis, but with a negligible effect against human tumoral cell lines (IC_{50}s from ~20–100 μM). Its anti-tumor mechanism is related to membrane permeabilization, and to DNA binding and prevention of vascular endothelial growth factor (VEGF) gene expression [93].

Different strategies have addressed structure-activity relationships of Bf-CATH to optimize its activity [86,87,96,97,124–127]. For instance, the 15-aa BF-15, mostly retains the antimicrobial activity of Bf-CATH, permeabilizes membranes, and is more stable in serum than the native peptide [96,97]. Cbf-K16, a Bf-CATH analog obtained from substitution of Glu16 $\rightarrow$ Lys, also showed antibacterial activity against a recombinant New Delhi metallo-beta-lactamase-1 (NMD-1)-carrying *E. coli* strain, as well as improved anti-tumor activity against human and mouse lung carcinoma cells, compared to Bf-CATH, both effects presumably achieved by membrane disruption and DNA binding [125,126]. Trp/Arg-rich analogs of Bf-CATH were also designed, from which ZY13, the most potent and least hemolytic analog, exhibited in vitro antibacterial and promising antifungal and anti-inflammatory properties both in vitro and in a mouse *C. albicans*-induced vaginitis model [87].

Finally, Bf-CATH production and delivery systems were investigated. Bf-CATH encapsulation was tested in poly(D,L-lactide-co-glycolide) (PLGA) microspheres and poly(ethylene glycol)-poly(lactic-acid-co-glycolic acid) block copolymers (4-arm-PEG-PLGA). Both systems retained the antibacterial activity of the free antimicrobial peptide and released Bf-CATH over >15 days [128,129]. Different recombinant DNA strategies were also reported to effectively produce Bf-CATH, such as small ubiquitin-related modifier (SUMO) technology or intein-based technology, both expressed in *Bacillus subtilis* (achieved peptide yields of ~3 mg/L and 0.5 mg/L, respectively) [130,131].

- **Cdt-CATH:** also named as crotalicidin (Ctn), is a 34-aa peptide from the South American rattlesnake (*Crotalus durissus terrificus*) venom. Among SV-CATHs identified in pit vipers by Falcão et al. [84] (lutzicidin, lachesicidin, batroxicidin, collectively named vipericidins), Ctn has been the most studied, showing potent bactericidal effects against Gram-negative and Gram-positive bacteria (MICs <10 μM) [84], anti-parasitic (anti-trypanosomatid) activity [109], and activity against opportunistic yeasts and dermatophytes, alone or in combination with conventional antifungals [108]. In addition, Ctn has shown potent anti-tumor activity against different leukemia cell lines (IC_{50}s <5 μM) [107].

The general mechanism of Ctn against bacteria or parasites is membrane-related and its ability to interfere and disrupt biological membranes have been extensively described [109,122]. Ctn toxicity to leukemia cells is also linked to its membranolytic effect, although it also seems able to interfere with key intracellular pathways, ultimately contributing to tumor cell death (Pérez-Peinado et al., submitted). In addition to the direct cytotoxic effect, the immunomodulatory properties of Ctn have been explored, showing an overall pro-inflammatory profile in the presence of heat-inactivated bacterial antigens and IFN-γ [110], contrasting with general anti-inflammatory behavior of other SV-CATHs.

From a structural viewpoint, circular dichroism and nuclear magnetic resonance (NMR) studies indicate that Ctn is fully in a random-coil conformation in aqueous solution, but may change its structure in membrane-like environments (i.e., dodecylphosphocholine micelles), displaying an α-helix conformation at the N-terminal end (residues 3 – 21) plus a C-terminal random coil tail (Figure 3a) [107]. The Ctn framework is quite similar to those proposed for Na-CATH and Bf-CATH (N-terminal α-helix plus a random coil C-term end), suggesting a common template shared by SV-CATHs.

A rational dissection of Ctn involving in silico enzymatic cleavage was performed by Falcao et al. in order to define a shorter, active motif [107]. Two fragments resulted from cleavage at Val14: Ctn [1–14] and Ctn [15–34]. Surprisingly, the former was inactive, regardless of its amphipathic α-helical conformation. In contrast, the latter preserved some of the activity of Ctn, despite its overall disordered structure. In fact, Ctn [15–34] presents lower hemolytic and cytotoxicity effects than its predecessor Ctn, despite improved selectivity for Gram-negative bacteria and enhanced stability in human serum, presumably due to serum protein binding and its preferred scaffold [107,132,133]. Ctn [15–34] lost activity against dermatophytes, but had enhanced activity against pathogenic yeasts, such as several (multi-resistant) *Candida* species, and acted synergistically with amphotericin B [108,134]. Although Ctn was also able to induce necrosis of all developmental forms of *T. cruzi* (the Chagas' disease agent), Ctn [15–34] only retained activity against the trypomastigote form [109]. The fragment also showed antiviral activity against the infectious myonecrosis virus, an epizootic agent that threatens shrimp production in Brazil, and for which no current treatment exists [135]. Like Ctn, Ctn [15–34] acts via membrane permeabilization and necrosis, as described for strains of *E.coli* and *C. albicans* [122,123].

Insights into the functionality of structural domains of SV-CATH were presented by Oliveira-Júnior et al. using Ctn [15–34] as a model. As described above, SV-CATHs contain an anionic region between the cathelin domain and the mature AMP domain, not usually found in other CATHs. Thus, by including a Glu decapeptide at the N-terminus of Ctn [15–34] to generate a "pro-peptide" model, Oliveira-Júnior et al. confirmed that this acidic moiety contributes to a more helical conformation and prevents peptide antimicrobial activity before its release [78].

- **Other snake-derived CATHs**: additional pit viper-derived CATHs were predicted by Falcao et al., from the venom gland of *Lachesis muta rhombeata* (lachesicidin), *Bothrops atrox* (batroxicidin) and *B. lutzi* (lutzicidin), as well as two clones from the elapid, *Pseudonaja textilis* (Pt-CATH1 and Pt-CATH2). Batroxicidin and Pt-CATH1 displayed antibacterial potency comparable to that of Ctn, but were more hemolytic [84]. Moreover, batroxicidin induced *T. cruzi* cell death by membrane disruption and showed an overall proinflammatory profile [110,111].

CATHs were similarly predicted/identified in the genome of the boid, *P. bivittatus*, both by Kim et al. and Cai et al. (Table 2), denominated Pb-CATH1-5 and CATHPb1-6, respectively [112,113]. From the set of Pb-CATHs identified by Kim et al., three (Pb-CATH1, Pb-CATH3 and Pb-CATH4) encode mature AMPs displaying powerful antibacterial activity against Gram-negative bacteria (MICs from 0.5 to 8 µg/mL) [113]. Pb-CATH4, for instance, induced bacterial death by toroidal pore formation and displayed low hemolysis and cytotoxicity, as well as considerable stability in serum [113]. In parallel, CATHPb1 exhibited protection in mice infected with MRSA and VRSA (vancomycin-resistant *S. aureus*), via neutrophil-mediated bacterial clearance and immunomodulation employing Mitogen-Activated Protein Kinases (MAPKs) and NF- κB pathways [112].

Finally, another CATH-related peptide, Hc-CATH (Table 2), was identified in the genome of the sea snake *Hydrophis cyanocinctus* (annulated sea snake) [105]. Unlike the overall preference of terrestrial snake-derived CATHs for Gram-negative bacteria, Hc-CATH displays more or less equal activity against both Gram-negative and Gram-positive bacteria as a result of membrane permeabilization [105]. Hc-CATH has intrinsic structural advantages compared to other CATHs, such as high stability in the presence of salts, high temperature (≤90 °C), and serum proteases, together with low toxicity against eukaryotic cells [105]. In a recent study, Carlile et al. demonstrated anti-inflammatory properties and bacterial load reduction by Hc-CATH in vivo, using wax moth and mouse models of intraperitoneal and respiratory infection induced by *P. aeruginosa* [106].

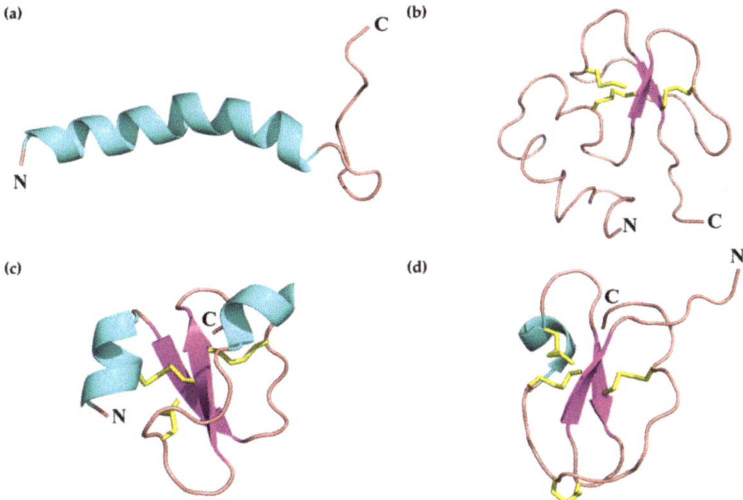

Figure 3. Three-dimensional structures adopted by SV-AMPs and -ACPs. (**a**) crotalicidin (Ctn), PDB 2MWT. (**b**) Cdt-defensin, 3D structure prediction obtained using the Iterative Threading ASSEmbly Refinement (I-TASSER) web server [136] (available at: https://zhanglab.ccmb.med.umich. edu/I-TASSER/). (**c**) Crotamine, PDB 4GV5. (**d**) Omwaprin, PDB 3NGG. Representation was performed using the PyMOL Molecular Graphical System, Version 2.0. Schrödinger, LLC [137]. Color code: blue for α-helix, magenta for β-sheet and pink for, loops. Disulfide pairing is also indicated in yellow.

2.1.2. SV-Defensins

Together with CATHs, defensins are a major group of host defense peptides found in vertebrates and invertebrates, exhibiting broad-spectrum activity against bacteria, fungi, and enveloped viruses [138]. Defensins are cationic, Cys-rich peptides of 3.5–6 kDa, with a typical β-sheet-rich folding stabilized by three disulfide bonds (Figure 3b). Defensins are divided into three major groups based on length and Cys-pairing: α-, β-, and θ-defensins [139]. Although defensins have been deeply studied in mammals, little information is available for those of SV origin. More than 20 β-defensin-like sequences have been described in snakes [140–142], half of them identified in the genus *Bothrops* (Table S2). However, no details regarding their antibacterial or antitumoral potency are available. Based on the sequence alignment presented in Figure 4, SV defensins present a highly conserved N-terminal region, as well as conserved Cys residues in disulfide bonds. Gly, Pro, and Asp residues are also highly conserved.

Structurally, crotamine is closely related to β-defensins, displaying an $\alpha_1\beta_1\alpha_2\beta_2$ arrangement, with the whole structure stabilized by three disulfide bonds [143] (Figure 3c). Crotamine is a 42-aa neuro- and myotoxic peptide, initially isolated from the South American rattlesnake (*C. durissus terrificus*). Crotamine's anti-infective, antitumoral, and cell-penetrating properties have been deeply characterized both in vitro and in vivo [144]. Crotamine displays modest antibacterial activity (MICs in the 25–100 μg/mL range against *E. coli* strains) and does not induce hemolysis at high concentrations (no hemolysis observed up to 1024 μg/mL) [145]. Additional studies revealed antifungal activity of crotamine (12.5–50 μg/mL) against *Candida* spp., including clinically resistant strains [146]. The anticancer potential of crotamine has also been studied in vitro and in vivo, showing selective cytotoxicity against tumor cell lines at low concentrations (~5 µg/mL) and significant inhibition of tumor growth and increased lifespan in a melanoma mouse model [147,148].

Derivatives of crotamine [CyLoP-1 (cytosol localizing peptide 1) and the NrTPs (nucleolar-targeting peptides)] have been designed, with cytosolic and nucleolar localization, respectively, instead of the nuclear distribution pattern of crotamine [149–151].

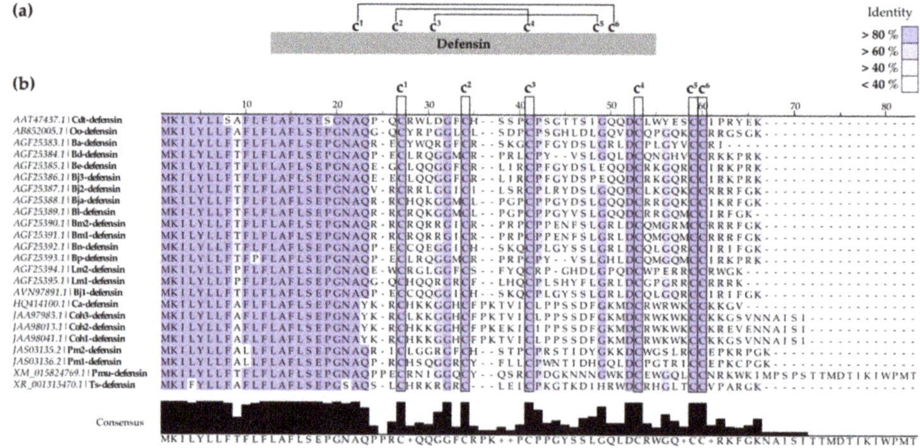

Figure 4. SV-defensins. (**a**) Schematic representation of general SV-defensin structure highlighting the Cys-pairing pattern. (**b**) Alignment of SV-defensin sequences. Multiple sequence alignment was performed using Clustal Omega [115] and represented with Jalview v2.11.0 software [116]. Background is colored according to the percentage of identity (residues in agreement with the consensus sequence depicted at the bottom).

2.1.3. Waprins

Waprins are a family of ~50-residue, Cys-rich peptides first isolated from the venom of *Naja nigricollis* (black-necked spitting cobra), named nawaprin (Naw), and subsequently in venom of the Indian taipan (*Oxyuranus microlepidotus*), named omwaprin (Omw) [152,153]. Omwaprin displays salt-resistant antibacterial activity against Gram-positive bacteria such as *Bacillus megaterium* and *Staphylococcus warneri* [153]. In vivo experiments in mice demonstrated that Omw is non-toxic at doses up to 10 µg/g after *i.p.* injection [153]. Tertiary structure of Omw and Naw has been studied by X-ray and NMR, respectively, showing a complex disc-like shape with four disulfide bonds [152,154] (Figure 3d).

Two derivatives, Omw1 and Omw2, also present modest antimicrobial, antifungal, and antibiofilm activity at concentrations from 16–500 µg/mL, presumably due to membrane disruption with low hemolysis induction in the same concentration range (~10 % hemolysis at 250–500 µg/mL) [155].

3. Conclusions

Despite major advances, important challenges remain in the anti-infective and anticancer drug discovery pipeline. The emergence of superbugs has created an antibiotic resistance crisis, evidencing the need for new anti-infective strategies. Paradoxically, investment and R&D by the pharmaceutical industry in this area is losing momentum [156,157]. In parallel, although meaningful progress has been made in cancer research, new therapeutic alternatives are still required in order to avoid undesired effects linked to current therapies and to target multidrug-resistant cells [158,159].

In this context, peptide-based therapeutics emerge as a feasible alternative to traditional anti-infective and cytotoxic drugs. Examples of promising natural AMPs and ACPs (e.g., cecropins, magainins, defensins) have been described since the 1980s. Study of SV-derived AMPs and ACPs, has been limited to the last decade, but thorough work has been done recently to search for ideal candidates. We expect to see SV-derived AMPs and ACPs in the pharmaceutical market in the near future. Although peptides still face important challenges (such as potential toxicity/immunogenicity, non-oral activity, limited bioavailability, cost, etc.) that hinder their penetration of the market, significant advances have been made to overcome these hurdles (e.g., rational design, cyclization, peptidomimetics, conjugation to macromolecules/polymers/delivery vectors, etc.).

Snake-derived AMPs and ACPs presented here are but a small number of the plethora of experimentally validated AMPs (~2800 sequences [160]) and ACPs (~500 [161]) reported to date. Throughout this manuscript, we have described their in vitro and in vivo validation and bioavailability, and have mentioned strategies for their optimization (downsizing, conjugation to polymers, etc.) and even some scale-up attempts (for Bf-CATH, for instance). Taken together, these advances testify to the significant potential of SV peptides as future anti-infective and antitumoral therapeutics and highlight the utility of snake venoms as a font for drug discovery. With so few snake venoms investigated to date and so many more waiting to be explored, the number of drug leads derived from snake venoms can only increase in the future.

Supplementary Materials: The following are available online at http://www.mdpi.com/2072-6651/12/4/255/s1: Table S1. SV-CATH precursors. SV-CATH protein precursors were collected, together with their sequences and domain information. Domains were highlighted in the protein sequence as: green, signal peptide; cathelicidin domain, blue; Glu-rich domain, yellow; mature peptide, red. Domain information (L, length; P, position) was extracted from NCBI (https://www.ncbi.nlm.nih.gov/protein). If no information was available (green background), domains were annotated based on bibliographic information and/or sequence similarities, Table S2. SV-defensins. Summary of SV-defensins identified to date, their sequences and accession numbers. A unified name was given to each SV-defensin according to the initial of the snake, Figure S1. Percentage identity matrix of SV-CATH precursors. Estimation of the homology/divergence between SV-CATH precursor proteins (sequences and NCBI accession numbers available in Table S1) illustrated by percentage identity matrix.

Author Contributions: C.P.-P. and S.D. revised literature. C.P.-P. and D.A. wrote the manuscript with contributions of all authors. All authors reviewed the final manuscript. All authors have read and agreed to the published version of the manuscript.

Funding: This research received no external funding.

Conflicts of Interest: The authors declare no conflicts of interest.

Abbreviations

ACE	angiotensin-converting enzyme
ACP	anticancer peptide
AMP	antimicrobial peptide
CATH	cathelicidin
Ctn	crotalicidin
EC_{50}	50% effective concentration
HC_{10}	10% hemolytic concentration;
HPLC	high-performance liquid chromatography
IC_{50}	50 % inhibitory concentration
L-AAO	L-amino acid oxidases
MRSA	methicillin-resistant *S. aureus*
MS	mass spectrometry
MSA	multiple sequence analysis
Naw	nawaprin
NMR	nuclear magnetic resonance
Omw	omwaprin
PLA_2	phospholipases A2
SV	snake venom

References

1. Waheed, H.; Moin, S.F.; Choudhary, M.I. Snake Venom: From Deadly Toxins to Life-saving Therapeutics. *Curr. Med. Chem.* **2017**, *24*, 1874–1891. [CrossRef] [PubMed]

2. Ferreira, S.H. A bradykinin-potentiating factor (BPF) present in the venom of bothrops jararaca. *Br. J. Pharmacol. Chemother.* **1965**, *24*, 163–169. [CrossRef] [PubMed]

3. Ondetti, M.A.; Rubin, B.; Cushman, D.W. Design of specific inhibitors of angiotensin-converting enzyme: New class of orally active antihypertensive agents. *Science* **1977**, *196*, 441–444. [CrossRef] [PubMed]

4. Michael, C.A.; Dominey-Howes, D.; Labbate, M. The antimicrobial resistance crisis: Causes, consequences, and management. *Front. Public Health* **2014**, *2*, 145. [CrossRef]

5. Ventola, C.L. The antibiotic resistance crisis: Part 1: Causes and threats. *P T Peer-Rev. J. Formul. Manag.* **2015**, *40*, 277–283.

6. O'Neill, J. Antimicrobial Resistance: Tackling a crisis for the health and wealth of nations. *Rev. Antimicrob. Resist* **2014**.

7. O'Neill, J. Tackling drug-resistant infection globally: Final report and recommendations. *Rev. Antimicrob. Resist* **2016**.

8. WHO. 2018. Cancer Fact Sheets. Available online: https://www.who.int/news-room/fact-sheets/detail/cancer (accessed on 6 August 2019).

9. Ferlay, J.E.M.; Lam, F.; Colombet, M.; Mery, L.; Piñeros, M.; Znaor, A.; Soerjomataram, I.; Bray, F. *Global Cancer Observatory: Cancer Today*; International Agency for Research on Cancer: Lyon, France, 2018. Available online: https://gco.iarc.fr/today (accessed on 6 August 2019).

10. Bray, F.; Ferlay, J.; Soerjomataram, I.; Siegel, R.L.; Torre, L.A.; Jemal, A. Global cancer statistics 2018: Globocan estimates of incidence and mortality worldwide for 36 cancers in 185 countries. *CA Cancer J. Clin.* **2018**, *68*, 394–424. [CrossRef]

11. Ferlay, J.; Colombet, M.; Soerjomataram, I.; Mathers, C.; Parkin, D.M.; Pineros, M.; Znaor, A.; Bray, F. Estimating the global cancer incidence and mortality in 2018: Globocan sources and methods. *Int. J. Cancer* **2019**, *144*, 1941–1953. [CrossRef]

12. Francischetti, I.M.B.; Reyes Gil, M. Chapter 164—Diagnostic Use of Venoms. In *Transfusion Medicine and Hemostasis*, 3rd ed.; Shaz, B.H., Hillyer, C.D., Reyes Gil, M., Eds.; Elsevier: Cambridge, MA, USA, 2019; pp. 969–975.

13. Koh, D.C.; Armugam, A.; Jeyaseelan, K. Snake venom components and their applications in biomedicine. *Cell. Mol. Life Sci.* **2006**, *63*, 3030–3041. [CrossRef]

14. Takacs, Z.; Nathan, S. Animal Venoms in Medicine. In *Encyclopedia of Toxicology*, 3rd ed.; Wexler, P., Ed.; Academic Press: Oxford, UK, 2014; pp. 252–259.

15. King, G.F. Venoms as a platform for human drugs: Translating toxins into therapeutics. *Expert Opin. Biol. Ther.* **2011**, *11*, 1469–1484. [CrossRef] [PubMed]

16. Fox, J.W.; Serrano, S.M. Approaching the golden age of natural product pharmaceuticals from venom libraries: An overview of toxins and toxin-derivatives currently involved in therapeutic or diagnostic applications. *Curr. Pharm. Des.* **2007**, *13*, 2927–2934. [CrossRef] [PubMed]

17. Vonk, F.J.; Jackson, K.; Doley, R.; Madaras, F.; Mirtschin, P.J.; Vidal, N. Snake venom: From fieldwork to the clinic: Recent insights into snake biology, together with new technology allowing high-throughput screening of venom, bring new hope for drug discovery. *Bioessays* **2011**, *33*, 269–279. [CrossRef] [PubMed]

18. Carr, A.; Schultz, J. Prospective evaluation of the incidence of wound infection in rattlesnake envenomation in dogs. *J. Vet. Emerg. Crit. Care* **2015**, *25*, 546–551. [CrossRef]

19. De Lima, D.C.; Alvarez Abreu, P.; de Freitas, C.C.; Santos, D.O.; Borges, R.O.; Dos Santos, T.C.; Mendes Cabral, L.; Rodrigues, C.R.; Castro, H.C. Snake Venom: Any Clue for Antibiotics and CAM? *Evid. Based Complement. Altern. Med.* **2005**, *2*, 39–47. [CrossRef]

20. Talan, D.A.; Citron, D.M.; Overturf, G.D.; Singer, B.; Froman, P.; Goldstein, E.J. Antibacterial activity of crotalid venoms against oral snake flora and other clinical bacteria. *J. Infect. Dis.* **1991**, *164*, 195–198. [CrossRef]

21. Perumal Samy, R.; Gopalakrishnakone, P.; Ho, B.; Chow, V.T. Purification, characterization and bactericidal activities of basic phospholipase A2 from the venom of Agkistrodon halys (Chinese pallas). *Biochimie* **2008**, *90*, 1372–1388. [CrossRef]

22. Stiles, B.G.; Sexton, F.W.; Weinstein, S.A. Antibacterial effects of different snake venoms: Purification and characterization of antibacterial proteins from Pseudechis australis (Australian king brown or mulga snake) venom. *Toxicon* **1991**, *29*, 1129–1141. [CrossRef]

23. Perumal Samy, R.; Pachiappan, A.; Gopalakrishnakone, P.; Thwin, M.M.; Hian, Y.E.; Chow, V.T.; Bow, H.; Weng, J.T. In vitro antimicrobial activity of natural toxins and animal venoms tested against Burkholderia pseudomallei. *BMC Infect. Dis.* **2006**, *6*, 100. [CrossRef]

24. Chaim-Matyas, A.; Ovadia, M. Cytotoxic activity of various snake venoms on melanoma, B16F10 and chondrosarcoma. *Life Sci.* **1987**, *40*, 1601–1607. [CrossRef]

25. Da Silva, R.J.; da Silva, M.G.; Vilela, L.C.; Fecchio, D. Antitumor effect of Bothrops jararaca venom. *Mediat. Inflamm.* **2002**, *11*, 99–104. [CrossRef] [PubMed]

26. Gomes, A.; Bhattacharjee, P.; Mishra, R.; Biswas, A.K.; Dasgupta, S.C.; Giri, B. Anticancer potential of animal venoms and toxins. *Indian J. Exp. Biol.* **2010**, *48*, 93–103. [PubMed]

27. Kerkkamp, H.; Bagowski, C.; Kool, J.; van Soolingen, B.; Vonk, F.J.; Vlecken, D. Whole snake venoms: Cytotoxic, anti-metastatic and antiangiogenic properties. *Toxicon* **2018**, *150*, 39–49. [CrossRef] [PubMed]

28. Izidoro, L.F.; Sobrinho, J.C.; Mendes, M.M.; Costa, T.R.; Grabner, A.N.; Rodrigues, V.M.; da Silva, S.L.; Zanchi, F.B.; Zuliani, J.P.; Fernandes, C.F.; et al. Snake venom L-amino acid oxidases: Trends in pharmacology and biochemistry. *BioMed Res. Int.* **2014**, *2014*, 196754. [CrossRef] [PubMed]

29. Almeida, J.R.; Palacios, A.L.V.; Patino, R.S.P.; Mendes, B.; Teixeira, C.A.S.; Gomes, P.; da Silva, S.L. Harnessing snake venom phospholipases A2 to novel approaches for overcoming antibiotic resistance. *Drug Dev. Res.* **2019**, *80*, 68–85. [CrossRef] [PubMed]

30. Modahl, C.M.; Frietze, S.; Mackessy, S.P. Transcriptome-facilitated proteomic characterization of rear-fanged snake venoms reveal abundant metalloproteinases with enhanced activity. *J. Proteom.* **2018**, *187*, 223–234. [CrossRef] [PubMed]

31. Samy, R.P.; Gopalakrishnakone, P.; Chow, V.T.; Ho, B. Viper metalloproteinase (Agkistrodon halys pallas) with antimicrobial activity against multi-drug resistant human pathogens. *J. Cell. Physiol.* **2008**, *216*, 54–68. [CrossRef]

32. Markland, F.S., Jr.; Swenson, S. Snake venom metalloproteinases. *Toxicon* **2013**, *62*, 3–18. [CrossRef]

33. Sala, A.; Cabassi, C.S.; Santospirito, D.; Polverini, E.; Flisi, S.; Cavirani, S.; Taddei, S. Novel Naja atra cardiotoxin 1 (CTX-1) derived antimicrobial peptides with broad spectrum activity. *PLoS ONE* **2018**, *13*, e0190778. [CrossRef]

34. Samy, R.P.; Stiles, B.G.; Chinnathambi, A.; Zayed, M.E.; Alharbi, S.A.; Franco, O.L.; Rowan, E.G.; Kumar, A.P.; Lim, L.H.; Sethi, G. Viperatoxin-II: A novel viper venom protein as an effective bactericidal agent. *FEBS Open Bio* **2015**, *5*, 928–941. [CrossRef]

35. Sampaio, S.C.; Hyslop, S.; Fontes, M.R.; Prado-Franceschi, J.; Zambelli, V.O.; Magro, A.J.; Brigatte, P.; Gutierrez, V.P.; Cury, Y. Crotoxin: Novel activities for a classic beta-neurotoxin. *Toxicon* **2010**, *55*, 1045–1060. [CrossRef] [PubMed]

36. Samy, R.P.; Kandasamy, M.; Gopalakrishnakone, P.; Stiles, B.G.; Rowan, E.G.; Becker, D.; Shanmugam, M.K.; Sethi, G.; Chow, V.T. Wound healing activity and mechanisms of action of an antibacterial protein from the venom of the eastern diamondback rattlesnake (Crotalus adamanteus). *PLoS ONE* **2014**, *9*, e80199. [CrossRef] [PubMed]

37. Allane, D.; Oussedik-Oumehdi, H.; Harrat, Z.; Seve, M.; Laraba-Djebari, F. Isolation and characterization of an anti-leishmanial disintegrin from Cerastes cerastes venom. *J. Biochem. Mol. Toxicol.* **2018**, *32*, e22018. [CrossRef] [PubMed]

38. Sulca, M.A.; Remuzgo, C.; Cardenas, J.; Kiyota, S.; Cheng, E.; Bemquerer, M.P.; Machini, M.T. Venom of the Peruvian snake Bothriopsis oligolepis: Detection of antibacterial activity and involvement of proteolytic enzymes and C-type lectins in growth inhibition of Staphylococcus aureus. *Toxicon* **2017**, *134*, 30–40. [CrossRef]

39. Nolte, S.; de Castro Damasio, D.; Barea, A.C.; Gomes, J.; Magalhaes, A.; Mello Zischler, L.F.; Stuelp-Campelo, P.M.; Elifio-Esposito, S.L.; Roque-Barreira, M.C.; Reis, C.A.; et al. BJcuL, a lectin purified from Bothrops jararacussu venom, induces apoptosis in human gastric carcinoma cells accompanied by inhibition of cell adhesion and actin cytoskeleton disassembly. *Toxicon* **2012**, *59*, 81–85. [CrossRef]

40. Nunes Edos, S.; de Souza, M.A.; Vaz, A.F.; Santana, G.M.; Gomes, F.S.; Coelho, L.C.; Paiva, P.M.; da Silva, R.M.; Silva-Lucca, R.A.; Oliva, M.L.; et al. Purification of a lectin with antibacterial activity from Bothrops leucurus snake venom. *Comp. Biochem. Physiol. B Biochem. Mol. Biol.* **2011**, *159*, 57–63. [CrossRef]

41. Calderon, L.A.; Sobrinho, J.C.; Zaqueo, K.D.; de Moura, A.A.; Grabner, A.N.; Mazzi, M.V.; Marcussi, S.; Nomizo, A.; Fernandes, C.F.; Zuliani, J.P.; et al. Antitumoral activity of snake venom proteins: New trends in cancer therapy. *BioMed Res. Int.* **2014**, *2014*, 203639. [CrossRef]

42. Gomes, V.M.; Carvalho, A.O.; Da Cunha, M.; Keller, M.N.; Bloch, C., Jr.; Deolindo, P.; Alves, E.W. Purification and characterization of a novel peptide with antifungal activity from Bothrops jararaca venom. *Toxicon* **2005**, *45*, 817–827. [CrossRef]

43. Sarzaeem, A.; Zare Mirakabadi, A.; Moradhaseli, S.; Morovvati, H.; Lotfi, M. Cytotoxic effect of ICD-85 (venom-derived peptides) on HeLa cancer cell line and normal LK cells using MTT Assay. *Arch. Iran. Med.* **2012**, *15*, 696–701.

44. Júnior, N.; Cardoso, M.H.; Franco, O.L. Snake venoms: Attractive antimicrobial proteinaceous compounds for therapeutic purposes. *Cell. Mol. Life Sci.* **2013**, *70*, 4645–4658. [CrossRef]

45. Ma, R.; Mahadevappa, R.; Kwok, H.F. Venom-based peptide therapy: Insights into anti-cancer mechanism. *Oncotarget* **2017**, *8*, 100908–100930. [CrossRef] [PubMed]

46. Yeaman, M.R.; Yount, N.Y. Mechanisms of antimicrobial peptide action and resistance. *Pharmacol. Rev.* **2003**, *55*, 27–55. [CrossRef] [PubMed]

47. Swain, S.S.; Paidesetty, S.K.; Padhy, R.N. Antibacterial, antifungal and antimycobacterial compounds from cyanobacteria. *Biomed. Pharmacother.* **2017**, *90*, 760–776. [CrossRef] [PubMed]

48. Bahar, A.A.; Ren, D. Antimicrobial peptides. *Pharmaceuticals* **2013**, *6*, 1543–1575. [CrossRef]

49. Yeung, A.T.; Gellatly, S.L.; Hancock, R.E. Multifunctional cationic host defence peptides and their clinical applications. *Cell. Mol. Life Sci.* **2011**, *68*, 2161–2176. [CrossRef]

50. Gomes, B.; Augusto, M.T.; Felicio, M.R.; Hollmann, A.; Franco, O.L.; Goncalves, S.; Santos, N.C. Designing improved active peptides for therapeutic approaches against infectious diseases. *Biotechnol. Adv.* **2018**, *36*, 415–429. [CrossRef]

51. Hancock, R.E.; Diamond, G. The role of cationic antimicrobial peptides in innate host defences. *Trends Microbiol.* **2000**, *8*, 402–410. [CrossRef]

52. Hilchie, A.L.; Wuerth, K.; Hancock, R.E. Immune modulation by multifaceted cationic host defense (antimicrobial) peptides. *Nat. Chem. Biol.* **2013**, *9*, 761–768. [CrossRef]

53. Epand, R.M.; Walker, C.; Epand, R.F.; Magarvey, N.A. Molecular mechanisms of membrane targeting antibiotics. *Biochim. Biophys. Acta* **2016**, *1858*, 980–987. [CrossRef]

54. Xu, N.; Wang, Y.S.; Pan, W.B.; Xiao, B.; Wen, Y.J.; Chen, X.C.; Chen, L.J.; Deng, H.X.; You, J.; Kan, B.; et al. Human alpha-defensin-1 inhibits growth of human lung adenocarcinoma xenograft in nude mice. *Mol. Cancer Ther.* **2008**, *7*, 1588–1597. [CrossRef]

55. Al-Benna, S.; Shai, Y.; Jacobsen, F.; Steinstraesser, L. Oncolytic activities of host defense peptides. *Int. J. Mol. Sci.* **2011**, *12*, 8027–8051. [CrossRef] [PubMed]

56. Baxter, A.A.; Lay, F.T.; Poon, I.K.H.; Kvansakul, M.; Hulett, M.D. Tumor cell membrane-targeting cationic antimicrobial peptides: Novel insights into mechanisms of action and therapeutic prospects. *Cell. Mol. Life Sci.* **2017**, *74*, 3809–3825. [CrossRef] [PubMed]

57. Rosca, E.V.; Koskimaki, J.E.; Rivera, C.G.; Pandey, N.B.; Tamiz, A.P.; Popel, A.S. Anti-angiogenic peptides for cancer therapeutics. *Curr. Pharm. Biotechnol.* **2011**, *12*, 1101–1116. [CrossRef] [PubMed]

58. Chavakis, T.; Cines, D.B.; Rhee, J.S.; Liang, O.D.; Schubert, U.; Hammes, H.P.; Higazi, A.A.; Nawroth, P.P.; Preissner, K.T.; Bdeir, K. Regulation of neovascularization by human neutrophil peptides (alpha-defensins): A link between inflammation and angiogenesis. *FASEB J.* **2004**, *18*, 1306–1308. [CrossRef] [PubMed]

59. Yoo, Y.C.; Watanabe, S.; Watanabe, R.; Hata, K.; Shimazaki, K.; Azuma, I. Bovine lactoferrin and lactoferricin, a peptide derived from bovine lactoferrin, inhibit tumor metastasis in mice. *Jpn. J. Cancer Res.* **1997**, *88*, 184–190. [CrossRef] [PubMed]

60. Tam, J.P.; Wang, S.; Wong, K.H.; Tan, W.L. Antimicrobial Peptides from Plants. *Pharmaceuticals* **2015**, *8*, 711–757. [CrossRef]

61. Brogden, K.A.; Ackermann, M.; McCray, P.B., Jr.; Tack, B.F. Antimicrobial peptides in animals and their role in host defences. *Int. J. Antimicrob. Agents* **2003**, *22*, 465–478. [CrossRef]

62. Yazici, A.; Ortucu, S.; Taskin, M.; Marinelli, L. Natural-based Antibiofilm and Antimicrobial Peptides from Microorganisms. *Curr. Top. Med. Chem.* **2018**, *18*, 2102–2107. [CrossRef]

63. Hassan, M.; Kjos, M.; Nes, I.F.; Diep, D.B.; Lotfipour, F. Natural antimicrobial peptides from bacteria: Characteristics and potential applications to fight against antibiotic resistance. *J. Appl. Microbiol.* **2012**, *113*, 723–736. [CrossRef]

64. Wong, J.H.; Ye, X.J.; Ng, T.B. Cathelicidins: Peptides with antimicrobial, immunomodulatory, anti-inflammatory, angiogenic, anticancer and procancer activities. *Curr. Protein Pept. Sci.* **2013**, *14*, 504–514. [CrossRef]

65. Agier, J.; Efenberger, M.; Brzezinska-Blaszczyk, E. Cathelicidin impact on inflammatory cells. *Cent. Eur. J. Immunol.* **2015**, *40*, 225–235. [CrossRef] [PubMed]
66. Braff, M.H.; Hawkins, M.A.; Di Nardo, A.; Lopez-Garcia, B.; Howell, M.D.; Wong, C.; Lin, K.; Streib, J.E.; Dorschner, R.; Leung, D.Y.; et al. Structure-function relationships among human cathelicidin peptides: Dissociation of antimicrobial properties from host immunostimulatory activities. *J. Immunol.* **2005**, *174*, 4271–4278. [CrossRef] [PubMed]
67. Tomasinsig, L.; Zanetti, M. The cathelicidins—Structure, function and evolution. *Curr. Protein Pept. Sci.* **2005**, *6*, 23–34. [CrossRef]
68. Gennaro, R.; Zanetti, M. Structural features and biological activities of the cathelicidin-derived antimicrobial peptides. *Biopolymers* **2000**, *55*, 31–49. [CrossRef]
69. Das, H.; Sharma, B.; Kumar, A. Cloning and characterization of novel cathelicidin cDNA sequence of Bubalus bubalis homologous to Bos taurus cathelicidin-4. *DNA Seq.* **2006**, *17*, 407–414. [CrossRef] [PubMed]
70. Brogden, K.A.; Kalfa, V.C.; Ackermann, M.R.; Palmquist, D.E.; McCray, P.B., Jr.; Tack, B.F. The ovine cathelicidin SMAP29 kills ovine respiratory pathogens in vitro and in an ovine model of pulmonary infection. *Antimicrob. Agents Chemother.* **2001**, *45*, 331–334. [CrossRef] [PubMed]
71. Zhao, C.; Nguyen, T.; Boo, L.M.; Hong, T.; Espiritu, C.; Orlov, D.; Wang, W.; Waring, A.; Lehrer, R.I. RL-37, an alpha-helical antimicrobial peptide of the rhesus monkey. *Antimicrob. Agents Chemother.* **2001**, *45*, 2695–2702. [CrossRef]
72. Nagaoka, I.; Tsutsumi-Ishii, Y.; Yomogida, S.; Yamashita, T. Isolation of cDNA encoding guinea pig neutrophil cationic antibacterial polypeptide of 11 kDa (CAP11) and evaluation of CAP11 mRNA expression during neutrophil maturation. *J. Biol. Chem.* **1997**, *272*, 22742–22750. [CrossRef]
73. Gallo, R.L.; Kim, K.J.; Bernfield, M.; Kozak, C.A.; Zanetti, M.; Merluzzi, L.; Gennaro, R. Identification of CRAMP, a cathelin-related antimicrobial peptide expressed in the embryonic and adult mouse. *J. Biol. Chem.* **1997**, *272*, 13088–13093. [CrossRef]
74. Termen, S.; Tollin, M.; Olsson, B.; Svenberg, T.; Agerberth, B.; Gudmundsson, G.H. Phylogeny, processing and expression of the rat cathelicidin rCRAMP: A model for innate antimicrobial peptides. *Cell. Mol. Life Sci.* **2003**, *60*, 536–549. [CrossRef]
75. Xiao, Y.; Cai, Y.; Bommineni, Y.R.; Fernando, S.C.; Prakash, O.; Gilliland, S.E.; Zhang, G. Identification and functional characterization of three chicken cathelicidins with potent antimicrobial activity. *J. Biol. Chem.* **2006**, *281*, 2858–2867. [CrossRef] [PubMed]
76. Zhao, H.; Gan, T.X.; Liu, X.D.; Jin, Y.; Lee, W.H.; Shen, J.H.; Zhang, Y. Identification and characterization of novel reptile cathelicidins from elapid snakes. *Peptides* **2008**, *29*, 1685–1691. [CrossRef] [PubMed]
77. Uzzell, T.; Stolzenberg, E.D.; Shinnar, A.E.; Zasloff, M. Hagfish intestinal antimicrobial peptides are ancient cathelicidins. *Peptides* **2003**, *24*, 1655–1667. [CrossRef] [PubMed]
78. Junior, N.G.O.; Cardoso, M.H.; Candido, E.S.; van den Broek, D.; de Lange, N.; Velikova, N.; Kleijn, J.M.; Wells, J.M.; Rezende, T.M.B.; Franco, O.L.; et al. An acidic model pro-peptide affects the secondary structure, membrane interactions and antimicrobial activity of a crotalicidin fragment. *Sci. Rep.* **2018**, *8*, 11127. [CrossRef]
79. Zhang, Y.; Zhao, H.; Yu, G.Y.; Liu, X.D.; Shen, J.H.; Lee, W.H.; Zhang, Y. Structure-function relationship of king cobra cathelicidin. *Peptides* **2010**, *31*, 1488–1493. [CrossRef]
80. Azim, S.; McDowell, D.; Cartagena, A.; Rodriguez, R.; Laughlin, T.F.; Ahmad, Z. Venom peptides cathelicidin and lycotoxin cause strong inhibition of Escherichia coli ATP synthase. *Int. J. Biol. Macromol.* **2016**, *87*, 246–251. [CrossRef]
81. Li, S.A.; Lee, W.H.; Zhang, Y. Efficacy of OH-CATH30 and its analogs against drug-resistant bacteria in vitro and in mouse models. *Antimicrob. Agents Chemother.* **2012**, *56*, 3309–3317. [CrossRef]
82. Zhao, F.; Lan, X.Q.; Du, Y.; Chen, P.Y.; Zhao, J.; Zhao, F.; Lee, W.H.; Zhang, Y. King cobra peptide OH-CATH30 as a potential candidate drug through clinic drug-resistant isolates. *Zool. Res.* **2018**, *39*, 87–96.
83. Chen, X.-x.; Yu, G.-y.; Zhan, Y.; Zhang, Y.; Shen, J.-h.; Lee, W.-h. Effects of the Antimicrobial Peptide OH-CATH on Escherichia coli. *Zool. Res.* **2009**, *30*, 171–177. [CrossRef]
84. Falcao, C.B.; de La Torre, B.G.; Perez-Peinado, C.; Barron, A.E.; Andreu, D.; Radis-Baptista, G. Vipericidins: A novel family of cathelicidin-related peptides from the venom gland of South American pit vipers. *Amino Acids* **2014**, *46*, 2561–2571. [CrossRef]

85. Tajbakhsh, M.; Karimi, A.; Tohidpour, A.; Abbasi, N.; Fallah, F.; Akhavan, M.M. The antimicrobial potential of a new derivative of cathelicidin from Bungarus fasciatus against methicillin-resistant Staphylococcus aureus. *J. Microbiol.* **2018**, *56*, 128–137. [CrossRef] [PubMed]

86. Tajbakhsh, M.; Akhavan, M.M.; Fallah, F.; Karimi, A. A Recombinant Snake Cathelicidin Derivative Peptide: Antibiofilm Properties and Expression in Escherichia coli. *Biomolecules* **2018**, *8*, 118. [CrossRef] [PubMed]

87. Jin, L.; Bai, X.; Luan, N.; Yao, H.; Zhang, Z.; Liu, W.; Chen, Y.; Yan, X.; Rong, M.; Lai, R.; et al. A Designed Tryptophan- and Lysine/Arginine-Rich Antimicrobial Peptide with Therapeutic Potential for Clinical Antibiotic-Resistant Candida albicans Vaginitis. *J. Med. Chem.* **2016**, *59*, 1791–1799. [CrossRef] [PubMed]

88. Liu, C.; Qi, J.; Shan, B.; Gao, R.; Gao, F.; Xie, H.; Yuan, M.; Liu, H.; Jin, S.; Wu, F.; et al. Pretreatment with cathelicidin-BF ameliorates Pseudomonas aeruginosa pneumonia in mice by enhancing NETosis and the autophagy of recruited neutrophils and macrophages. *Int. Immunopharmacol.* **2018**, *65*, 382–391. [CrossRef]

89. Zhang, H.; Zhang, B.; Zhang, X.; Wang, X.; Wu, K.; Guan, Q. Effects of cathelicidin-derived peptide from reptiles on lipopolysaccharide-induced intestinal inflammation in weaned piglets. *Vet. Immunol. Immunopathol.* **2017**, *192*, 41–53. [CrossRef]

90. Zhang, H.; Xia, X.; Han, F.; Jiang, Q.; Rong, Y.; Song, D.; Wang, Y. Cathelicidin-BF, a Novel Antimicrobial Peptide from Bungarus fasciatus, Attenuates Disease in a Dextran Sulfate Sodium Model of Colitis. *Mol. Pharm.* **2015**, *12*, 1648–1661. [CrossRef]

91. Yi, H.; Yu, C.; Zhang, H.; Song, D.; Jiang, D.; Du, H.; Wang, Y. Cathelicidin-BF suppresses intestinal inflammation by inhibiting the nuclear factor-kappaB signaling pathway and enhancing the phagocytosis of immune cells via STAT-1 in weanling piglets. *Int. Immunopharmacol.* **2015**, *28*, 61–69. [CrossRef]

92. Song, D.; Zong, X.; Zhang, H.; Wang, T.; Yi, H.; Luan, C.; Wang, Y. Antimicrobial peptide Cathelicidin-BF prevents intestinal barrier dysfunction in a mouse model of endotoxemia. *Int. Immunopharmacol.* **2015**, *25*, 141–147. [CrossRef]

93. Wang, H.; Ke, M.; Tian, Y.; Wang, J.; Li, B.; Wang, Y.; Dou, J.; Zhou, C. BF-30 selectively inhibits melanoma cell proliferation via cytoplasmic membrane permeabilization and DNA-binding in vitro and in B16F10-bearing mice. *Eur. J. Pharmacol.* **2013**, *707*, 1–10. [CrossRef]

94. Zhou, H.; Dou, J.; Wang, J.; Chen, L.; Wang, H.; Zhou, W.; Li, Y.; Zhou, C. The antibacterial activity of BF-30 in vitro and in infected burned rats is through interference with cytoplasmic membrane integrity. *Peptides* **2011**, *32*, 1131–1138. [CrossRef]

95. Wang, Y.; Zhang, Z.; Chen, L.; Guang, H.; Li, Z.; Yang, H.; Li, J.; You, D.; Yu, H.; Lai, R. Cathelicidin-BF, a snake cathelicidin-derived antimicrobial peptide, could be an excellent therapeutic agent for acne vulgaris. *PLoS ONE* **2011**, *6*, e22120. [CrossRef] [PubMed]

96. Chen, W.; Yang, B.; Zhou, H.; Sun, L.; Dou, J.; Qian, H.; Huang, W.; Mei, Y.; Han, J. Structure-activity relationships of a snake cathelicidin-related peptide, BF-15. *Peptides* **2011**, *32*, 2497–2503. [CrossRef] [PubMed]

97. Wang, Y.; Hong, J.; Liu, X.; Yang, H.; Liu, R.; Wu, J.; Wang, A.; Lin, D.; Lai, R. Snake cathelicidin from Bungarus fasciatus is a potent peptide antibiotics. *PLoS ONE* **2008**, *3*, e3217. [CrossRef]

98. Amer, L.S.; Bishop, B.M.; van Hoek, M.L. Antimicrobial and antibiofilm activity of cathelicidins and short, synthetic peptides against Francisella. *Biochem. Biophys. Res. Commun.* **2010**, *396*, 246–251. [CrossRef]

99. de Latour, F.A.; Amer, L.S.; Papanstasiou, E.A.; Bishop, B.M.; van Hoek, M.L. Antimicrobial activity of the Naja atra cathelicidin and related small peptides. *Biochem. Biophys. Res. Commun.* **2010**, *396*, 825–830. [CrossRef]

100. Dean, S.N.; Bishop, B.M.; van Hoek, M.L. Susceptibility of Pseudomonas aeruginosa Biofilm to Alpha-Helical Peptides: D-enantiomer of LL-37. *Front. Microbiol.* **2011**, *2*, 128. [CrossRef]

101. Dean, S.N.; Bishop, B.M.; van Hoek, M.L. Natural and synthetic cathelicidin peptides with anti-microbial and anti-biofilm activity against Staphylococcus aureus. *BMC Microbiol.* **2011**, *11*, 114. [CrossRef]

102. Gupta, K.; Singh, S.; van Hoek, M.L. Short, Synthetic Cationic Peptides Have Antibacterial Activity against Mycobacterium smegmatis by Forming Pores in Membrane and Synergizing with Antibiotics. *Antibiotics* **2015**, *4*, 358–378. [CrossRef]

103. Blower, R.J.; Barksdale, S.M.; van Hoek, M.L. Snake Cathelicidin NA-CATH and Smaller Helical Antimicrobial Peptides Are Effective against Burkholderia thailandensis. *PLoS Negl. Trop. Dis.* **2015**, *9*, e0003862. [CrossRef]

104. Blower, R.J.; Popov, S.G.; van Hoek, M.L. Cathelicidin peptide rescues G. mellonella infected with B. anthracis. *Virulence* **2017**, *9*, 287–293. [CrossRef]

105. Wei, L.; Gao, J.; Zhang, S.; Wu, S.; Xie, Z.; Ling, G.; Kuang, Y.Q.; Yang, Y.; Yu, H.; Wang, Y. Identification and Characterization of the First Cathelicidin from Sea Snakes with Potent Antimicrobial and Anti-inflammatory Activity and Special Mechanism. *J. Biol. Chem.* **2015**, *290*, 16633–16652. [CrossRef]

106. Carlile, S.R.; Shiels, J.; Kerrigan, L.; Delaney, R.; Megaw, J.; Gilmore, B.F.; Weldon, S.; Dalton, J.P.; Taggart, C.C. Sea snake cathelicidin (Hc-cath) exerts a protective effect in mouse models of lung inflammation and infection. *Sci. Rep.* **2019**, *9*, 6071. [CrossRef]

107. Falcao, C.B.; Perez-Peinado, C.; de la Torre, B.G.; Mayol, X.; Zamora-Carreras, H.; Jimenez, M.A.; Radis-Baptista, G.; Andreu, D. Structural Dissection of Crotalicidin, a Rattlesnake Venom Cathelicidin, Retrieves a Fragment with Antimicrobial and Antitumor Activity. *J. Med. Chem.* **2015**, *58*, 8553–8563. [CrossRef]

108. Cavalcante, C.S.; Falcao, C.B.; Fontenelle, R.O.; Andreu, D.; Radis-Baptista, G. Anti-fungal activity of Ctn[15–34], the C-terminal peptide fragment of crotalicidin, a rattlesnake venom gland cathelicidin. *J. Antibiot.* **2016**, *70*, 231–237. [CrossRef]

109. Bandeira, I.C.J.; Bandeira-Lima, D.; Mello, C.P.; Pereira, T.P.; De Menezes, R.; Sampaio, T.L.; Falcao, C.B.; Radis-Baptista, G.; Martins, A.M.C. Antichagasic effect of crotalicidin, a cathelicidin-like vipericidin, found in Crotalus durissus terrificus rattlesnake's venom gland. *Parasitology* **2017**, *145*, 1059–1064. [CrossRef]

110. Oliveira-Junior, N.G.; Freire, M.S.; Almeida, J.A.; Rezende, T.M.B.; Franco, O.L. Antimicrobial and proinflammatory effects of two vipericidins. *Cytokine* **2018**, *111*, 309–316. [CrossRef]

111. Mello, C.P.; Lima, D.B.; Menezes, R.R.; Bandeira, I.C.; Tessarolo, L.D.; Sampaio, T.L.; Falcao, C.B.; Radis-Baptista, G.; Martins, A.M. Evaluation of the antichagasic activity of batroxicidin, a cathelicidin-related antimicrobial peptide found in Bothrops atrox venom gland. *Toxicon* **2017**, *130*, 56–62. [CrossRef]

112. Cai, S.; Qiao, X.; Feng, L.; Shi, N.; Wang, H.; Yang, H.; Guo, Z.; Wang, M.; Chen, Y.; Wang, Y.; et al. Python Cathelicidin CATHPb1 Protects against Multidrug-Resistant Staphylococcal Infections by Antimicrobial-Immunomodulatory Duality. *J. Med. Chem.* **2018**, *61*, 2075–2086. [CrossRef]

113. Kim, D.; Soundrarajan, N.; Lee, J.; Cho, H.S.; Choi, M.; Cha, S.Y.; Ahn, B.; Jeon, H.; Le, M.T.; Song, H.; et al. Genome wide analysis of the antimicrobial peptides in Python bivittatus and characterization of cathelicidins with potent antimicrobial activity and low cytotoxicity. *Antimicrob. Agents Chemother.* **2017**, *61*, e00530-17. [CrossRef]

114. Wang, A.; Zhang, F.; Guo, Z.; Chen, Y.; Zhang, M.; Yu, H.; Wang, Y. Characterization of a Cathelicidin from the Colubrinae Snake, Sinonatrix annularis. *Zool. Sci.* **2019**, *36*, 68–76. [CrossRef]

115. Madeira, F.; Park, Y.M.; Lee, J.; Buso, N.; Gur, T.; Madhusoodanan, N.; Basutkar, P.; Tivey, A.R.N.; Potter, S.C.; Finn, R.D.; et al. The EMBL-EBI search and sequence analysis tools APIs in 2019. *Nucleic Acids Res.* **2019**, *47*, W636–W641. [CrossRef]

116. Waterhouse, A.M.; Procter, J.B.; Martin, D.M.; Clamp, M.; Barton, G.J. Jalview Version 2–a multiple sequence alignment editor and analysis workbench. *Bioinformatics* **2009**, *25*, 1189–1191. [CrossRef]

117. UniProt, C. UniProt: A worldwide hub of protein knowledge. *Nucleic Acids Res.* **2019**, *47*, D506–D515.

118. Du, H.; Samuel, R.L.; Massiah, M.A.; Gillmor, S.D. The structure and behavior of the NA-CATH antimicrobial peptide with liposomes. *Biochim. Biophys. Acta* **2015**, *1848*, 2394–2405. [CrossRef]

119. Juba, M.; Porter, D.; Dean, S.; Gillmor, S.; Bishop, B. Characterization and performance of short cationic antimicrobial peptide isomers. *Biopolymers* **2013**, *100*, 387–401. [CrossRef]

120. Samuel, R.; Gillmor, S. Membrane phase characteristics control NA-CATH activity. *Biochim. Biophys. Acta* **2016**, *1858*, 1974–1982. [CrossRef]

121. Xia, X.; Zhang, L.; Wang, Y. The antimicrobial peptide cathelicidin-BF could be a potential therapeutic for Salmonella typhimurium infection. *Microbiol. Res.* **2015**, *171*, 45–51. [CrossRef]

122. Perez-Peinado, C.; Dias, S.A.; Domingues, M.M.; Benfield, A.H.; Freire, J.M.; Radis-Baptista, G.; Gaspar, D.; Castanho, M.; Craik, D.J.; Henriques, S.T.; et al. Mechanisms of bacterial membrane permeabilization by crotalicidin (Ctn) and its fragment Ctn(15-34), antimicrobial peptides from rattlesnake venom. *J. Biol. Chem.* **2018**, *293*, 1536–1549. [CrossRef]

123. Cavalcante, C.S.P.; de Aguiar, F.L.L.; Fontenelle, R.O.S.; de Menezes, R.; Martins, A.M.C.; Falcao, C.B.; Andreu, D.; Radis-Baptista, G. Insights into the candidacidal mechanism of Ctn[15–34]—A carboxyl-terminal, crotalicidin-derived peptide related to cathelicidins. *J. Med Microbiol.* **2018**, *67*, 129–138. [CrossRef]

124. Zhang, Z.; Mu, L.; Tang, J.; Duan, Z.; Wang, F.; Wei, L.; Rong, M.; Lai, R. A small peptide with therapeutic potential for inflammatory acne vulgaris. *PLoS ONE* **2013**, *8*, e72923. [CrossRef]

125. Hao, Q.; Wang, H.; Wang, J.; Dou, J.; Zhang, M.; Zhou, W.; Zhou, C. Effective antimicrobial activity of Cbf-K16 and Cbf-A7 A13 against NDM-1-carrying Escherichia coli by DNA binding after penetrating the cytoplasmic membrane in vitro. *J. Pept. Sci.* **2013**, *19*, 173–180. [CrossRef] [PubMed]

126. Tian, Y.; Wang, H.; Li, B.; Ke, M.; Wang, J.; Dou, J.; Zhou, C. The cathelicidin-BF Lys16 mutant Cbf-K16 selectively inhibits non-small cell lung cancer proliferation in vitro. *Oncol. Rep.* **2013**, *30*, 2502–2510. [CrossRef] [PubMed]

127. Fang, Y.; He, X.; Zhang, P.; Shen, C.; Mwangi, J.; Xu, C.; Mo, G.; Lai, R.; Zhang, Z. In Vitro and In Vivo Antimalarial Activity of LZ1, a Peptide Derived from Snake Cathelicidin. *Toxins* **2019**, *11*, 379. [CrossRef] [PubMed]

128. Li, L.; Wang, Q.; Li, H.; Yuan, M.; Yuan, M. Preparation, characterization, in vitro release and degradation of cathelicidin-BF-30-PLGA microspheres. *PLoS ONE* **2014**, *9*, e100809. [CrossRef]

129. Bao, Y.; Wang, S.; Li, H.; Wang, Y.; Chen, H.; Yuan, M. Characterization, Stability and Biological Activity In Vitro of Cathelicidin-BF-30 Loaded 4-Arm Star-Shaped PEG-PLGA Microspheres. *Molecules* **2018**, *23*, 497. [CrossRef]

130. Luan, C.; Zhang, H.W.; Song, D.G.; Xie, Y.G.; Feng, J.; Wang, Y.Z. Expressing antimicrobial peptide cathelicidin-BF in Bacillus subtilis using SUMO technology. *Appl. Microbiol. Biotechnol.* **2014**, *98*, 3651–3658. [CrossRef]

131. He, Q.; Fu, A.Y.; Li, T.J. Expression and one-step purification of the antimicrobial peptide cathelicidin-BF using the intein system in Bacillus subtilis. *J. Ind. Microbiol. Biotechnol.* **2015**, *42*, 647–653. [CrossRef]

132. Perez-Peinado, C.; Defaus, S.; Sans-Comerma, L.; Valle, J.; Andreu, D. Decoding the human serum interactome of snake-derived antimicrobial peptide Ctn[15–34]: Toward an explanation for unusually long half-life. *J. Proteom.* **2019**, *204*, 103372. [CrossRef]

133. Perez-Peinado, C.; Dias, S.A.; Mendonca, D.A.; Castanho, M.; Veiga, A.S.; Andreu, D. Structural determinants conferring unusual long life in human serum to rattlesnake-derived antimicrobial peptide Ctn[15–34]. *J. Pept. Sci.* **2019**, *25*, e3195. [CrossRef]

134. De Aguiar, F.L.L.; de Paula Cavalcante, C.S.; Dos Santos Fontenelle, R.O.; Falcao, C.B.; Andreu, D.; Radis-Baptista, G. The antiproliferative peptide Ctn[15–34] is active against multidrug-resistant yeasts Candida albicans and Cryptococcus neoformans. *J. Appl. Microbiol.* **2019**, *2*, 414–425. [CrossRef]

135. Vieira-Girao, P.R.N.; Falcao, C.B.; Rocha, I.; Lucena, H.M.R.; Costa, F.H.F.; Radis-Baptista, G. Antiviral Activity of Ctn[15–34], A Cathelicidin-Derived Eicosapeptide, Against Infectious Myonecrosis Virus in Litopenaeus vannamei Primary Hemocyte Cultures. *Food Environ. Virol.* **2017**, *9*, 277–286. [CrossRef] [PubMed]

136. Yang, J.; Zhang, Y. I-TASSER server: New development for protein structure and function predictions. *Nucleic Acids Res.* **2015**, *43*, W174–W181. [CrossRef] [PubMed]

137. DeLano, W.L. *The PyMOL Molecular Graphics System*; DeLano Scientific: San Carlos, CA, USA, 2002.

138. De Smet, K.; Contreras, R. Human antimicrobial peptides: Defensins, cathelicidins and histatins. *Biotechnol. Lett.* **2005**, *27*, 1337–1347. [CrossRef] [PubMed]

139. Ganz, T. Defensins: Antimicrobial peptides of innate immunity. *Nat. Rev. Immunol.* **2003**, *3*, 710–720. [CrossRef]

140. Correa, P.G.; Oguiura, N. Phylogenetic analysis of beta-defensin-like genes of Bothrops, Crotalus and Lachesis snakes. *Toxicon* **2013**, *69*, 65–74. [CrossRef]

141. Van Hoek, M.L. Antimicrobial peptides in reptiles. *Pharmaceuticals* **2014**, *7*, 723–753. [CrossRef]

142. De Oliveira, Y.S.; Correa, P.G.; Oguiura, N. Beta-defensin genes of the Colubridae snakes Phalotris mertensi, Thamnodynastes hypoconia, and T. strigatus. *Toxicon* **2018**, *146*, 124–128. [CrossRef]

143. Coronado, M.A.; Gabdulkhakov, A.; Georgieva, D.; Sankaran, B.; Murakami, M.T.; Arni, R.K.; Betzel, C. Structure of the polypeptide crotamine from the Brazilian rattlesnake Crotalus durissus terrificus. *Acta Crystallogr. Sect. D Biol. Crystallogr.* **2013**, *69*, 1958–1964. [CrossRef]

144. Kerkis, I.; Hayashi, M.A.; Prieto da Silva, A.R.; Pereira, A.; De Sa Junior, P.L.; Zaharenko, A.J.; Radis-Baptista, G.; Kerkis, A.; Yamane, T. State of the art in the studies on crotamine, a cell penetrating peptide from South American rattlesnake. *BioMed Res. Int.* **2014**, *2014*, 675985. [CrossRef]

145. Oguiura, N.; Boni-Mitake, M.; Affonso, R.; Zhang, G. In vitro antibacterial and hemolytic activities of crotamine, a small basic myotoxin from rattlesnake Crotalus durissus. *J. Antibiot.* **2011**, *64*, 327–331. [CrossRef]

146. Yamane, E.S.; Bizerra, F.C.; Oliveira, E.B.; Moreira, J.T.; Rajabi, M.; Nunes, G.L.; de Souza, A.O.; da Silva, I.D.; Yamane, T.; Karpel, R.L.; et al. Unraveling the antifungal activity of a South American rattlesnake toxin crotamine. *Biochimie* **2013**, *95*, 231–240. [CrossRef] [PubMed]

147. Pereira, A.; Kerkis, A.; Hayashi, M.A.; Pereira, A.S.; Silva, F.S.; Oliveira, E.B.; Prieto da Silva, A.R.; Yamane, T.; Radis-Baptista, G.; Kerkis, I. Crotamine toxicity and efficacy in mouse models of melanoma. *Expert Opin. Investig. Drugs* **2011**, *20*, 1189–1200. [CrossRef]

148. Nascimento, F.D.; Sancey, L.; Pereira, A.; Rome, C.; Oliveira, V.; Oliveira, E.B.; Nader, H.B.; Yamane, T.; Kerkis, I.; Tersariol, I.L.; et al. The natural cell-penetrating peptide crotamine targets tumor tissue in vivo and triggers a lethal calcium-dependent pathway in cultured cells. *Mol. Pharm.* **2012**, *9*, 211–221. [CrossRef] [PubMed]

149. Ponnappan, N.; Budagavi, D.P.; Chugh, A. CyLoP-1: Membrane-active peptide with cell-penetrating and antimicrobial properties. *Biochim. Biophys. Acta* **2017**, *1859*, 167–176. [CrossRef] [PubMed]

150. Radis-Baptista, G.; de la Torre, B.G.; Andreu, D. Insights into the uptake mechanism of NrTP, a cell-penetrating peptide preferentially targeting the nucleolus of tumour cells. *Chem. Biol. Drug Des.* **2012**, *79*, 907–915. [CrossRef] [PubMed]

151. Rodrigues, M.; Andreu, D.; Santos, N.C. Uptake and cellular distribution of nucleolar targeting peptides (NrTPs) in different cell types. *Biopolymers* **2015**, *104*, 101–109. [CrossRef]

152. Torres, A.M.; Wong, H.Y.; Desai, M.; Moochhala, S.; Kuchel, P.W.; Kini, R.M. Identification of a novel family of proteins in snake venoms. Purification and structural characterization of nawaprin from Naja nigricollis snake venom. *J. Biol. Chem.* **2003**, *278*, 40097–40104. [CrossRef]

153. Nair, D.G.; Fry, B.G.; Alewood, P.; Kumar, P.P.; Kini, R.M. Antimicrobial activity of omwaprin, a new member of the waprin family of snake venom proteins. *Biochem. J.* **2007**, *402*, 93–104. [CrossRef]

154. Banigan, J.R.; Mandal, K.; Sawaya, M.R.; Thammavongsa, V.; Hendrickx, A.P.; Schneewind, O.; Yeates, T.O.; Kent, S.B. Determination of the X-ray structure of the snake venom protein omwaprin by total chemical synthesis and racemic protein crystallography. *Protein Sci.* **2010**, *19*, 1840–1849. [CrossRef]

155. Thankappan, B.; Angayarkanni, J. Biological characterization of omw1 and omw2: Antimicrobial peptides derived from omwaprin. *3 Biotech* **2019**, *9*, 295. [CrossRef]

156. Cooper, M.A.; Shlaes, D. Fix the antibiotics pipeline. *Nature* **2011**, *472*, 32. [CrossRef] [PubMed]

157. Tomaras, A.P.; Dunman, P.M. In the midst of the antimicrobial discovery conundrum: An overview. *Curr. Opin. Microbiol.* **2015**, *27*, 103–107. [CrossRef] [PubMed]

158. Chabner, B.A.; Roberts, T.G., Jr. Timeline: Chemotherapy and the war on cancer. *Nat. Rev. Cancer* **2005**, *5*, 65–72. [CrossRef] [PubMed]

159. Kelderman, S.; Schumacher, T.N.; Haanen, J.B. Acquired and intrinsic resistance in cancer immunotherapy. *Mol. Oncol.* **2014**, *8*, 1132–1139. [CrossRef]

160. Waghu, F.H.; Barai, R.S.; Gurung, P.; Idicula-Thomas, S. CAMPR3: A database on sequences, structures and signatures of antimicrobial peptides. *Nucleic Acids Res.* **2015**, *44*, D1094–D1097. [CrossRef]

161. Tyagi, A.; Tuknait, A.; Anand, P.; Gupta, S.; Sharma, M.; Mathur, D.; Joshi, A.; Singh, S.; Gautam, A.; Raghava, G.P. CancerPPD: A database of anticancer peptides and proteins. *Nucleic Acids Res.* **2014**, *43*, D837–D843. [CrossRef]

Article

Hadrurid Scorpion Toxins: Evolutionary Conservation and Selective Pressures

Carlos E. Santibáñez-López [1], Matthew R. Graham [1], Prashant P. Sharma [2], Ernesto Ortiz [3] and Lourival D. Possani [3,*]

[1] Department of Biology, Eastern Connecticut State University, 83 Windham St. Willimantic, CT 06266, USA; santibanezlopezc@easternct.edu (C.E.S.-L.); grahamm@easternct.edu (M.R.G.)

[2] Department of Integrative Biology, University of Wisconsin-Madison, 430 Lincoln Drive, Madison, WI 53706, USA; prashant.sharma@wisc.edu

[3] Departamento de Medicina Molecular y Bioprocesos, Instituto de Biotecnología, Universidad Nacional Autónoma de México, Avenida Universidad 2001, Cuernavaca, Morelos 62210, Mexico; erne@ibt.unam.mx

* Correspondence: possani@ibt.unam.mx

Received: 20 September 2019; Accepted: 30 October 2019; Published: 1 November 2019

Abstract: Scorpion toxins are thought to have originated from ancestral housekeeping genes that underwent diversification and neofunctionalization, as a result of positive selection. Our understanding of the evolutionary origin of these peptides is hindered by the patchiness of existing taxonomic sampling. While recent studies have shown phylogenetic inertia in some scorpion toxins at higher systematic levels, evolutionary dynamics of toxins among closely related taxa remain unexplored. In this study, we used new and previously published transcriptomic resources to assess evolutionary relationships of closely related scorpions from the family Hadruridae and their toxins. In addition, we surveyed the incidence of scorpine-like peptides (SLP, a type of potassium channel toxin), which were previously known from 21 scorpion species. We demonstrate that scorpine-like peptides exhibit gene duplications. Our molecular analyses demonstrate that only eight sites of two SLP copies found in scorpions are evolving under positive selection, with more sites evolving under negative selection, in contrast to previous findings. These results show evolutionary conservation in toxin diversity at shallow taxonomic scale.

Keywords: evolutionary shifts; Hadruridae; negative selection; phylogenomics; venom transcriptome

Key Contribution: Multiple gene copies encoding scorpine-like peptides occur in iurid scorpions. We show that these paralogs evolved under negative selection.

---

1. Introduction

Animal venoms are outstanding evolutionary innovations used to subdue and digest prey, with other important functions, such as self-defense or intraspecific conflicts [1–5]. Venoms are secretions consisting of organic and inorganic components that alter cellular processes in the target organism. Several arthropod groups are well known for their venoms, such as bees and wasps [6,7], centipedes [8–10], remipedes [11], and arachnids (e.g., see [12,13]). Some of the most efficient venoms occur in scorpions, which possess venom cocktails rich in peptides that affect ion channels [14–16]. Toxins are thought to have arisen from diversification of paralogs of housekeeping genes, followed by neofunctionalization driven by positive selection in coding regions [15,17–20].

Our knowledge of scorpion toxins has benefitted from high-throughput sequencing technologies, though taxonomically, sampling favors the diverse scorpion family Buthidae. In recent years, however, transcriptomic analyses have expanded beyond the Buthidae to the other 19 scorpion families (e.g., see [21–26]). These efforts have demonstrated remarkable phylogenetic inertia in scorpion venom components at higher-level categories, but little is known about how venom composition and toxin evolution vary among species below the family level.

The North American scorpion family Hadruridae is represented by nine species in two genera, including several of the largest species worldwide [27]. Of these, the diversity and effects of venom components from *Hoffmannihadrurus gertschi* have been extensively studied [28–30]. More recently, a transcriptome analysis, conducted on the venom gland of *Hadrurus spadix*, was published [31]. Though both species showed similar venom composition, a comparative analysis can shed light on the evolution of toxin components among these and other closely related species.

In this study, we focused on five hadrurid species: *Hadrurus arizonensis*, *Hadrurus spadix*, *Hadrurus concolorous*, *Hoffmannihadrurus aztecus*, and *Hoffmannihadrurus gertschi*. We used RNA-Seq to sequence de novo the venom gland transcriptome of *Hoffmannihadrurus aztecus* and *Hadrurus concolorous*.

We assessed the commonality of venom composition by comparing the number of putative transcripts found in the five species. Specifically, we focused on the evolutionary history of scorpine-like peptides (SLP, sensu [22]). Originally, SLPs were called 'orphan peptides' due to the presence of two domains with dual functionality. The N-terminal region has cytolytic activity, whereas the C-terminus blocks potassium channel activity [32–34]. SLPs were first isolated from the venom of *Pandinus imperator* [35]. They have now been described in one or two gene copies, from 21 scorpion species, including nine buthids and 12 species from 10 other, non-buthid families [22]. Intriguingly, buthid SLPs (with the exception of one sequence) and non-buthid SLPs form two mutually monophyletic clusters that are phylogenetically restricted to the parvorders Buthida and Iurida, respectively [22], suggesting that the origin of SLPs could predate the diversification of scorpions. To assess SLP evolution in the context of comprehensive hadrurid phylogeny, we inferred internal hadrurid relationships and branch lengths using phylogenomic datasets, and surveyed venom gland transcriptomes to discover and map the distribution of non-buthid SLP homologs. We then inferred the direction of selection acting on the codon sequences of non-buthid SLP genes and evaluated the evolutionary dynamics of paralogs of SLPs within the Hadruridae.

2. Results

2.1. Hadrurid Phylogenomic Tree and Divergence Time Estimate

Maximum likelihood (ML) analysis and species tree reconciliation of a 1982-locus matrix (727,571 amino acid sites, 35.6% missing data) yielded a topology that was largely congruent with phylogenomic trees previously published by us [27,36]. The monophyly of the two basal clades Buthidae and Iuridae, along with relationships of all superfamilies within the Iuridae, were recovered with 100% nodal support (Figure 1 and Figure S1). Monophyly of the Hadrurinae, *Hadrurus*, and *Hoffmannihadrurus* were all strongly supported (Figure 1). *Cercophonius* and *Urodacus* (two distantly related lineages with sampling of multiple congeners and patristic distances comparable to those within *Hadrurus*) were both monophyletic with strong support (Figure 1). Estimates of the time to the most recent common ancestor (tMRCA) were 7–11 Myr for the Hadrurinae, 8–34 Myr for the Urodacidae, and 86–196 Myr for the Bothriuridae (Figures S2 and S3), in agreement with previous hypotheses [37].

2.2. The Repertoire of Venom-Specific Transcripts in the H. concolorous and H. aztecus Libraries

After sequencing, quality assessment, and rRNA and adapter removal, 92,004,852 reads were assembled into 122,574 putative transcripts from *H. concolorous* (N50:1467 bp) and 44,355,818 reads were assembled into 83,643 putative transcripts from *H. aztecus* (N50:709 bp). Of these, 22,189 *H. concolorous* sequences and 28,101 *H. aztecus* sequences were identified, matching gene sequences in databases. We recorded 74 putative coding transcripts for *H. concolorous* and 96 for *H. aztecus*, representing the following venom components: potassium-channel toxins (KTx), sodium-channel toxins (NaTx), enzymes, protease inhibitors, La1-like peptides, host-defense peptides (HDPs), members of the cysteine-rich secretory protein, antigen 5, and pathogenesis-related 1 protein (CAP) superfamily, scorpine-like peptides (SLP), and other venom components (Figure 2A, Tables S1 and S2). Comparative venom transcriptomic analysis showed similar numbers of putative transcripts in the venom repertoires of *H. gertschi* and *H. spadix* (except for KTx), and among *Urodacus* and *Cercophonius* species (Figure 2B).

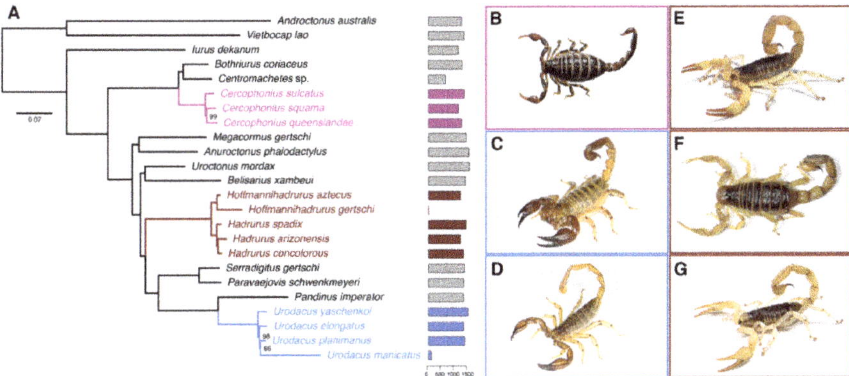

Figure 1. (**A**) Maximum likelihood tree topology recovered from the analysis of 1982 genes, with 24 scorpion species. Bars to the right of termini indicate numbers of orthologs. Numbers on nodes indicate ultrafast bootstrap support values under 100. (**B–G**) Representative species of scorpions studied here: (**B**) *Cercophonius squama* (Gervais, 1843); (**C**) *Urodacus yaschenkoi* (Birula, 1903); (**D**) *Urodacus elongatus* Koch, 1977; (**E**) *Hadrurus spadix* Stahnke, 1940; (**F**) *Hadrurus concolorous* Stahnke, 1969; (**G**) *Hadrurus arizonensis* Ewing, 1928. Photos by Nick Volpe (**C,D**), Matthew Graham (**B,E,G**) and Carlos Santibañez (**F**).

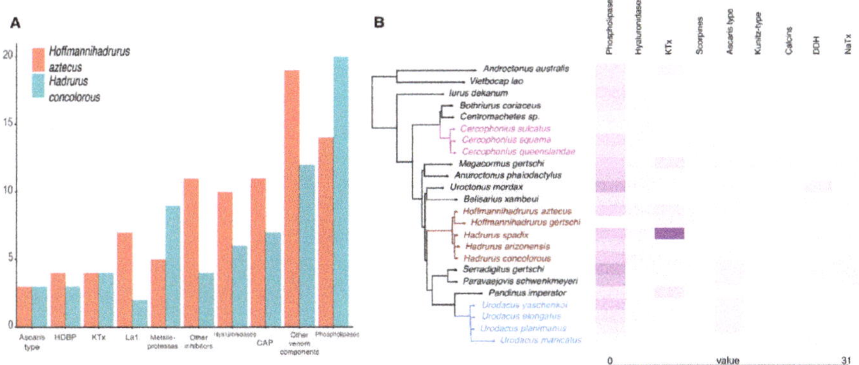

Figure 2. Comparative transcriptomic analyses. (**A**) Distribution of the annotated transcripts from the venom gland transcriptomes of *Hadrurus concolorous* and *Hoffmannihadrurus aztecus*, according to protein families. (**B**) Comparison of selected venom protein families in the libraries of the scorpion species studied herein.

2.3. Molecular Evolution of Scorpine-Like Peptides (SLP)

Eighty-four scorpine-like sequences were retrieved from the libraries generated herein, with others from GenBank, UniProt, and previous studies. Maximum Likelihood (ML) and Bayesian Inference (BI) gene trees recovered SLPs as a monophyletic cluster with strong support (Figure 3A). Among SLPs, three clades including only buthid sequences and three clades including mainly non-buthid sequences were recovered, in agreement with previous hypotheses (in [22]; Figure 3A and Figure S4). We observed a distal duplication of SLP (SLP1 and SLP2, Figure 3 and Figure S4) that is retained by a subset of iurid scorpions (the clade excluding the Iuridae and Bothriuridae). Our results showed that *H. concolorous* and *H. aztecus* have three putative SLPs, whereas *H. gertschi* has two and *H. spadix* one (as opposed to four, as reported by [31], missing in our assembly).

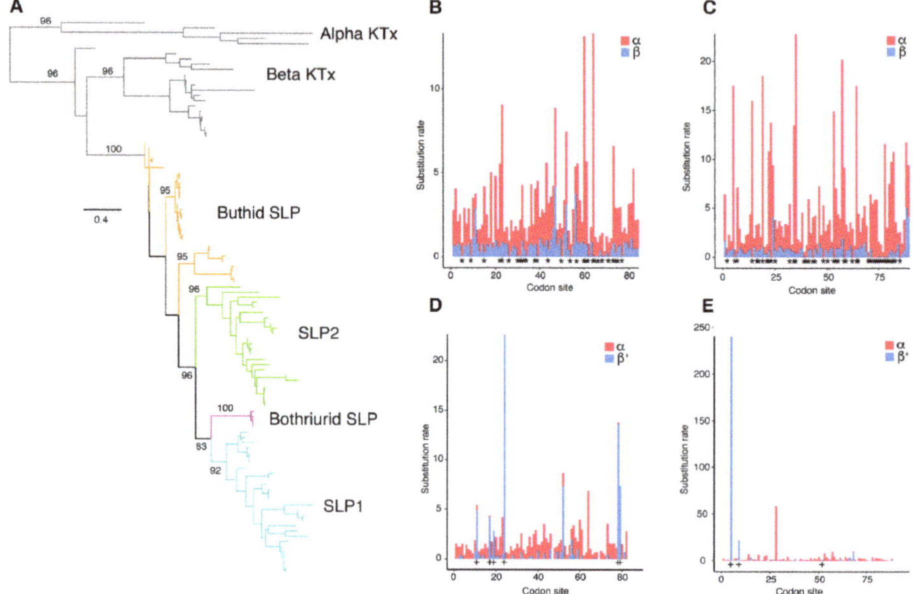

Figure 3. Evolutionary analyses of scorpine-like peptides. (**A**). Maximum likelihood (ML) gene tree topology of scorpine-like peptides (SLPs) and related toxins (αKTx and other βKTx). Ultrafast bootstrap values are shown above selected basal nodes (>75%). (**B,C**) Site selection analyses of SLP1 (**B**) and SLP2 (**C**) sequences with FUBAR. Visualization of synonymous substitution rates (α, red) and non-synonymous substitution rates (β, blue) for sites exhibiting positive/neutral evolution. Asterisks indicate sites with α greater than β, $p > 0.95$. (**D,E**) Site selection analyses of SLP1 (**D**) and SLP2 (**E**) sequences with MEME. + signs indicate sites with β^+ values greater than α, $p > 0.95$.

Evidence of positively selected sites in these peptides was assessed using MEME and FUBAR. The MEME results suggested that four sites have experienced episodic positive selection ($p < 0.05$; Figure S5). In contrast, the FUBAR analysis identified episodic negative selection at 44 sites, but none evolving under positive selection ($p < 0.05$; Figure S5).

To analyze evolution of scorpine paralogs, MEME and FUBAR analyses were conducted on each gene copy (SLP1 and SLP2). FUBAR analysis detected 30 sites evolving under negative selection in both copies. Three sites in SLP1 and six sites in SLP2 were considered to be evolving under positive selection by MEME analysis (Figure 3C–E). Multiple sequence alignments (MSA) of each SLP (SLP1 and 2) showed consistency among hadrurid toxins (Figure 4A,B). For downstream analyses, we selected only pairs of copies that were recovered as monophyletic with SLPs described previously for *H. gertschi* (Hge-scorpine 1-2).

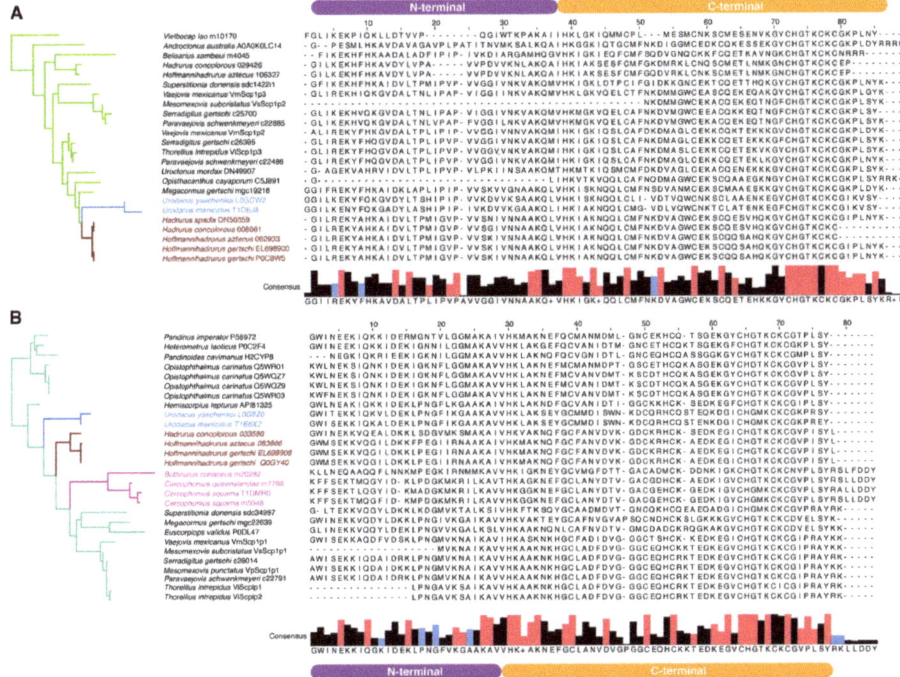

Figure 4. Multiple sequence alignment (MSA) of the mature peptide of SLP2 (**A**) and SLP1 (**B**) in order of appearance in the phylogeny (left, color coded as in previous figure). Consensus sequence histograms of each clade below the MSA (+ sign indicates a highly variable site). In red, sites evolving under negative selection, as detected with FUBAR, and in blue, sites evolving under positive selection as detected with MEME (see Figure 3).

The effect of molecular weight, molecular volume, net charge, and isoelectric point were evaluated for both copies of SLPs (combined and independently) using a principal component analysis (*prcomp* in R). Combined PCAs of both SLPs showed that 99.63% of the variation was explained by PC1, which segregated the SLP1 from SLP2 by molecular weight and volume, whereas PC2 (0.67%) split them by net charge (Figure S6). Three SLP2s copies were grouped with most of the SLP1s, due to their low molecular weight and volume. Similarly, almost 100% of the variation (99.72%) was explained by PC1 in both independent analyses. By molecular weight and volume, it segregated SLP2s of *Vietbocap lao*, *H. aztecus*, and *H. concolorous* from the rest of the sequences, and the SLP1 from *P. imperator* from other SLP1 sequences. In this analysis, PC2 also separated the sequences by net charge in both analyses. Superimposition of the phylogram onto the principal component space of SLP1 and SLP2 showed that SLP1 more strongly retains a phylogenetic signal (Figure 5A,C). We tested the correlation between these biochemical properties and found that molecular weight and volume are strongly correlated ($p < 0.05$; $r^2 = 1.00$), as are the net charge and the isoelectric point ($p < 0.05$, $r^2 = 0.92$; Figure S6).

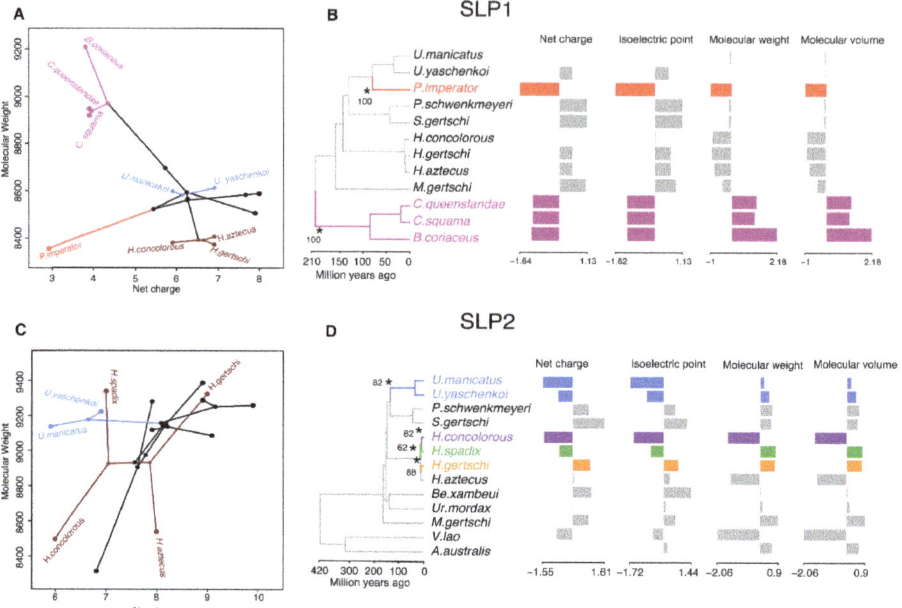

Figure 5. Evolutionary analyses of the chemical properties of SLP homologs. (**A**) Visualization of the phylogenomic tree on the morphospace of the molecular weight and net charge of SLP1 (**A**) and SLP2 (**C**). Hadrurids, urodacids, and *Cercophonius* are colored accordingly to phylogenomic topology. (**B–D**) Two (SLP1) and four (SLP2) evolutionary shifts in the optimum chemical properties of SLPs under an Ornstein–Uhlenbeck (OU) process. Edges with a major shift are annotated with asterisk and bootstrap support values.

We then inferred changes in evolutionary dynamics of the SLP gene family using 84 sequences. Two evolutionary shifts were detected in SLP1 with bootstrap support (BS) above 50% (Figure 5B). These occurred in the branch subtending *P. imperator* and the branch subtending the Bothriuridae, both with BS = 100%. In contrast, four major shifts were detected with BS > 50% across the evolutionary history of SLP2 (Figure 5D). These subtended *Uroctonus* (BS = 82%), *H. concolorous* (BS = 82%), *H. aztecus* (BS = 88%), and *H. spadix* (BS = 62%). However, when only two variables were considered (molecular weight and net charge), no major evolutionary shifts were detected in the gene tree of SLP2, with the same two shifts detected in evolution of SLP1.

3. Discussion

The study of the early evolutionary origins of scorpion toxins has been complicated by issues of homology inference and taxonomic sampling. Recently, the use of transcriptomics to resolve the scorpion tree of life has yielded sequence data that can be used to assess venom composition concomitantly with reconstruction of phylogenetic relationships. Our analyses, based on nearly 2000 loci, provide the first resolution of relationships within the scorpion family Hadruridae. Most notably, we provide evidence of the conservation of venom composition up to the family level. Furthermore, our results detected different evolutionary selective pressures on the two domains of scorpine-like peptides. These results strongly contrast with previous hypotheses that scorpion cystine-stabilized alpha/beta fold toxins (CSαβ) peptides are evolving under positive selection to increase potency. Instead, more sites are evolving under negative selection, suggesting an evolutionary conservation in function for these peptides since the estimated Permian diversification of scorpions [37], continuing through the estimated Neogene period of diversification of the Hadruridae. This study provides parameters to test the significance of toxic evolutionary dynamics in the extraordinary diversification of this culturally iconic group of arachnids.

Maximum likelihood and species tree analyses supported the monophyly of the family Hadruridae and its two constituent genera. Our results are congruent with those presented in a more sparsely sampled phylogenomic study [27]. The molecular phylogeny presented here demonstrates the monophyly of *Hoffmannihadrurus* as a lineage distinct from *Hadrurus*, in agreement with previous hypotheses based on morphology [38,39]. Our results suggest that the tradeoff between missing data and number of genes does not adversely impact reconstruction of the major scorpion clades (compare results in [27,36,37,40]). Molecular dating supported the divergence of the Hadruridae 7–11 Mya, in partial agreement with formation of the Trans-Mexican Volcanic Belt, which could have been a vicariance event that isolated *Hoffmannihadrurus* in the south from *Hadrurus* in the north [41–43].

Of marked interest to us in the transcriptomic analyses of the Hadruridae were the evolutionary dynamics of, and selective pressures on, scorpine-like peptides, which exhibit paralogous copies in a subset of hadrurid species. Buthid scorpine-like peptides differ from non-buthid SLPs with respect to the size of the mature peptide (<75 sites in buthids vs. >80 sites in non-buthids). Gene tree topologies, however, do not reflect a phylogenetic signal due to additional duplications in two hadrurid species (*H. concolorous* and *H. aztecus*) or other scorpions (e.g., in vaejovids). Only SLP2 was recovered from buthid scorpions.

Comparative transcriptomic analyses of hadrurid species highlights the evolutionary process that contributes to the diversity of CSαβ scorpion toxins. Results suggest that the SLP gene underwent duplication, which is concordant with the traditional model of venom toxin gene evolution [17–20]. However, evolutionary analyses of scorpine-like peptides showed that few non-cysteine amino acid substitutions occurred between the cysteines of the C-terminus, including insertion/deletions. Consequently, our results do not support a previous hypothesis [15] that CSαβ toxins have evolved under the influence of positive selection. Fewer than five sites in each SLP copy have experienced diversifying selection. Moreover, the majority of these sites were found at the N-terminus (a putative alpha helix), suggesting that positive selection might play an important role in evolution of this domain, but not at the C-terminus (Figure 4). Thus, this highly conserved conformation of both SLP copies, along with the incidence of several sites evolving under negative selection to reduce alternate states, is consistent with the tendency to preserve function (e.g., venom potency or target specificity [19,36,44]).

Parametric analyses of the predicted peptides also suggest that both SLP copies possess similar biochemical properties. For example, SLP1 homologs have shorter sequences (<84 amino acids) with lower molecular weights and volumes (with two exceptions) and more negative net charges (with three exceptions, $p < 0.01$). In contrast, SLP2 homologs have longer sequences (>84 amino acids) with correspondingly greater molecular weights and volumes (with three exceptions) and more positive net charges (with three exceptions; $p < 0.01$). To test the evolutionary dynamics of these protein traits, we assessed the role of selection during hadrurid evolution by modeling shifts in SLP biochemical traits under an Ornstein–Uhlenbeck (OU) process. Several studies have used this process to model evolution of continuous traits by detecting shifts to different regimes with different adaptive optima, along a time calibrated phylogeny (e.g., in [45–47], with a recent application to 3D structures of toxins [32]). The incidence of shifts in hadrurid SLP2s may be biased by sequence lengths from *H. aztecus* and *H. concolorous* (smaller than other SLP2) and the divergence time of these species (between 7 and 11 Myr). In contrast, the incidence of shifts in *Cercophonius* and *Bothriurus* suggests a unique evolutionary regime in the family Bothriuridae.

4. Conclusions

Reconstructing scorpion venom evolution is challenging due to the diversity of scorpions, the diversity of venom components, and the relative paucity of genome-scale sampling for various lineages. Application of RNA-Seq in the past decade has transformed both our understanding of scorpion phylogeny and the potential for gene discovery in venom gland transcriptomes. However, while transcriptomics provides distinct advantages, like scalability and rapidly deployable analytical toolkits, care must be taken to validate predictions of transcriptomic approaches using genomic, proteomic, and functional datasets (e.g., see [48,49]). In the absence of high-quality genomes for all surveyed species, conclusions on the number of toxin paralogs should be treated cautiously, as gene absence cannot be equated with gene loss. Similarly, proteomic and functional data are needed to bridge the gap between bioinformatic predictions and validation of venom components. Future efforts should therefore use transcriptomics as a guide for selection of high-priority targets (either specific peptides or in poorly studied species) for assessments using proteomes or functional experiments.

5. Materials and Methods

5.1. Specimen Collection and RNA Sequencing

Three specimens of *Hadrurus concolorous* were collected at night with the aid of ultraviolet lamps at Guerrero Negro in Baja California Sur (27°59′1.81″N, 114°1′5.12″W) on August 9, 2015 (Col. CESL and G. Contreras-Félix). By the same means, two specimens of *Hoffmannihadrurus aztecus* were collected from Zapotitlán de Salinas in Puebla (18°19′30.68″N, 97°28′37.09″W) on March 9, 2016 (Col. CESL and R. Paredes). One female specimen from each species was selected for RNAseq, while the others were kept under laboratory conditions for venom collection. Telsons were dissected and placed directly into RNA denaturing solution. An SV Total RNA Isolation System (Promega, Madison, WI, USA) was used for Total RNA extraction and purification. RNA quality was assessed using a Bioanalyzer 2100 (Agilent, Santa Clara, CA, USA). Samples lacked a 28S rRNA peak, as previously reported [21], but there was no indication of RNA degradation. Paired-end cDNA libraries were prepared with an Illumina TruSeq Stranded mRNA Sample Preparation Kit (Illumina, Inc., San Diego, CA, USA). A Genome Analyzer IIx (Illumina, Inc., San Diego, CA, USA) at the Massive DNA Sequencing Facility in the Institute of Biotechnology (Cuernavaca, Mexico) was employed, with a 72-bp paired-end sequencing scheme over 200–400-bp cDNA fragments. Raw sequences were deposited in the ENA SRA database under project number PRJEB34464. After sequence cleaning and adapter clipping, reads were assembled de novo into contigs using Trinity (v. 2.8.5, https://trinityrnaseq.github.io [50]) as reported before [51], and annotated with Trinotate (v3.1.1, https://trinotate.github.io, [52]). Transcriptomic data from all other species in this study were retrieved from GenBank, with most of them (19/22) previously sequenced by our team. Protein coding regions within these assemblies were identified and predicted using TransDecoder v. 5.3.0 [50].

5.2. Orthology Inference and Phylogenetic Methods

Our phylogenomic dataset consisted of 24 scorpion taxa (five ingroup species, 19 outgroups; Table S3), and our methodology followed recent scorpion phylogenomic studies (i.e., in [53]). Briefly, transcriptomes were combined, and a de novo homology search was conducted using the informed orthology criterion implemented in UPhO v1.0 [54]. Libraries of four representative species, *Cercophonius sulcatus*, *Hoffmannihadrurus aztecus*, *Serradigitus gertschi*, and *Urodacus yaschenkoi*, were combined and used as queries against the database containing all species using Mmseqs2 [55]. Subsequently, resulting sequences were clustered into gene families using mcl [56,57] with the inflation parameter $i = 6$ selected. In total, 13,317 clusters were produced with at least six species, with downstream analyses parallelized implemented through gnu-parallel [58]. A multiple sequence alignment of these gene clusters was performed with MAFFT 7.0 [59], gap masked with trimAl v 1.2 [60], and sanitized by removing sequences having fewer than 50 amino acids or <25% unambiguous sites with the scripts *Al2Phylo.py* (-m 50 -p 0.25) [52] and *paMATRAX+.sh* [54]. Gene family trees (GFT) were calculated using IQ-TREE v. 1.6.10 [61] and the LG+R4 substitution model, with the resulting GFT analyzed in search of groups of orthologs with at least 12 species using UPhO. In-paralogs, duplicates and/or isoforms among retained orthogroups were resolved in favor of the longest sequence, with a total of 1982 orthologs retained.

Similarly, these orthologs were aligned and sanitized as described above, but with only one sequence retained per species (option *-r* in the script *Al2Phylo.py*).

Phylogenetic inference of orthologous gene trees (OGT) was computed using 1000 ultrafast bootstrap resampling replicates [62,63] and substitution models and heterogeneity suggested by ModelFinder [64], based on the Bayesian Information Criterion (BIC). Cleaned sequences recovered from the collection of OGTs were concatenated in a supermatrix partitioned by locus, using the script *geneStitcher.py* [54]. Maximum likelihood (ML) analysis of this supermatrix was also conducted with IQ-TREE using the precomputed best substitution models from the collection of OGTs (-spp partition.nex). We estimated a species tree with ASTRAL-II [65] using the collection of OGTs to account for potentially deleterious effect of concatenating loci. The resulting ML tree was calibrated for downstream analyses using the penalized likelihood [66] as implemented in the chronos function of the R package 'ape' [67,68], under relaxed and correlated models, with lambda = 1.0. We set the crown age of the Scorpiones to 430 Mya, the stem age of the Buthidae to a minimum of 120 Mya [37], and the split of *Hadrurus-Hoffmannihadrurus* to a maximum of 11 Mya, based on calibrations from our previous studies [37,69]. To compare venom evolutionary regimes within the Hadrurinae against other families, we aimed to test the monophyly of two more scorpion families (Bothriuridae and Urodacidae) by the inclusion of more than one representative species. Our selection included all species of *Cercophonius* (family Bothriuridae) used in our previous studies, and four species of *Urodacus* (family Urodacidae).

5.3. Venom Transcriptome Analyses of Hadrurus Concolorous and Hoffmannihadrurus Aztecus

Transcriptomes of *H. concolorous* and *H. aztecus* were annotated with the Trinotate pipeline [50]. Venom components reported from the transcriptome, proteome, or isolated from the venom of *Megacormus gertschi* [21], *Hoffmannihadrurus gertschi* [29], and *Hadrurus spadix* [31] were used as queries to search our two transcriptomes with the *blastp* algorithm. Matching sequences with low e-values ($<1 \times 10^{-15}$) were selected. Subsequently, these sequences were used as queries to search the GenBank and UniProt databases, keeping only hits with low e-values ($<1 \times 10^{-10}$), high query cover values (>70%), and percentages of identity (>50%) as definitive matches (Tables S2 and S3 and Figure S7). To assess the diversity and composition of Hadrurinae venoms, we selected sequences encoding, or putatively encoding enzymes (phospholipases, hyaluronidases), protease inhibitors (Ascaris-type, Kunitz-type), ryanodine receptor ligands (calcins, DDH), and sodium channel toxins (NaTx) from 24 scorpion libraries. To test our recent hypothesis that calcins retain a phylogenetic signal above the familial level [36], we retrieved calcin sequences from each species used in the phylogenomic analysis to generate an ML gene tree.

5.4. Gene Tree Analysis and Molecular Evolution of Scorpines

Scorpine (long scorpion toxin) homologs were retrieved from the complete dataset generated here, or from UniProt and GenBank (Table S4). Signal and mature peptides were predicted using SpiderP from Arachnoserver [70]. Outgroup taxa for gene tree analysis consisted of scorpion potassium channel toxins (KTx) α and β, with αKTx used to root the tree. Phylogenetic relationships of scorpion KTxs have been addressed elsewhere [16,22]. Multiple sequence alignments for the full precursor (ca 97–124 amino acids) were generated using MAFFT and subsequently trimmed with trimAl, resulting in a matrix of 107 terminals and 77 amino acid sites. ML analysis was performed with IQ-TREE as stated above. Additionally, a phylogeny was generated using Bayesian inference (BI) for the same matrix, with the LG model [71] in MrBayes v 3.2.2 [72]. The analysis was run four times, each with four Markov chains that sampled every 1000 generations for 1×10^7 generations, using default priors and discarding 2×10^6 generations as burn in.

Multiple sequence alignments of nucleotide sequences encoding mature peptides were generated based on their corresponding amino acid sequences using PAL2NAL v. 14 [73]. The resulting codon alignment was used to calculate synonymous and non-synonymous substitution rates under different models implemented in HyPhy 2.6 [74]. Detection of sites evolving under positive or negative selection was accomplished with FUBAR [75] using default parameters, whereas MEME [76] was used to detect episodic or diversifying selection at individual sites in amino acid sequences.

Lastly, molecular weights, net charges, and isoelectric points of SLPs were calculated for mature peptide sequences using the R package '*Peptides*' [77]. We investigated evolutionary dynamics of these protein traits for shifts in trait regimes using a continuous multivariate Ornstein–Uhlenbeck (OU) approach implemented in the R package, *l1ou* [45], as in other studies on venom evolution [36]. These data were mapped to the phylogeny and the best shift configuration was estimated and selected using the phylogenetic Bayesian Information Criterion (pBIC). Edges were painted according to their corresponding regime with statistical support for each regime shift assessed with 100 bootstrap resampling replicates [36].

Supplementary Materials: The following are available online at http://www.mdpi.com/2072-6651/11/11/637/s1, Table S1: Putative venom sequences encoded by 96 transcripts of the *Hoffmannihadrurus aztecus* transcriptome, Table S2: Putative venom sequences encoded by 74 transcripts of the *Hadrurus concolorus* transcriptome, Table S3: List of the 24 scorpion species used in the phylogenomic analyses, Table S4: One hundred and seven sequences of Scorpine-like Peptides (SLP) isolated from venoms, or deduced from cDNA or transcriptome analyses of 44 scorpion species, Figure S1: ASTRAL-II topology, Figure S2: Chronogram under the correlated model, Figure S3: Chronogram under the relaxed model, Figure S4: ML gene tree SLPs, Figure S5: Site selection analyses of both SLP1-2 combined, Figure S6: Visualization of PCA and Correlation tests, Figure S7: Multiple sequence alignments of calcins, Kunitz-type inhibitors, and La1-like peptides.

Author Contributions: Conceptualization: C.E.S.-L.; methodology, C.E.S.-L., M.R.G., P.P.S., E.O.; software, C.E.S.-L.; validation, C.E.S.-L., M.R.G., P.P.S., E.O., L.D.P.; formal analysis, C.E.S.-L.; investigation, C.E.S.-L.; resources, M.R.G., P.P.S., L.D.P.; data curation, C.E.S.-L.; writing—original draft preparation, C.E.S.-L.; writing—review and editing, M.R.G., P.P.S., E.O., L.D.P.; visualization, C.E.S.-L.; supervision, M.R.G., P.P.S.; project administration, C.E.S.-L.; funding acquisition, L.D.P.

Funding: C.E.S.L. was supported by postdoctoral CONACyT (reg. 207146/454834) and CONACyT (No. 237864) grants. M.R.G. was supported by NSF grant DEB-1754030. P.P.S. was supported by NSF IOS-1552610. L.D.P. was funded by grant IN202619 from Dirección General del Personal Académico, UNAM.

Acknowledgments: We are greatly indebted to Nick Volpe who kindly gave us permission to use his photographs. We thank G. Contreras, M. Perzabal, and R. Paredes for their help during the collecting field trip. The authors are grateful to Oscar Francke from the Biology Institute of UNAM for support during collection of scorpions used in this work (Scientific Permit FAUT-0175, from SEMARNAT). We are indebted to Jimena I. Cid-Uribe and Maria T. Romero-Gutiérrez for their help with transcriptomic analysis. E.W. Setton made valuable comments on early drafts of this manuscript.

Conflicts of Interest: The authors declare no conflict of interest. The funders had no role in the design of the study, collection, analyses, or interpretation of data, in writing the manuscript, or in the decision to publish the results.

References

1. Fry, B.G.; Roelants, K.; Champagne, D.E.; Scheib, H.; Tyndall, J.D.A.; King, G.F.; Nevalainen, T.J.; Norman, J.A.; Lewis, R.J.; Norton, R.S.; et al. The Toxicogenomic Multiverse: Convergent Recruitment of Proteins Into Animal Venoms. *Annu. Rev. Genom. Hum. Genet.* **2009**, *10*, 483–511. [CrossRef]

2. Nelsen, D.R.; Nisani, Z.; Cooper, A.M.; Fox, G.A.; Gren, E.C.K.; Corbit, A.G.; Hayes, W.K. Poisons, toxungens, and venoms: Redefining and classifying toxic biological secretions and the organisms that employ them. *Biol. Rev.* **2013**, *89*, 450–465. [CrossRef]

3. Morgenstern, D.; King, G.F. The venom optimization hypothesis revisited. *Toxicon* **2013**, *63*, 120–128. [CrossRef]

4. Casewell, N.R.; Wüster, W.; Vonk, F.J.; Harrison, R.A.; Fry, B.G. Complex cocktails: The evolutionary novelty of venoms. *Trends Ecol. Evol.* **2013**, *28*, 219–229. [CrossRef]

5. King, G.F.; Hardy, M.C. Spider-Venom Peptides: Structure, Pharmacology, and Potential for Control of Insect Pests. *Annu. Rev. Entomol.* **2013**, *58*, 475–496. [CrossRef]

6. Torres, A.F.C.; Huang, C.; Chong, C.-M.; Leung, S.W.; Prieto-da-Silva, Á.R.B.; Havt, A.; Quinet, Y.P.; Martins, A.M.C.; Lee, S.M.Y.; Rádis-Baptista, G. Transcriptome Analysis in Venom Gland of the Predatory Giant Ant *Dinoponera quadriceps*: Insights into the Polypeptide Toxin Arsenal of Hymenopterans. *PLoS ONE* **2014**, *9*, e87556. [CrossRef]

7. Asgari, S.; Rivers, D.B. Venom Proteins from Endoparasitoid Wasps and Their Role in Host-Parasite Interactions. *Annu. Rev. Entomol.* **2011**, *56*, 313–335. [CrossRef]

8. Undheim, E.A.B.; King, G.F. On the venom system of centipedes (Chilopoda), a neglected group of venomous animals. *Toxicon* **2011**, *57*, 512–524. [CrossRef]

9. Undheim, E.A.B.; Jones, A.; Clauser, K.R.; Holland, J.W.; Pineda, S.S.; King, G.F.; Fry, B.G. Clawing through Evolution: Toxin Diversification and Convergence in the Ancient Lineage Chilopoda (Centipedes). *Mol. Biol. Evol.* **2014**, *31*, 2124–2148. [CrossRef]

10. Undheim, E.; Fry, B.; King, G. Centipede Venom: Recent Discoveries and Current State of Knowledge. *Toxins* **2015**, *7*, 679–704. [CrossRef]

11. von Reumont, B.M.; Blanke, A.; Richter, S.; Alvarez, F.; Bleidorn, C.; Jenner, R.A. The First Venomous Crustacean Revealed by Transcriptomics and Functional Morphology: Remipede Venom Glands Express a Unique Toxin Cocktail Dominated by Enzymes and a Neurotoxin. *Mol. Biol. Evol.* **2013**, *31*, 48–58. [CrossRef] [PubMed]

12. Gopalakrishnakone, P.; Possani, L.D.; Schwartz, E.F.; Rodriguez de la Vega, R.C. *Scorpion Venoms*; Springer: Amsterdam, The Netherlands, 2015.

13. Gopalakrishnakone, P.; Corzo, G.; de Lima, M.E.; Diego-Garcia, E. *Spider Venoms*; Springer: Amsterdam, The Netherlands, 2016.

14. Possani, L.D.; Becerril, B.; Delepierre, M.; Tytgat, J. Scorpion toxins specific for Na+-channels. *FEBS J.* **1999**, *264*, 287–300. [CrossRef]

15. Sunagar, K.; Undheim, E.; Chan, A.; Koludarov, I.; Muñoz-Gómez, S.; Antunes, A.; Fry, B. Evolution Stings: The Origin and Diversification of Scorpion Toxin Peptide Scaffolds. *Toxins* **2013**, *5*, 2456–2487. [CrossRef]

16. Santibáñez-López, C.E.; Possani, L.D. Overview of the Knottin scorpion toxin-like peptides in scorpion venoms: Insights on their classification and evolution. *Toxicon* **2015**, *107*, 317–326. [CrossRef]

17. Juárez, P.; Comas, I.; González-Candelas, F.; Calvete, J.J. Evolution of snake venom disintegrins by positive Darwinian selection. *Mol. Biol. Evol.* **2008**, *25*, 2391–2407. [CrossRef]

18. Dowell, N.L.; Giorgianni, M.W.; Kassner, V.A.; Selegue, J.E.; Sanchez, E.E.; Carroll, S.B. The deep origin and recent loss of venom toxin genes in rattlesnakes. *Curr. Bio.* **2016**, *26*, 2434–2445. [CrossRef]

19. Weinberger, H.; Moran, Y.; Gordon, D.; Turkov, M.; Kahn, R.; Gurevitz, M. Positions under positive selection—Key for selectivity and potency of scorpion alpha toxins. *Mol. Biol. Evol.* **2010**, *27*, 1025–1034. [CrossRef]

20. Haney, R.A.; Clarke, T.H.; Gadgil, R.; Fitzpatrick, R.; Hayashi, C.Y.; Ayoub, N.A.; Garb, J.E. Effects of gene duplication, positive selection, and shifts in gene expression on the evolution of the venom gland transcriptome in widow spiders. *Genome Biol. Evol.* **2016**, *8*, 228–242. [CrossRef]

21. Santibáñez-López, C.E.; Cid-Uribe, J.I.; Zamudio, F.Z.; Batista, C.V.F.; Ortiz, E.; Possani, L.D. Venom gland transcriptomic and venom proteomic analyses of the scorpion *Megacormus gertschi* Díaz-Najera, 1966 (Scorpiones: Euscorpiidae: Megacorminae). *Toxicon* **2017**, *133*, 95–109. [CrossRef]

22. Santibáñez-López, C.; Cid-Uribe, J.; Batista, C.; Ortiz, E.; Possani, L. Venom Gland Transcriptomic and Proteomic Analyses of the Enigmatic Scorpion *Superstitionia donensis* (Scorpiones: Superstitioniidae), with Insights on the Evolution of Its Venom Components. *Toxins* **2016**, *8*, 367. [CrossRef]

23. Romero-Gutiérrez, M.; Santibáñez-López, C.; Jiménez-Vargas, J.; Batista, C.; Ortiz, E.; Possani, L. Transcriptomic and Proteomic Analyses Reveal the Diversity of Venom Components from the Vaejovid Scorpion *Serradigitus gertschi*. *Toxins* **2018**, *10*, 359. [CrossRef] [PubMed]

24. Luna-Ramírez, K.; Quintero-Hernández, V.; Juárez-González, V.R.; Possani, L.D. Whole Transcriptome of the Venom Gland from *Urodacus yaschenkoi* Scorpion. *PLoS ONE* **2015**, *10*, e0127883. [CrossRef] [PubMed]

25. Kazemi-Lomedasht, F.; Khalaj, V.; Bagheri, K.P.; Behdani, M.; Shahbazzadeh, D. The first report on transcriptome analysis of the venom gland of Iranian scorpion, *Hemiscorpius lepturus*. *Toxicon* **2017**, *125*, 123–130. [CrossRef] [PubMed]

26. Zhong, J.; Zeng, X.-C.; Zeng, X.; Nie, Y.; Zhang, L.; Wu, S.; Bao, A. Transcriptomic analysis of the venom glands from the scorpion *Hadogenes troglodytes* revealed unique and extremely high diversity of the venom peptides. *J. Proteom.* **2017**, *150*, 40–62. [CrossRef] [PubMed]

27. Santibáñez-López, C.E.; Ojanguren-Affilastro, A.; Sharma, P.P. Another one bites the dust: Taxonomic sampling of a key genus in phylogenomic datasets reveals more non-monophyletic groups in traditional scorpion classification. *Inv. Syst.* **2019**, in press. [CrossRef]

28. Schwartz, E.F.; Capes, E.M.; Diego-Garcia, E.; Zamudio, F.Z.; Fuentes, O.; Possani, L.D.; Valdivia, H.H. Characterization of hadrucalcin, a peptide from *Hadrurus gertschi* scorpion venom with pharmacological activity on ryanodine receptors. *Br. J. Pharmacol.* **2009**, *157*, 392–403. [CrossRef] [PubMed]
29. Schwartz, E.F.; Diego-Garcia, E.; Rodríguez de la Vega, R.C.; Possani, L.D. Transcriptome analysis of the venom gland of the Mexican scorpion *Hadrurus gertschi* (Arachnida: Scorpiones). *BMC Genom.* **2007**, *8*, 119. [CrossRef]
30. Chen, Z.-Y.; Hu, Y.-T.; Yang, W.-S.; He, Y.-W.; Feng, J.; Wang, B.; Zhao, R.-M.; Ding, J.-P.; Cao, Z.-J.; Li, W.-X.; et al. Hg1, Novel Peptide Inhibitor Specific for Kv1.3 Channels from First Scorpion Kunitz-type Potassium Channel Toxin Family. *J. Biol. Chem.* **2012**, *287*, 13813–13821. [CrossRef]
31. Rokyta, D.R.; Ward, M.J. Venom-gland transcriptomics and venom proteomics of the black- back scorpion (*Hadrurus spadix*) reveal detectability challenges and an unexplored realm of animal toxin diversity. *Toxicon* **2017**, *128*, 23–37. [CrossRef]
32. Diego-García, E.; Schwartz, E.F.; D'Suze, G.; González, S.A.R.; Batista, C.V.F.; García, B.I.; de la Vega, R.C.R.; Possani, L.D. Wide phylogenetic distribution of Scorpine and long-chain β-KTx-like peptides in scorpion venoms: Identification of "orphan" components. *Peptides* **2007**, *28*, 31–37. [CrossRef]
33. Diego-García, E.; Abdel-Mottaleb, Y.; Schwartz, E.F.; de la Vega, R.C.R.; Tytgat, J.; Possani, L.D. Cytolytic and K+ channel blocking activities of β-KTx and scorpine-like peptides purified from scorpion venoms. *CMLS Cell. Mol. Life Sci.* **2007**, *65*, 187–200. [CrossRef] [PubMed]
34. Zhu, S.; Tytgat, J. The scorpine family of defensins: Gene structure, alternative polyadenylation and fold recognition. *CMLS Cell. Mol. Life Sci.* **2004**, *61*, 1–13. [CrossRef] [PubMed]
35. Conde, R.; Zamudio, F.Z.; Rodríguez, M.H.; Possani, L.D. Scorpine, an anti-malaria and anti-bacterial agent purified from scorpion venom. *FEBS Lett.* **2000**, *471*, 165–168. [CrossRef]
36. Santibáñez-López, C.E.; Kriebel, R.; Ballesteros, J.A.; Rush, N.; Witter, Z.; Williams, J.; Janies, D.A.; Sharma, P.P. Integration of phylogenomics and molecular modeling reveals lineage-specific diversification of toxins in scorpions. *PeerJ* **2018**, *6*, e5902. [CrossRef]
37. Sharma, P.P.; Baker, C.M.; Cosgrove, J.G.; Johnson, J.E.; Oberski, J.T.; Raven, R.J.; Harvey, M.S.; Boyer, S.L.; Giribet, G. A revised dated phylogeny of scorpions—Phylogenomic support for ancient divergence of the temperate Gondwanan family Bothriuridae. *Mol. Phylogen. Evol.* **2018**, *122*, 37–45. [CrossRef]
38. Fet, V.; Soleglad, M.E. Cladistic analysis of superfamily Iuroidea, with emphasis on subfamily Hadrurinae (Scorpiones: Iurida). *Bol. Soc. Entomol. Aragonesa* **2008**, *43*, 255–281.
39. Soleglad, M.; Fet, V. Further observations on scorpion genera *Hadrurus* and *Hoffmannihadrurus* (Scorpiones, Caraboctonidae). *Zookeys* **2010**, *59*, 1–13. [CrossRef]
40. Sharma, P.P.; Fernández, R.; Esposito, L.A.; González-Santillán, E.; Monod, L. Phylogenomic resolution of scorpions reveals multilevel discordance with morphological phylogenetic signal. *Proc. Biol. Sci.* **2015**, *282*, 20142953. [CrossRef]
41. Bryson, R.W., Jr.; Murphy, R.W.; Lathrop, A.; Lazcano-Villareal, D. Evolutionary drivers of phylogeographical diversity in the highlands of Mexico: A case study of the *Crotalus triseriatus* species group of montane rattlesnakes. *J. Biogeogr.* **2010**, *38*, 697–710. [CrossRef]
42. Becerra, J.X. Timing the origin and expansion of the Mexican tropical dry forest. *Proc. Natl. Acad. Sci. USA* **2005**, *102*, 10919–10923. [CrossRef]
43. Ferrusquia-Villafranca, I.; González-Guzmán, L.I. Northern Mexico's landscape, part II: The biotic setting across time. In *Biodiversity, Ecosystems, and Conservation in Northern Mexico*; Oxford University Press: Oxford, UK, 2005; pp. 39–51.
44. Undheim, E.A.B.; Mobli, M.; King, G.F. Toxin structures as evolutionary tools: Using conserved 3D folds to study the evolution of rapidly evolving peptides. *BioEssays* **2016**, *38*, 539–548. [CrossRef] [PubMed]
45. Khabbazian, M.; Kriebel, R.; Rohe, K.; Ané, C. Fast and accurate detection of evolutionary shifts in Ornstein-Uhlenbeck models. *Methods Ecol. Evol.* **2016**, *7*, 811–824. [CrossRef]
46. Cressler, C.E.; Butler, M.A.; King, A.A. Detecting Adaptive Evolution in Phylogenetic Comparative Analysis Using the Ornstein–Uhlenbeck Model. *Syst. Biol.* **2015**, *64*, 953–968. [CrossRef] [PubMed]
47. Kriebel, R.; Khabbazian, M.; Sytsma, K.J. A continuous morphological approach to study the evolution of pollen in a phylogenetic context: An example with the order Myrtales. *PLoS ONE* **2017**, *12*, e0187228. [CrossRef] [PubMed]

48. Von Reumont, B.M. Studying smaller and neglected organisms in modern evolutionary venomics implementing rnaseq (transcriptomics)—A critical guide. *Toxins* **2018**, *10*, 292. [CrossRef] [PubMed]

49. Hofmann, E.P.; Rautsaw, R.M.; Strickland, J.L.; Holding, M.L.; Hogan, M.P.; Mason, A.J.; Rokyta, D.R.; Parkinson, C.L. Comparative venom-gland transcriptomics and venom proteomics of four sidewinder rattlesnake (*Crotalus cerastes*) lineages reveal little differential expression despite individual variation. *Sci. Rep.* **2018**, *8*, 15534. [CrossRef]

50. Haas, B.J.; Papanicolaou, A.; Yassour, M.; Grabherr, M.; Blood, P.D.; Bowden, J.; Couger, M.B.; Eccles, D.; Li, B.; Lieber, M.; et al. De novo transcript sequence reconstruction from RNA-seq using the Trinity platform for reference generation and analysis. *Nat. Protoc.* **2013**, *8*, 1494–1512. [CrossRef]

51. Cid-Uribe, J.I.; Santibáñez-López, C.E.; Meneses, E.P.; Batista, C.V.F.; Jiménez-Vargas, J.M.; Ortiz, E.; Possani, L.D. The diversity of venom components of the scorpion species *Paravaejovis schwenkmeyeri* (Scorpiones: Vaejovidae) revealed by transcriptome and proteome analyses. *Toxicon* **2018**, *151*, 47–62. [CrossRef]

52. Bryant, D.M.; Johnson, K.; DiTommaso, T.; Tickle, T.; Couger, M.B.; Payzin-Dogru, D.; Lee, T.J.; Leigh, N.D.; Kuo, T.H.; Davis, F.G.; et al. A Tissue-Mapped Axolotl De Novo Transcriptome Enables Identification of Limb Regeneration Factors. *Cell Rep.* **2017**, *18*, 762–776. [CrossRef]

53. Santibáñez-López, C.E.; González-Santillán, E.; Monod, L.; Sharma, P.P. Phylogenomics facilitates stable scorpion systematics_ Reassessing the relationships of Vaejovidae and a new higher-level classification of Scorpiones (Arachnida). *Mol. Phylogen. Evol.* **2019**, *135*, 22–30. [CrossRef]

54. Ballesteros, J.A.; Hormiga, G. A New Orthology Assessment Method for Phylogenomic Data: Unrooted Phylogenetic Orthology. *Mol. Biol. Evol.* **2016**, *33*, 2117–2134. [CrossRef] [PubMed]

55. Steinegger, M.; Söding, J. MMseqs2 enables sensitive protein sequence searching for the analysis of massive data sets. *Nat. Biotech.* **2017**, *35*, 1026–1028. [CrossRef] [PubMed]

56. Dongen, S. Graph Clustering by Low Simulation. Ph.D. Thesis, University of Utrecht, Utrecht, The Netherlands, 2000.

57. Enright, A.J.; Van Dongen, S.; Ouzounis, C.A. An efficient algorithm for large-scale detection of protein families. *Nucleic Acids Res.* **2002**, *30*, 1575–1584. [CrossRef] [PubMed]

58. Tange, O. Gnu parallel-the command-line power tool. *USENIX Mag.* **2011**, *36*, 42–47.

59. Katoh, K.; Standley, D.M. MAFFT Multiple Sequence Alignment Software Version 7: Improvements in Performance and Usability. *Mol. Biol. Evol.* **2013**, *30*, 772–780. [CrossRef]

60. Capella-Gutiérrez, S.; Silla-Martínez, J.M.; Gabaldón, T. trimAl: A tool for automated alignment trimming in large-scale phylogenetic analyses. *Bioinformatics* **2009**, *25*, 1972–1973. [CrossRef]

61. Nguyen, L.-T.; Schmidt, H.A.; von Haeseler, A.; Minh, B.Q. IQ-TREE: A Fast and Effective Stochastic Algorithm for Estimating Maximum-Likelihood Phylogenies. *Mol. Biol. Evol.* **2014**, *32*, 268–274. [CrossRef]

62. Hoang, D.T.; Chernomor, O.; von Haeseler, A.; Minh, B.Q.; Vinh, L.S. UFBoot2: Improving the ultrafast bootstrap approximation. *Mol. Biol. Evol.* **2018**, *35*, 518–522. [CrossRef]

63. Minh, B.Q.; Nguyen, M.A.T.; von Haeseler, A. Ultrafast Approximation for Phylogenetic Bootstrap. *Mol. Biol. Evol.* **2013**, *30*, 1188–1195. [CrossRef]

64. Kalyaanamoorthy, S.; Minh, B.Q.; Wong, T.K.F.; von Haeseler, A.; Jermiin, L.S. ModelFinder: Fast model selection for accurate phylogenetic estimates. *Nat. Methods* **2017**, *14*, 587–589. [CrossRef]

65. Mirarab, S.; Warnow, T. ASTRAL-II: Coalescent-based species tree estimation with many hundreds of taxa and thousands of genes. *Bioinformatics* **2015**, *31*, 44–52. [CrossRef] [PubMed]

66. Sanderson, M.J. Estimating absolute rates of molecular evolution and divergence times: A penalized likelihood approach. *Mol. Biol. Evol.* **2002**, *19*, 101–109. [CrossRef] [PubMed]

67. Paradis, E.; Claude, J.; Strimmer, K. APE: Analyses of Phylogenetics and Evolution in R language. *Bioinformatics* **2004**, *20*, 289–290. [CrossRef] [PubMed]

68. Paradis, E. Molecular dating of phylogenies by likelihood methods: A comparison of models and a new information criterion. *Mol. Phylogen. Evol.* **2013**, *67*, 436–444. [CrossRef] [PubMed]

69. Santibáñez-López, C.E.; Kriebel, R.; Sharma, P.P. eadem figura manet: Measuring morphological convergence in diplocentrid scorpions (Arachnida: Scorpiones: Diplocentridae) under a multilocus phylogenetic framework. *Inv. Syst.* **2017**, *31*, 233–248. [CrossRef]

70. Herzig, V.; Wood, D.L.A.; Newell, F.; Chaumeil, P.A.; Kaas, Q.; Binford, G.J.; Nicholson, G.M.; Gorse, D.; King, G.F. ArachnoServer 2.0, an updated online resource for spider toxin sequences and structures. *Nucleic Acids Res.* **2010**, *39*, D653–D657. [CrossRef]

71. Le, S.Q.; Gascuel, O. An improved general amino acid replacement matrix. *Mol. Biol. Evol.* **2008**, *25*, 1307–1320. [CrossRef]

72. Ronquist, F.; Teslenko, M.; van der Mark, P.; Ayres, D.L.; Darling, A.; Höhna, S.; Larget, B.; Liu, L.; Suchard, M.A.; Huelsenbeck, J.P. MrBayes 3.2: Efficient Bayesian Phylogenetic Inference and Model Choice Across a Large Model Space. *Syst. Biol.* **2012**, *61*, 539–542. [CrossRef]

73. Suyama, M.; Torrents, D.; Bork, P. PAL2NAL: Robust conversion of protein sequence alignments into the corresponding codon alignments. *Nucleic Acids Res.* **2006**, *34*, W609–W612. [CrossRef]

74. Pond, S.L.K.; Frost, S.D.W.; Muse, S.V. HyPhy: Hypothesis testing using phylogenies. *Bioinformatics* **2005**, *21*, 676–679. [CrossRef]

75. Murrell, B.; Moola, S.; Mabona, A.; Weighill, T.; Sheward, D.; Kosakovsky Pond, S.L.; Scheffler, K. FUBAR: A Fast, Unconstrained Bayesian AppRoximation for Inferring Selection. *Mol. Biol. Evol.* **2013**, *30*, 1196–1205. [CrossRef] [PubMed]

76. Murrell, B.; Wertheim, J.O.; Moola, S.; Weighill, T.; Scheffler, K.; Kosakovsky Pond, S.L. Detecting individual sites subject to episodic diversifying selection. *PLoS Genet.* **2012**, *8*, e1002764. [CrossRef] [PubMed]

77. Osorio, D.; Rondón-Villarrea, P.; Torres, R. Peptides: A package for data mining of antimicrobial peptides. *R J.* **2015**, *7*, 4–14. [CrossRef]

Article

The Dual α-Amidation System in Scorpion Venom Glands

Gustavo Delgado-Prudencio, Lourival D. Possani, Baltazar Becerril and Ernesto Ortiz *

Departamento de Medicina Molecular y Bioprocesos, Instituto de Biotecnología, Universidad Nacional Autónoma de México, Avenida Universidad 2001, Colonia Chamilpa, Cuernavaca, Morelos 62210, Mexico
* Correspondence: erne@ibt.unam.mx

Received: 4 June 2019; Accepted: 18 July 2019; Published: 20 July 2019

Abstract: Many peptides in scorpion venoms are amidated at their C-termini. This post-translational modification is paramount for the correct biological function of ion channel toxins and antimicrobial peptides, among others. The discovery of canonical amidation sequences in transcriptome-derived scorpion proproteins suggests that a conserved enzymatic α-amidation system must be responsible for this modification of scorpion peptides. A transcriptomic approach was employed to identify sequences putatively encoding enzymes of the α-amidation pathway. A dual enzymatic α-amidation system was found, consisting of the membrane-anchored, bifunctional, peptidylglycine α-amidating monooxygenase (PAM) and its paralogs, soluble monofunctional peptidylglycine α-hydroxylating monooxygenase (PHM*m*) and peptidyl-α-hydroxyglycine α-amidating lyase (PAL*m*). Independent genes encode these three enzymes. Amino acid residues responsible for ion coordination and enzymatic activity are conserved in these sequences, suggesting that the enzymes are functional. Potential endoproteolytic recognition sites for proprotein convertases in the PAM sequence indicate that PAM-derived soluble isoforms may also be expressed. Sequences potentially encoding proprotein convertases (PC1 and PC2), carboxypeptidase E (CPE), and other enzymes of the α-amidation pathway, were also found, confirming the presence of this pathway in scorpions.

Keywords: amidation; evolution; posttranslational modifications; scorpion; transcriptomics

Key Contribution: A dual enzymatic system responsible for α-amidation of scorpion venom peptides is described. Independent genes encode a bifunctional PAM enzyme and the monofunctional PHMm and PALm enzymes.

---

1. Introduction

The order Scorpiones constitutes one of the most ancient lineages within the phylum Arthropoda [1,2]. The key to the ecological success of these arachnids resides in the production of potent venoms used for feeding, defense, and deterring competitors [3,4]. Scorpion venoms are complex mixtures of components, including bioactive peptides with potential therapeutic applications [4], enzymes, metabolites, and most importantly, an arsenal of toxins active on Na^+, K^+, Ca^{2+}, and Cl^- channels [5–10]. By altering the normal mechanics of these channels, scorpion toxins unleash systemic havoc in their victims, which can lead to severe envenomation symptoms, including death [11,12]. The venom is produced and secreted by two symmetrical glands located in the last segment of the metasoma, the telson [12]. In these glands, the peptidyl venom components undergo synthesis and maturation, a complex process involving a series of post-translational modifications (PTMs) that result in the biologically active molecules [3]. The most common PTMs found in scorpion venom peptides are the formation of disulfide bridges from pairs of cysteines, proteolytic cleavage, and C-terminal amidation (α-amidation). Amidated toxins and peptides without disulfide bonds (NDBP) are well known in scorpion venoms (Table 1).

Table 1. Diversity of amidated peptides from scorpion venoms. Signal peptides are underlined. Mature peptides are in bold upper-case letters. Propeptides are italicized and amidation signals are shown in red.

Type	Peptide	Uniprot	Precursor Sequence	C-terminus
Sodium toxins	AaH2	P01484	<u>MNYLVMISLALLFVTGVES</u>**VKDGYIVDDVNCTYFCGRNAYCNEECTKLKGESGYCQWASPYGNAC YCYKLPDHVRTKGPGRCH**_GR_	H-NH$_2$
	LqhIT2	Q26292	<u>MKLLLLLIVSASMLIESLVNA</u>**DGYIKRRDGCKVACLIGNEGCDKECKAYGGSYGYCWTWGLACWCE GLPDDKTWKSETNTCG**_GKK_	G-NH$_2$
	BmKITa	Q9XY87	<u>MKLFLLLLISASMLIDGLVNA</u>**DGYIRGSNGCKVSCLWGNEGCNKECRAYGASYGYCWTWGLACWC QGLPDDKTWKSESNTCG**_GKK_	G-NH$_2$
	Cn2	P01495	<u>LLIITACLALIGTVWA</u>**KEGYLVDKNTGCKYECLKLGDNDYCLRECKQQYGKGAGGYCYAFACWC THLYEQAIVWPLPNKRCS**_GK_	S-NH$_2$
	Css4	P60266	<u>MNSLLMITACLALVGTVWA</u>**KEGYLVNSYTGCKFECFKLGDNDYCLRECRQQYGKGSGGYCYAFG CWCTHLYEQAVVWPLPNKTCN**_GK_	N-NH$_2$
	CsEI	P01491	<u>MNSLLMITACLVLIGTVWA</u>**KDGYLVEKTGCKKTCYKLGENDFCNRECKWKHIGGSYGYCYGFGC YCEGLPDSTQTWPLPNKTC**_GKK_	C-NH$_2$
	CsEv3	P01494	<u>MNSLLMITACLFLIGTVWA</u>**KEGYLVNKSTGCKYGCLKLGENEGCDKECKAKNQGGSYGYCYAFA CWCEGLPESTPTYPLPNKSC**_GKK_	C-NH$_2$
	Ts1	P15226	<u>MKGMILFISCLLLIGIVVEC</u>**KEGYLMDHEGCKLSCFIRPSGYCGRECGIKKGSSGYCAWPACYCY GLPNWVKVWDRATNKC**_GKK_	C-NH$_2$
	Ts3	P01496	<u>LVVVCLLTAGTEG</u>**KKDGYPVEYDNCAYICWNYDNAYCDKLCKDKKADSGYCYWVHILCYCY GLPDSEPTKTNGKCKS**_GKK_	S-NH$_2$
Potassium toxins	NTx	P08815	<u>MKAFYGILIILLFCSMFNLNES</u>**TIINVKCTSPKQCSKPCKELYGSSAGAKCMNGKCKCYNN**_G_	N-NH$_2$
	BmKTX	Q9NII7	<u>MKVFFAVLITLFICSMIIGIHG</u>**VGINVKCKHSGQCLKPCKDAGMRFGKCINGKCDCTPK**_G_	K-NH$_2$
	CoTx1	O46028	<u>MEGIAKITLILLFLFVTMHTFANWNTEA</u>**AVCVYRTCDKDCKRRGYRSGKCINNACKCYPY**_GK_	Y-NH$_2$
	OcKTx5	Q6XLL5	<u>MNAKFILLLVLTTMMLLPDTKG</u>**AEVIRCSGSKQCYGPCKQQTGCTNSKCMNKVCKCYGC**_G_	C-NH$_2$
	OcKTx1	Q6XLL9	<u>MNAKFILLLLVVATTMLLPDTQG</u>**AEVIKCRTPKDCAGPCRKQTGCPHGKCMNRTCRCNRC**_G_	C-NH$_2$
Non disulfide bridged peptides	IsCT	Q8MMJ7	<u>MKTQFAILLVALVLFQMFAQSDA</u>**ILGKIWEGIKSLF**_GKR_GLSDLDGLDELFDGEISKADRDFLRELMR	F-NH$_2$
	BmKb1	Q718F4	<u>MEIKYLLTVFLVLLIVSDHCQA</u>**FLFSLIPSAISGLISAFK**_GRRKR_DLNGYIDHFKNFRKRDAELEELLSKLPIY	K-NH$_2$
	Hp1090	P0DJ02	<u>MKTQFAIFLITLVLFQMFSQSDA</u>**IFKAIWSGIKSLF**_GKR_GLSDLDDLDESFDGEVSQADIDFLKELMQ	F-NH$_2$
	IsCT2	Q8MTX2	<u>MKTQFAILLVALVLFQMFAQSEA</u>**IFGAIWNGIKSLF**_GRR_ALNNDLDLDGLDELFDGEISQADVDFLKELMR	F-NH$_2$
	VAMP-2	E4VP07	<u>MKSQTFFLLFLVVFLLAITQSEA</u>**IFGAIAGLLKNIF**_GKR_SLRDMDTMKYLYDPSLSAADLKTLQKLMENY	F-NH$_2$

Post-translational α-amidation is characteristic of bioactive peptides from many eukaryotic taxa [13]. C-terminal amidation confers on peptides enhanced resilience to degradation by carboxypeptidases, thus increasing their half-lives and decreasing their turnover rates [14]. Moreover, amidation is essential for correct functioning of many mammalian neuropeptides and hormones [15]. Several scorpion toxins have also been shown to require C-terminal amidation for full biological activity, without which, potency is severely reduced [16,17]. C-terminal amidation results in a change with two possible functional implications: the amidated terminal amino acid could be directly involved in molecular recognition events, or the amidation could simply reduce the negative charge of the carboxyl moiety and the peptide as a whole [18].

In general, metazoan amidated peptides are translated as larger polypeptidyl precursors, which contain an amidation signal, a glycine typically followed by one or two basic residues (R-X-Gly-Basic or R-X-Gly-Basic-Basic) and the rest of the propeptide sequence. This signal is first targeted by endoproteolytic proprotein-processing enzymes of the protein convertase family (PCs), resulting in peptides terminated with basic residues, which are substrates for carboxypeptidases that remove those residues from the processing intermediate and expose the C-terminal glycine. This glycine is then further subjected to sequential reactions that amidate the peptide [19,20] (Figure 1A). Two enzymatic activities catalyze these reactions. First, a peptidylglycine α-hydroxylating monooxygenase (PHM, EC 1.14.17.3) catalyzes the hydroxylation of the glycine residue, using ascorbate and molecular oxygen as co-substrates. Then, a peptidyl-α-hydroxyglycine α-amidating lyase (PAL, EC 4.3.2.5) cleaves the hydroxyglycine residue, yielding the amidated product and glyoxylate [21,22] (Figure 1B).

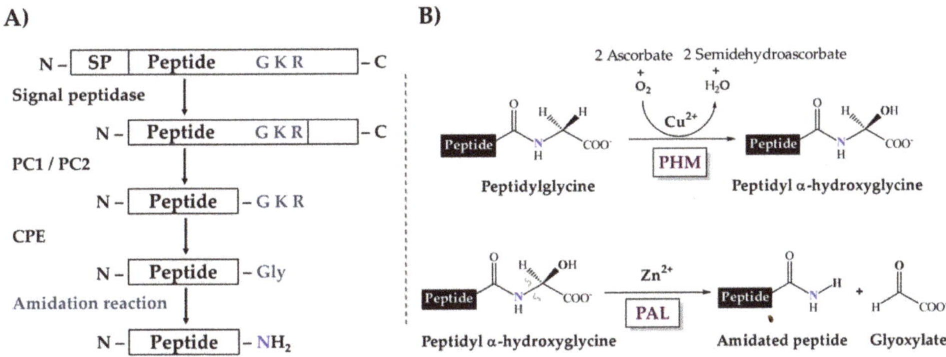

Figure 1. Enzymatic processing of the amidation signal. (**A**) The complete pathway leading to amidation. SP, Signal Peptide; PC1/PC2, proprotein convertases 1/2; CPE, carboxypeptidase E. (**B**) Sequential amidation reaction catalyzed by the PHM and PAL domains.

Peptide amidation seems to be common to all metazoans, and PHM and PAL are assumed to have monophyletic origins [23]. However, the way these activities are expressed differs among taxa. For example, in insects *Apis mellifera*, *Drosophila melanogaster* and others, PHM and PAL are encoded by independent genes [24–26]. In vertebrates, such as *Bos taurus*, *Rattus norvegicus*, *Xenopus laevis*, *Homo sapiens* and others, a single gene encodes both activities in a bifunctional enzyme comprising a single polypeptide, peptidylglycine α-amidating monooxygenase (PAM) [27–30]. The same two-domain PAM structure was reported for the gastropod *Aplysia californica* [31]. Curiously, another gastropod, *Lymnaea stagnalis*, produces a zymogen comprising four different PHM domains and a single PAL domain, which is endogenously converted to a mixture of monofunctional isoenzymes [32]. PAM isoforms have been reported in *R. norvegicus*, with up to seven isoforms generated by alternative splicing. These isoforms include configurations with and without internal proteolytic sites, resulting in both independent PHM and PAL, and the bifunctional PAM [27,33]. Among arthropods, independent PHM and PAL, but not the bifunctional PAM, are expressed in insects, as indicated above, whereas both independent and bifunctional enzymes are expressed in crustaceans [26]. No information is available on the α-amidating system of other subphyla, e.g., the chelicerates, and in particular, the arachnids.

Amidated peptides are common in venoms produced by various animals. For example, marine snails of the genus *Conus*, produce a large array of peptidyl toxins (conotoxins), a significant fraction of which are amidated. From the venom ducts of cone snails, cDNAs were cloned that encode bifunctional PAMs. Heterologously expressed PAMs were demonstrated to be active [34]. Although no amidating system has been described in arachnids, the abundance of amidated peptides in their venoms, in particular scorpion venoms, suggests that amidating enzymes are active in their venom glands. The correlation between canonical α-amidation signals in transcripts from different transcriptomic analyses and amidation of the mature encoded peptides, confirmed by biochemical and proteomic analyses [35–39], indicates that the classical PHM plus PAL and/or PAM systems must be present in scorpion venom glands. In this work, the enzymatic amidation system of Old and New World scorpions is assessed by transcriptomic analysis.

2. Results

2.1. The Dual Enzymatic System for α-Amidation in the Order Scorpiones

We investigated venom gland transcriptomes of 21 scorpion species and the genome of *Centruroides sculpturatus* and identified sequences encoding orthologs of the bifunctional PAM enzyme in 13 of them. Partial sequences for PAM were found in the remaining eight transcriptomes. A 14th complete PAM-coding sequence was recovered by RT-PCR from venom-gland total RNA from the scorpion *Centruroides noxius* (Table 2 and Supplementary Table S2). Complete coding sequences (CDS) from those transcripts translate into proteins of 861–887 amino acids (Supplementary Figure S1). The deduced topology of the scorpion PAM precursor is similar to that of the PAM-2 isoform described for *Rattus norvegicus* (Figure 2A,B). A signal peptide sequence (SP) for secretion is followed by a short propeptide (PP) region, a PHM domain, a linker sequence (Linker 1), a PAL domain, a second linker sequence (Linker 2), a membrane spanning domain (MSD), and a cytosolic domain (CD) (Figure 2A). The rat PAM-2 isoform lacks the Exon A-encoded linker region with respect to the rat PAM-1 isoform. This extra region contains an endoproteolytic site which, after processing, cleaves the PHM and PAL monofunctional enzymes into separate polypeptides. This Exon-A-encoded region has been described only for vertebrates [23], and has no equivalent sequence in the scorpion PAM (Supplementary Figure S2). It is notable that although the scorpion PAM lacks this region, two putative endoproteolytic sites are still present in the scorpion PAM sequence (Figure 2A). The first site, defined by a lysine dyad (KK), is located between the PHM and PAL domains, and is proposed to delimit the PHM domain. The second site, located between the PAL sequence and the MSD, is also defined by a KK dyad, and if subjected to post-translational processing, would liberate a soluble PAL enzyme from the MSD and CD domains. Thus, the scorpion bifunctional PAM enzyme could be post-translationally processed to generate independent, soluble PHM and PAL enzymes.

Shorter transcripts encoding the monofunctional PHM and PAL enzymes (PHM*m* and PAL*m*) were also identified in most of the analyzed scorpion transcriptomes/genome (Table 2 and Supplementary Table S2). The encoded proteins are 345–350 amino acids long (PHM*m*) and 356–366 (PAL*m*) (Supplementary Figures S3 and S4). Topologies of the monofunctional enzymes are similar to those of the PHM and PAL-2 isoforms from *D. melanogaster* (Figure 2C). The proproteins include a SP and the catalytic domain. No MSD and CD domains are detected; therefore, the monofunctional enzymes are predicted to be soluble.

Table 2. Enzymes of the α-amidation pathway detected in scorpions.

Family	Species	PAM	PHM	PAL	PC1	PC2	CPE
Buthidae	*Centruroides sculpturatus*	■	■	■	○	●	●
	Centruroides hentzi	■	■	■	●		●
	Centruroides noxius [a]	■	■	■	○		○
	Centruroides limpidus [b]	■	■	■	●	●	●
	Centruroides orizaba	■	■	■	●	●	●
	Centruroides ochraceus	■	■	■	○	○	●
	Centruroides hirsutipalpus	✓	■	■	○		●
	Tityus trivittatus	■	■	■	●	●	●
	Leiurus abdullahbayrami *	■	■	■	●	●	●
	Mesobuthus martensii *	■	■	■			
Vaejovidae	*Thorellius cristimanus*	▨	■	■	●	●	●
	Paravaejovis schwenkmeyeri	✓	■	■		●	●
	Chihuahuanus coahuilae	▨	■	■	●	○	●
	Serradigitus gertschi	✓	■	■			
Caraboctonidae	*Hoffmannihadrurus aztecus*	■	■	■			
	Hadrurus concolorus	■	■	■		✓	●
Euscorpiidae	*Megacormus gertschi*	■	■	■			
Chactidae	*Anuroctonus pococki bajae*	✓	■	■			
Superstitionidae	*Superstitionia donensis*	✓	■	✓		✓	
Diplocentridae	*Diplocentrus melici*	▨	■	■		●	
Urodacidae	*Urodacus yaschenkoi* *	■	■	■	●	✓	●
Scorpionidae	*Pandinus imperator* *	■	■			●	

(■-■, ■ and ■): Complete PAM, PHM*m* and PAL*m* sequences; (▨ -): Partial PAM sequences with 93% or more of the sequence determined; (●, ●, ●): Complete PC1, PC2 and CPE sequences; (○, ○, ○): PC1, PC2 and CPE sequences with more than 50% of the sequence determined; (✓): Partial sequences with less than 50% of the estimated total sequence determined; [a] PAM sequence amplified by PCR; [b] PAM sequence verified by DNA sequencing; * Old World scorpion. The tblast and blastn algorithms were used to identify sequences in the local scorpion transcriptomic databases, with an e-value of 1×10^{-6}. Empty spaces indicate that no sequences were identified in those transcriptomes.

Key residues involved in catalysis and metal coordination are conserved in both scorpion amidation systems (Figure 2A and Supplementary Figures S5 and S6), suggesting that those enzymes are probably functional. The percentage of sequence identity between homologous domains of the bifunctional and independent enzymes for each species are indicated in Supplementary Table S3. As an example, for *C. noxius*, the percentage of identity between the PAM subdomains and the PHM*m* and PAL*m* are 29.8% and 32.5%, respectively.

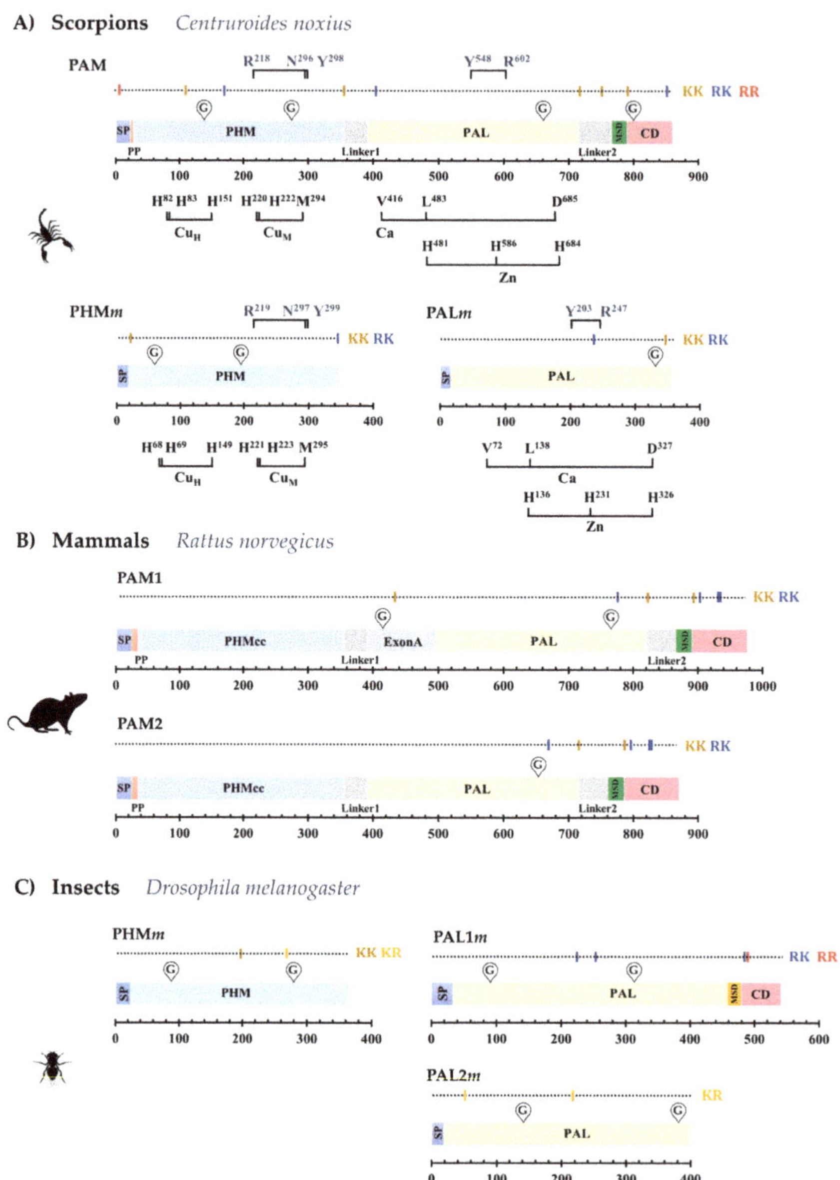

Figure 2. Representative structures of the precursors of (**A**) scorpion (*C. noxius*) bifunctional PAM and mono-functional PHM*m* and PAL*m*; (**B**) mammalian (*R. norvegicus*) PAM-1 and PAM-2 isoforms; (**C**) monofunctional PHM*m*, PAL1*m* and PAL2*m* in *D. melanogaster*. Structural and sequence features are indicated as: SP, Signal peptide; PP, Propeptide; MSD, Membrane Spanning Domain; CD, Cytosolic Domain; KK, RK, KR and RR, putative proprotein convertase cleavage sites at basic dyads; G, predicted glycosylation site; PHMcc, catalytic core of the PHM domain.

Sequences encoding other components of the α-amidation pathway were also sought among available scorpion transcriptomic/genomic sequences. Transcripts encoding orthologs of proprotein convertases 1 and 2 (PC1 and PC2) and carboxypeptidase E (CPE), enzymes that operate upstream in the α-amidation pathway (Figure 1A), were also found, as well as their genes in the *Centruroides sculpturatus* genome (Table 2, Supplementary Table S2), reinforcing the notion of a conserved α-amidation pathway in scorpions.

These results indicate that in scorpions, a dual enzymatic system for α-amidation is responsible for the amidation of venom peptides. Transcripts for both the bifunctional PAM and the monofunctional PHMm and PALm are present in scorpion venom glands. Among arthropods, a similar dual system is present in crustaceans, but not in insects [26].

2.2. The PAM-, PHMm- and PALm-Coding Genes

The search for genomic sequences in *C. sculpturatus* using blastn showed that separate genes encode the bifunctional and monofunctional enzymes, demonstrating that they are encoded by paralogous genes and are not the result of alternative splicing, a phenomenon reported in the generation of isoforms in *R. norvegicus* [27,33]. Figure 3A shows the structure of the genes for the PAM, PHMm and PALm enzymes in *C. sculpturatus*, including their sizes, exon numbers and distributions. The structures of the rat PAM and fruit fly PHM and PAL genes are also shown for comparison (Figure 3B,C).

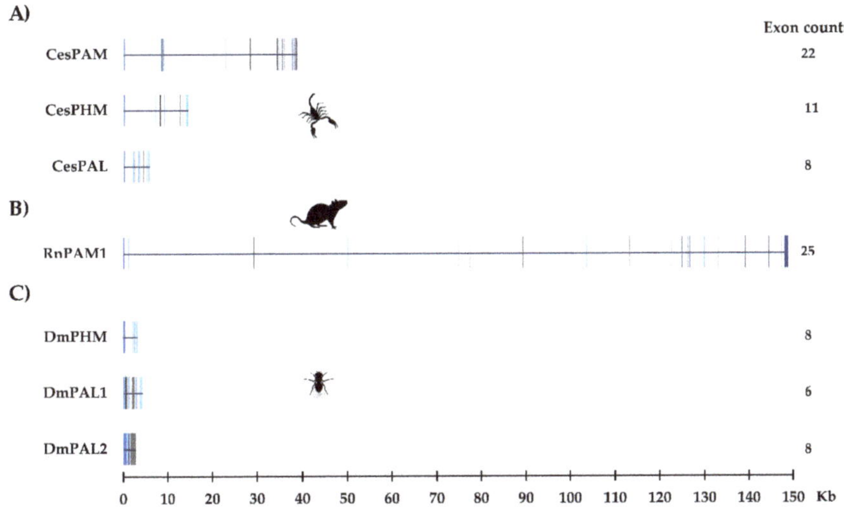

Figure 3. The structure of genes encoding the amidating enzymes of (**A**) *C. sculpturatus* (CesPAM, CesPHM and CesPAL); (**B**) *R. norvegicus* (PAM1); (**C**) *D. melanogaster* (DmPHM, DmPAL1 and DmPAL2). Exons are indicated as vertical blocks.

2.3. Phylogenetic Reconstruction of Amidating Enzymes of Arachnids

Phylogenomic analyses have proposed two basal branches from which all scorpions have descended (parvorders Buthida and Iurida) [1,40,41]. Maximum likelihood analyses with the nucleotide sequences of the PHM and PAL domains from the PAM (designated as *phm*-PAM and *pal*-PAM in these analyses, respectively) and the PHMm and PALm enzymes, show a correlation between the phylogeny of these enzymes and the phylogeny of the scorpion families from which they originate [40]. Figure 4; Figure 5 show the two main clades in which the sequences of the *phm*-PAM and *pal*-PAM are separated from the sequences PHMm and PALm, respectively. Within those clades, there is a clear divergence between sequences from species belonging to the family Buthidae (parvorder Buthida) and sequences from species belonging to families of the parvorder Iurida (Vaejovidae, Caraboctonidae, Euscorpiidae, Chactidae, Superstitionidae, Diplocentridae, Urodacidae, Scorpionidae). Within the family Buthidae, sequences from Old World scorpions *Leiurus abdullahbayrami* (Turkey) and *Mesobuthus martensii* (Eastern Asian countries) are placed in an independent, supported clade that precedes the clade of New World species. The sequences from *Tityus trivittatus* (from the south-central part of South America, Argentina, and Brazil) are in independent supported clades with respect to those of the genus *Centruroides* (distributed in Central America, the Caribbean, and North America). The variable numbers of sequences recovered from different transcriptomes, limits comparative analyses of the catalytic domains, though a consistent topology for the phylogenetic trees is observed. Sequences putatively encoding a dual amidation system, as in scorpions, were also found in other arachnids, including members of the orders Araneae (*Liphistius malayanus*, *Frontinella communis*, *Parasteatoda tepidariorum*, *Leucauge venusta*), Opiliones (*Siro boyerae*, *Trogulus martensi*) Ricinulei (*Ricinoides atewa*), and the xiphosuran, *Limulus polyphemus* (recently placed within the class Arachnida [42]), among others (Supplementary Table S1). This indicates that the same dual α-amidation system is also employed by other arachnids.

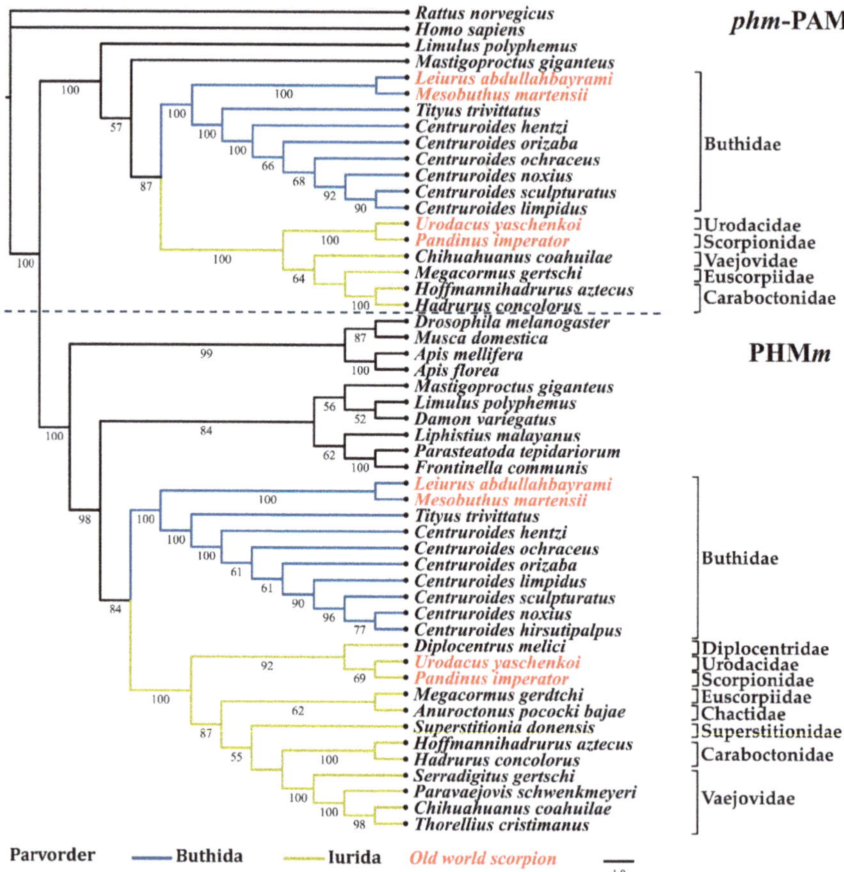

Figure 4. Reconstructed evolutionary history of the phm-PAM and PHM*m* domains. Maximum likelihood analyses were performed with nucleotide sequences corresponding to the respective domains. Numbers under the nodes indicate the values of ultrafast bootstrap (UFBoot) (only branches with values higher that 50 are shown).

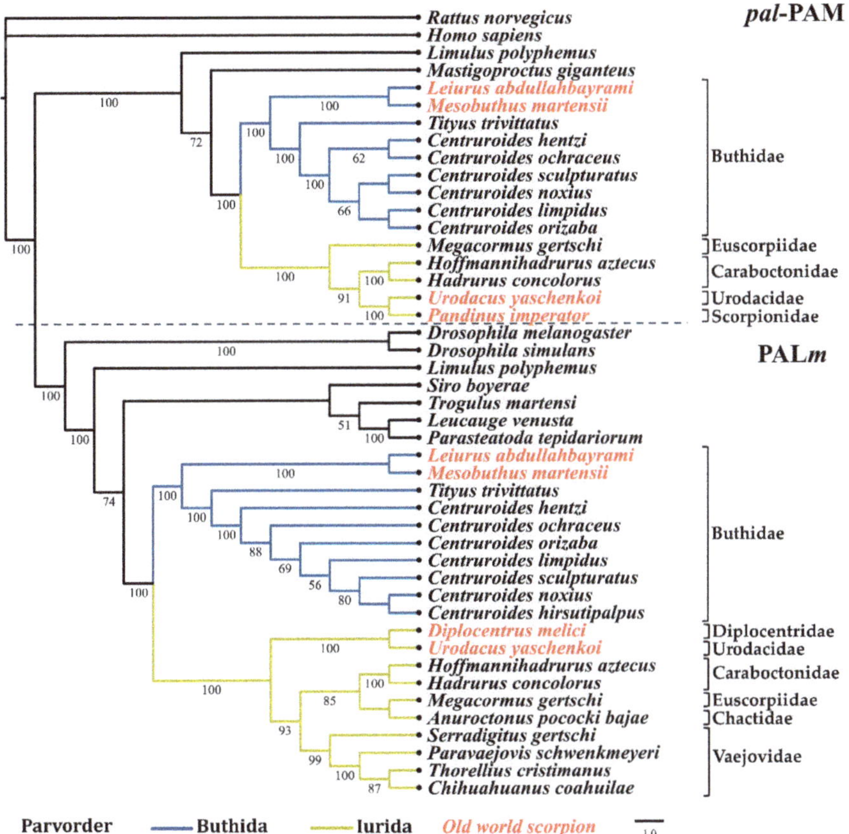

Figure 5. Reconstructed evolutionary history of the pal-PAM and PALm domains. Maximum likelihood analyses were performed with the nucleotide sequences corresponding to the respective domains. Number under the nodes indicate the values of ultrafast bootstrap (UFBoot) (only branches with values higher that 50 are shown).

3. Discussion

Venom gland transcriptomic analyses performed with representative scorpion families from both the Old and New Worlds have shown the enormous diversity of compounds that comprise these important biofluids [4]. Together with available biochemical information on scorpion venom components, sequences of many transcripts indicate that amidation is one of the most common PTMs of scorpion venom peptides. The discovery of canonical amidation signals in the translated sequences suggested that a conserved α-amidation system might be present in scorpion venom glands to convert propeptides into shorter, amidated, mature peptides. In this work, transcripts encoding the relevant components of this pathway are described, confirming that a dual amidation system, including a bifunctional PAM enzyme and individual non-membrane bound PHMm and PALm is employed. Genes for this dual system were found in the genome of *C. sculpturatus*, demonstrating than the bifunctional and the monofunctional enzymes are encoded by independent genes and are not the result of alternative splicing. Paralogs involved in various developmental processes and cellular functions within the orders Scorpiones and Araneae arose as a consequence of a genome duplication in the common ancestor of scorpions and spiders [43,44]. Given the importance of amidation in peptide signaling and the functionality of toxins and other amidated venom peptides, it is not surprising that both amidation enzyme systems were retained in this lineage of venomous arachnids, where they evolved to target specific substrates, or to be expressed in particular cell types or physiological conditions.

Together with conserved functional residues for cation coordination and enzymatic activity, the scorpion PAM sequence contains all the structural elements for generation of a membrane-anchored protein. However, the sequence of the bifunctional PAM contains putative endoprotease cleavage sites (dyads of basic amino acids), which are normally targeted by proprotein convertases, flanking the catalytic domains. This means that the PAM proprotein could in principle be processed to the complete membrane-bound two-domain enzyme or it could be post-translationally cleaved by convertases to render soluble monofunctional domains. The presence of transcripts encoding convertases in the scorpion venom glands, also described in this work, reinforces this possibility. Whether both the two-domain PAM and the PAM-derived monofunctional enzymes coexist in the venom gland remains to be established. We expect that the soluble PHM*m* and PAL*m*, as well as the putative PAM-derived soluble isoforms, are secreted by the venom glands into the venom. This has been confirmed, at least for PHM*m* with liquid chromatography-mass spectrometry (LC-MS/MS) in scorpion venom proteomic analyses. Although it is not clear what additional functions they might have in scorpion venom, it is known that the bovine PAM enzyme is capable of catalyzing three alternative reactions: sulfoxidation, N-dealkylation of amines and O-dealkylation [45]. This raises the possibility of finding new natural substrates for this set of enzymes and taking advantage of their catalytic capacities for synthesis or chemical modification of molecules of biotechnological interest.

Other proteomic analyses have confirmed the presence of putative amidating enzymes in arachnid venoms. One of these sequences was reported as a PAM from the spider *Cupiennius salei* (annotated as PAM_CUPSA [MH766628]) [46]. However, a rigorous sequence analysis demonstrates that this sequence is not from a PAM ortholog, but a monofunctional PHM*m*. Similarly, for the scorpion *Tityus obscurus*, a sequence reported as a PAM (GenBank: JAT91064) [38], shares 87% sequence identity with the PHM*m* from *T. trivittatus*, as reported here, and is therefore also a PHM*m*. A third report found a PHM*m* sequence in transcriptomic and proteomic analyses of the scorpion *C. hentzi* (annotated as GFWZ01000197.1 TSA: *Centruroides hentzi* Chent_MonoO transcribed RNA sequence) [47]. Sequences encoding orthologs of PHM*m* were also identified in venoms of *Centruroides limpidus*, *Centruroides hirsutipalpus* and *Superstitionia donensis* (data not shown). Therefore, this constitutes the first report of the monofunctional PAL*m* and the bifunctional PAM enzymes from any arachnid, and demonstrates that a conserved, functional dual α-amidation system is present in scorpion venom glands, as well as in other arachnids.

4. Materials and Methods

4.1. Sequence Data and Transcriptome Assembly

Previously reported transcriptomic analyses from venom glands of the scorpion species C. limpidus, Paravaejovis schwenkmeyeri, Urodacus yaschenkoi, Thorellius cristimanus (reported as T. atrox), Serradigitus gertschi, S. donensis, and Megacormus gertschi [37,48–53] were used to obtain relevant sequence information. Complementary sequence information was obtained from other unpublished transcriptomes for the species Centruroides noxius, C. orizaba, C. ochraceus, C. hirsutipalpus, T. trivittatus, L. abdullahbayrami, Hoffmannihadrurus aztecus, Hadrurus concolorus, Anuroctonus pococki bajae, Chihuahuanus coahuilae and Diplocentrus melici. Publicly available reads from massive transcriptome analyses of other species were assembled de novo and also used (M. martensii SRR3061379, Pandinus imperator SRR1721600, C. hentzi SRR6041834/SRR6041835; external groups are shown in Supplementary Table S1). Assembly was performed using Trinity 2.0.3 [54] with previously reported parameters [37]. Genomic sequences from C. sculpturatus (BioProject: PRJNA168116) were obtained from NCBI. Sequence information from 22 different scorpion species, belonging to nine of the 20 recognized scorpion families [55] was used in this work.

4.2. Identification and Annotation of Amidating Enzymes in Scorpions and Related Organisms

Sequences putatively encoding PAM, PHM and PAL homologs were identified in transcriptomes using tBLASTn, with the sequence of the *R. norvegicus* PAM (Uniprot, P14925) as query. Recovered nucleotide sequences were translated with the ExPASy server [56]. The presence and organization of characteristic domains was evaluated with NCBI-CDART [57] in accordance with [23]. Other sequence hallmarks were identified: the signal peptide (SP) with SignalP 4.1 and Phobius [58,59], the propeptide region (Pp) with ArachnoServer v. 3.0 [60] and the transmembrane domain with the TMHMM server v. 2.0 [61]. Identification and delimitation of the catalytic domains and the residues involved in metal coordination and disulfide formation was manually performed by sequence alignment with the reference *R. norvegicus* PAM (Uniport P14925). Potential glycosylation sites were predicted with the NetNGlyc 1.0 Server (http://www.cbs.dtu.dk/services/NetNGlyc/). The annotation of each determined sequence can be found in Supplementary Table S1. The sequences were submitted to the European Nucleotide Archive (ENA) under project PRJEB32831.

4.3. Amplification and Cloning of the PAM Sequence from Centruroides noxius

Total RNA was extracted from the telson of a single female *C. noxius* using an SV Total RNA Isolation System kit (Promega Corporation, Madison, WI, USA). cDNA was amplified with a First Strand cDNA Synthesis Kit for RT-PCR (AMV) (Roche, Basel, Switzerland). Primers, Cen-Fw3 (5′-GAT CTT GTA AAC GGC GTA TTT CCC TT-3′) and Cen-Rv4 (5′-CCG ATA TCC TCC CAA CCA TCC TTT C-3′), were designed from the consensus of the PAM sequences from two scorpions of the genus *Centruroides* (*C. limpidus* and *C. orizaba*). Amplification conditions were 3 min at 96 °C, followed by 30 cycles of 3 sec at 96 °C, 1 min at 56 °C and 2 min at 68 °C, plus a final step of 5 min at 68 °C. A recombinant Pfu polymerase produced in-house was used. The PCR product was purified with the QIAQuick Gel extraction Kit (QIAGEN GmbH, Hilden, Germany), ligated into an EcoRV-digested pBluescript II KS(+) vector, and electroporated to electrocompetent DH5α *Escherichia coli* cells. Positive clones were selected with the blue/white system by growing the cells in X-Gal/IPTG-complemented LB/ampicillin medium. Plasmids were prepared by alkaline lysis and submitted to sequencing with the primers T7-Like (5′-GCG TAA TAC GAC TCA CTA TA-3′), T3-Like (5′-CTC ACT AAA GGG AAC AAA AGC-3′), Cen-In1 (5′-CTC GTT GCT TAG ATA TAG AGA-3′), Cen-In2 (5′-ACA TCA GTC AAC CAA ACA-3′) and Cnox-In3 (5′-ATT GAT GCT GAT GAT GCC TA-3′).

4.4. Multiple Alignments and Phylogeny Reconstruction of PAM, PHM, and PAL

Phylogenetic reconstruction of the PAM enzyme and its two catalytic domains *phm*-PAM and *pal*-PAM (with the suffix '-PAM' used to differentiate them from those of the monofunctional enzymes), and of the independent enzymes PHM*m* and PAL*m* (with the suffix '*m*', for 'monofunctional') was performed using the maximum likelihood (ML) method with nucleotide sequences. Additional sequences from phylogenetically related organisms (external groups) were obtained from NCBI or assembled from transcriptome raw reads deposited at SRA-NCBI. All sequences were aligned with MAFFT v7.407 [62]. The best substitution model (GTR+F+I+G4) and the ML analysis were evaluated with IQ-TREE v1.6.9 [63,64], using the ultrafast bootstrap method (UFBoot2) [65] with 10,000 replicates.

4.5. Genomic Organization of Scorpion PAM, PHM, and PAL

Genome sequences of *C. sculpturatus* (NCBI:txid218467) corresponding to the amidating enzymes were recovered from NCBI using BLASTn, with the nucleotide sequences for PAM, PHM, and PAL from *C. limpidus* as queries. Identification of introns and exons was performed with the Splign utility [66].

Supplementary Materials: The following are available online at http://www.mdpi.com/2072-6651/11/7/425/s1: Figure S1: Schematic alignment of PAM sequences with >90% of the estimated sequence determined, Figure S2: Schematic alignment of PAM1 and PAM2 isoforms from *R. norvegicus* and the completely sequenced scorpion PAM, Figure S3: Schematic alignment of the PHM*m* sequences found in 22 analyzed scorpion transcriptomes, Figure S4: Schematic alignment of the 20 PAL*m* sequences found in 22 analyzed scorpion transcriptomes, Figure S5: Sequence alignment of PHM domains, Figure S6: Sequence alignment of PAL domains, Table S1: Nomenclature of transcripts in various scorpion species, Table S2: Sequence conservation between catalytic domains of the bifunctional and monofunctional enzymes by species (% of identity), Table S3: External groups used for phylogenetic reconstruction of the evolutionary history of the functional domains.

Author Contributions: Conceptualization: E.O.; methodology: G.D.-P. Validation: E.O., L.D.P., B.B. Formal analysis: G.D.-P. Investigation: G.D.-P. Resources: L.D.P. Data curation: G.D.-P. Writing—original draft preparation: G.D.-P., E.O. Writing—review and editing: L.D.P., B.B. Visualization: G.D.-P., E.O. Supervision: E.O. Project administration: E.O. Funding acquisition: L.D.P.

Funding: This research was funded by grant IN202619 from Dirección General del Personal Académico, UNAM, awarded to L.D.P., G.D.-P. is the recipient of a PhD fellowship from CONACyT (540036).

Acknowledgments: We are indebted to Jorge L. Folch-Mallol and Marcela Ayala-Aceves for insightful discussions. We appreciate the support given by María T. Romero-Gutiérrez with transcriptomic analyses, and by Erika P. Meneses with proteomic analyses. The help of Edmundo González-Santillán with phylogenetic analysis is greatly appreciated. We thank Jérôme Verleyen for the bioinformatics tools in the cluster of the Instituto de Biotecnología-UNAM. The technical assistance of Arturo Ocádiz-Ramírez, David S. Castañeda-Carreón, Roberto P. Rodríguez-Bahena, Jesús O. Arriaga-Pérez, Juan M. Hurtado-Ramírez, and Servando Aguirre-Cruz is also greatly acknowledged.

Conflicts of Interest: The authors declare no conflict of interest. The funders had no role in the design of the study; in the collection, analyses, or interpretation of data; in the writing of the manuscript, or in the decision to publish the results.

References

1. Sharma, P.P.; Fernandez, R.; Esposito, L.A.; Gonzalez-Santillan, E.; Monod, L. Phylogenomic resolution of scorpions reveals multilevel discordance with morphological phylogenetic signal. *Proc. Biol. Sci.* **2015**, *282*, 20142953. [CrossRef] [PubMed]

2. Waddington, J.; Rudkin, D.M.; Dunlop, J.A. A new mid-Silurian aquatic scorpion-one step closer to land? *Biol. Lett.* **2015**, *11*, 20140815. [CrossRef] [PubMed]

3. Oldrati, V.; Arrell, M.; Violette, A.; Perret, F.; Sprungli, X.; Wolfender, J.L.; Stocklin, R. Advances in venomics. *Mol. Biosyst.* **2016**, *12*, 3530–3543. [CrossRef] [PubMed]

4. Ortiz, E.; Gurrola, G.B.; Schwartz, E.F.; Possani, L.D. Scorpion venom components as potential candidates for drug development. *Toxicon* **2015**, *93*, 125–135. [CrossRef] [PubMed]

5. Xiao, L.; Gurrola, G.B.; Zhang, J.; Valdivia, C.R.; SanMartin, M.; Zamudio, F.Z.; Zhang, L.; Possani, L.D.; Valdivia, H.H. Structure-function relationships of peptides forming the calcin family of ryanodine receptor ligands. *J. Gen. Physiol.* **2016**, *147*, 375–394. [CrossRef]

6. Fuller, M.D.; Thompson, C.H.; Zhang, Z.R.; Freeman, C.S.; Schay, E.; Szakacs, G.; Bakos, E.; Sarkadi, B.; McMaster, D.; French, R.J.; et al. State-dependent inhibition of cystic fibrosis transmembrane conductance regulator chloride channels by a novel peptide toxin. *J. Biol. Chem.* **2007**, *282*, 37545–37555. [CrossRef] [PubMed]

7. Gurevitz, M. Mapping of scorpion toxin receptor sites at voltage-gated sodium channels. *Toxicon* **2012**, *60*, 502–511. [CrossRef] [PubMed]

8. Possani, L.D.; Becerril, B.; Delepierre, M.; Tytgat, J. Scorpion toxins specific for Na+-channels. *Eur. J. Biochem.* **1999**, *264*, 287–300. [CrossRef]

9. Rodriguez de la Vega, R.C.; Possani, L.D. Overview of scorpion toxins specific for Na+ channels and related peptides: Biodiversity, structure-function relationships and evolution. *Toxicon* **2005**, *46*, 831–844. [CrossRef]

10. Rodriguez de la Vega, R.C.; Possani, L.D. Current views on scorpion toxins specific for K+-channels. *Toxicon* **2004**, *43*, 865–875. [CrossRef]

11. Chippaux, J.P. Emerging options for the management of scorpion stings. *Drug Des. Dev. Ther.* **2012**, *6*, 165–173. [CrossRef] [PubMed]

12. Chippaux, J.P.; Goyffon, M. Epidemiology of scorpionism: A global appraisal. *Acta Trop.* **2008**, *107*, 71–79. [CrossRef] [PubMed]

13. Jekely, G. Global view of the evolution and diversity of metazoan neuropeptide signaling. *Proc. Natl. Acad. Sci. USA* **2013**, *110*, 8702–8707. [CrossRef] [PubMed]

14. Gutte, B. *Peptides: Synthesis, Structures, and Applications*, 1st ed.; Academic Press: San Diego, CA, USA, 1995; pp. 288–289.

15. Merkler, D.J. C-terminal amidated peptides: Production by the in vitro enzymatic amidation of glycine-extended peptides and the importance of the amide to bioactivity. *Enz. Microb. Technol.* **1994**, *16*, 450–456. [CrossRef]

16. Coelho, V.A.; Cremonez, C.M.; Anjolette, F.A.; Aguiar, J.F.; Varanda, W.A.; Arantes, E.C. Functional and structural study comparing the C-terminal amidated beta-neurotoxin Ts1 with its isoform Ts1-G isolated from Tityus serrulatus venom. *Toxicon* **2014**, *83*, 15–21. [CrossRef] [PubMed]

17. Liu, Z.; Yang, G.; Li, B.; Chi, C.; Wu, X. Cloning, co-expression with an amidating enzyme, and activity of the scorpion toxin BmK ITa1 cDNA in insect cells. *Mol. Biotechnol.* **2003**, *24*, 21–26. [CrossRef]

18. Estrada, G.; Restano-Cassulini, R.; Ortiz, E.; Possani, L.D.; Corzo, G. Addition of positive charges at the C-terminal peptide region of CssII, a mammalian scorpion peptide toxin, improves its affinity for sodium channels Nav1.6. *Peptides* **2011**, *32*, 75–79. [CrossRef] [PubMed]

19. Eipper, B.A.; Stoffers, D.A.; Mains, R.E. The biosynthesis of neuropeptides: Peptide alpha-amidation. *Annu. Rev. Neurosci.* **1992**, *15*, 57–85. [CrossRef] [PubMed]

20. Fricker, L.D. Neuropeptide-processing enzymes: Applications for drug discovery. *AAPS J.* **2005**, *7*, E449–E455. [CrossRef] [PubMed]

21. Eipper, B.A.; Perkins, S.N.; Husten, E.J.; Johnson, R.C.; Keutmann, H.T.; Mains, R.E. Peptidyl-alpha-hydroxyglycine alpha- amidating lyase. Purification, characterization, and expression. *J. Biol. Chem.* **1991**, *266*, 7827–7833.

22. Perkins, S.N.; Husten, E.J.; Eipper, B.A. The 108-kDA peptidylglycine alpha-amidating monooxygenase precursor contains two separable enzymatic activities involved in peptide amidation. *Biochem. Biophys. Res. Commun.* **1990**, *171*, 926–932. [CrossRef]

23. Attenborough, R.M.; Hayward, D.C.; Kitahara, M.V.; Miller, D.J.; Ball, E.E. A "neural" enzyme in nonbilaterian animals and algae: Preneural origins for peptidylglycine alpha-amidating monooxygenase. *Mol. Biol. Evol.* **2012**, *29*, 3095–3109. [CrossRef] [PubMed]

24. Han, M.; Park, D.; Vanderzalm, P.J.; Mains, R.E.; Eipper, B.A.; Taghert, P.H. Drosophila uses two distinct neuropeptide amidating enzymes, dPAL1 and dPAL2. *J. Neurochem.* **2004**, *90*, 129–141. [CrossRef] [PubMed]

25. Kolhekar, A.S.; Roberts, M.S.; Jiang, N.; Johnson, R.C.; Mains, R.E.; Eipper, B.A.; Taghert, P.H. Neuropeptide amidation in Drosophila: Separate genes encode the two enzymes catalyzing amidation. *J. Neurosci.* **1997**, *17*, 1363–1376. [CrossRef] [PubMed]

26. Zabriskie, T.M.; Klinge, M.; Szymanski, C.M.; Cheng, H.; Vederas, J.C. Peptide amidation in an invertebrate: Purification, characterization, and inhibition of peptidylglycine alpha-hydroxylating monooxygenase from the heads of honeybees (Apis mellifera). *Arch. Insect Biochem. Physiol.* **1994**, *26*, 27–48. [CrossRef] [PubMed]

27. Eipper, B.A.; Green, C.B.; Campbell, T.A.; Stoffers, D.A.; Keutmann, H.T.; Mains, R.E.; Ouafik, L. Alternative splicing and endoproteolytic processing generate tissue-specific forms of pituitary peptidylglycine alpha-amidating monooxygenase (PAM). *J. Biol. Chem.* **1992**, *267*, 4008–4015. [PubMed]

28. Eipper, B.A.; Park, L.P.; Dickerson, I.M.; Keutmann, H.T.; Thiele, E.A.; Rodriguez, H.; Schofield, P.R.; Mains, R.E. Structure of the precursor to an enzyme mediating COOH-terminal amidation in peptide biosynthesis. *Mol. Endocrinol.* **1987**, *1*, 777–790. [CrossRef]

29. Glauder, J.; Ragg, H.; Rauch, J.; Engels, J.W. Human peptidylglycine alpha-amidating monooxygenase: cDNA, cloning and functional expression of a truncated form in COS cells. *Biochem. Biophys. Res. Commun.* **1990**, *169*, 551–558. [CrossRef]

30. Mizuno, K.; Ohsuye, K.; Wada, Y.; Fuchimura, K.; Tanaka, S.; Matsuo, H. Cloning and sequence of cDNA encoding a peptide C-terminal alpha-amidating enzyme from Xenopus laevis. *Biochem. Biophys. Res. Commun.* **1987**, *148*, 546–552. [CrossRef]

31. Fan, X.; Spijker, S.; Akalal, D.B.; Nagle, G.T. Neuropeptide amidation: Cloning of a bifunctional alpha-amidating enzyme from Aplysia. *Brain Res. Mol. Brain Res.* **2000**, *82*, 25–34. [CrossRef]

32. Spijker, S.; Smit, A.B.; Eipper, B.A.; Malik, A.; Mains, R.E.; Geraerts, W.P. A molluscan peptide alpha-amidating enzyme precursor that generates five distinct enzymes. *FASEB J.* **1999**, *13*, 735–748. [CrossRef]

33. Eipper, B.A.; Milgram, S.L.; Husten, E.J.; Yun, H.Y.; Mains, R.E. Peptidylglycine alpha-amidating monooxygenase: A multifunctional protein with catalytic, processing, and routing domains. *Protein Sci.* **1993**, *2*, 489–497. [CrossRef]

34. Ul-Hasan, S.; Burgess, D.M.; Gajewiak, J.; Li, Q.; Hu, H.; Yandell, M.; Olivera, B.M.; Bandyopadhyay, P.K. Characterization of the peptidylglycine alpha-amidating monooxygenase (PAM) from the venom ducts of neogastropods, Conus bullatus and Conus geographus. *Toxicon* **2013**, *74*, 215–224. [CrossRef]

35. Becerril, B.; Corona, M.; Garcia, C.; Bolivar, F.; Possani, L.D. Cloning of Genes Encoding Scorpion Toxins—An Interpretative Review. *J. Tox. Tox. Rev.* **1995**, *14*, 339–357. [CrossRef]

36. Bougis, P.E.; Rochat, H.; Smith, L.A. Precursors of Androctonus australis scorpion neurotoxins. Structures of precursors, processing outcomes, and expression of a functional recombinant toxin II. *J. Biol. Chem.* **1989**, *264*, 19259–19265.

37. Romero-Gutierrez, T.; Peguero-Sanchez, E.; Cevallos, M.A.; Batista, C.V.F.; Ortiz, E.; Possani, L.D. A Deeper Examination of Thorellius atrox Scorpion Venom Components with Omic Techonologies. *Toxins* **2017**, *9*, 3399. [CrossRef]

38. De Oliveira, U.C.; Nishiyama, M.Y., Jr.; Dos Santos, M.B.V.; Santos-da-Silva, A.P.; Chalkidis, H.M.; Souza-Imberg, A.; Candido, D.M.; Yamanouye, N.; Dorce, V.A.C.; Junqueira-de-Azevedo, I.L.M. Proteomic endorsed transcriptomic profiles of venom glands from Tityus obscurus and T. serrulatus scorpions. *PLoS ONE* **2018**, *13*, e0193739. [CrossRef]

39. Abdel-Rahman, M.A.; Quintero-Hernandez, V.; Possani, L.D. Venom proteomic and venomous glands transcriptomic analysis of the Egyptian scorpion Scorpio maurus palmatus (Arachnida: Scorpionidae). *Toxicon* **2013**, *74*, 193–207. [CrossRef]

40. Santibanez-Lopez, C.E.; Gonzalez-Santillan, E.; Monod, L.; Sharma, P.P. Phylogenomics facilitates stable scorpion systematics: Reassessing the relationships of Vaejovidae and a new higher-level classification of Scorpiones (Arachnida). *Mol. Phylogenet. Evol.* **2019**, *135*, 22–30. [CrossRef]

41. Sharma, P.P.; Baker, C.M.; Cosgrove, J.G.; Johnson, J.E.; Oberski, J.T.; Raven, R.J.; Harvey, M.S.; Boyer, S.L.; Giribet, G. A revised dated phylogeny of scorpions: Phylogenomic support for ancient divergence of the temperate Gondwanan family Bothriuridae. *Mol. Phylogenet. Evol.* **2018**, *122*, 37–45. [CrossRef]

42. Ballesteros, J.A.; Sharma, P.P. A Critical Appraisal of the Placement of Xiphosura (Chelicerata) with Account of Known Sources of Phylogenetic Error. *Syst. Biol.* **2019**, *10*. [CrossRef]

43. Leite, D.J.; Baudouin-Gonzalez, L.; Iwasaki-Yokozawa, S.; Lozano-Fernandez, J.; Turetzek, N.; Akiyama-Oda, Y.; Prpic, N.M.; Pisani, D.; Oda, H.; Sharma, P.P.; et al. Homeobox gene duplication and divergence in arachnids. *Mol. Biol. Evol.* **2018**, *35*. [CrossRef]

44. Schwager, E.E.; Sharma, P.P.; Clarke, T.; Leite, D.J.; Wierschin, T.; Pechmann, M.; Akiyama-Oda, Y.; Esposito, L.; Bechsgaard, J.; Bilde, T.; et al. The house spider genome reveals an ancient whole-genome duplication during arachnid evolution. *BMC Biol.* **2017**, *15*, 62. [CrossRef]

45. Katopodis, A.G.; May, S.W. Novel substrates and inhibitors of peptidylglycine alpha-amidating monooxygenase. *Biochemistry* **1990**, *29*, 4541–4548. [CrossRef]

46. Kuhn-Nentwig, L.; Langenegger, N.; Heller, M.; Koua, D.; Nentwig, W. The Dual Prey-Inactivation Strategy of Spiders-In-Depth Venomic Analysis of Cupiennius salei. *Toxins* **2019**, *11*, 167. [CrossRef]

47. Ward, M.J.; Ellsworth, S.A.; Rokyta, D.R. Venom-gland transcriptomics and venom proteomics of the Hentz striped scorpion (Centruroides hentzi; Buthidae) reveal high toxin diversity in a harmless member of a lethal family. *Toxicon* **2018**, *142*, 14–29. [CrossRef]

48. Cid-Uribe, J.I.; Meneses, E.P.; Batista, C.V.F.; Ortiz, E.; Possani, L.D. Dissecting Toxicity: The Venom Gland Transcriptome and the Venom Proteome of the Highly Venomous Scorpion Centruroides limpidus (Karsch, 1879). *Toxins* **2019**, *11*, 247. [CrossRef]

49. Cid-Uribe, J.I.; Santibanez-Lopez, C.E.; Meneses, E.P.; Batista, C.V.F.; Jimenez-Vargas, J.M.; Ortiz, E.; Possani, L.D. The diversity of venom components of the scorpion species Paravaejovis schwenkmeyeri (Scorpiones: Vaejovidae) revealed by transcriptome and proteome analyses. *Toxicon* **2018**, *151*, 47–62. [CrossRef]

50. Luna-Ramirez, K.; Quintero-Hernandez, V.; Juarez-Gonzalez, V.R.; Possani, L.D. Whole Transcriptome of the Venom Gland from Urodacus yaschenkoi Scorpion. *PLoS ONE* **2015**, *10*, e0127883. [CrossRef]

51. Romero-Gutierrez, M.T.; Santibanez-Lopez, C.E.; Jimenez-Vargas, J.M.; Batista, C.V.F.; Ortiz, E.; Possani, L.D. Transcriptomic and Proteomic Analyses Reveal the Diversity of Venom Components from the Vaejovid Scorpion Serradigitus gertschi. *Toxins* **2018**, *10*, 359. [CrossRef]

52. Santibanez-Lopez, C.E.; Cid-Uribe, J.I.; Batista, C.V.; Ortiz, E.; Possani, L.D. Venom Gland Transcriptomic and Proteomic Analyses of the Enigmatic Scorpion Superstitionia donensis (Scorpiones: Superstitioniidae), with Insights on the Evolution of Its Venom Components. *Toxins* **2016**, *8*, 367. [CrossRef]

53. Santibanez-Lopez, C.E.; Cid-Uribe, J.I.; Zamudio, F.Z.; Batista, C.V.F.; Ortiz, E.; Possani, L.D. Venom gland transcriptomic and venom proteomic analyses of the scorpion Megacormus gertschi Diaz-Najera, 1966 (Scorpiones: Euscorpiidae: Megacorminae). *Toxicon* **2017**, *133*, 95–109. [CrossRef]

54. Grabherr, M.G.; Haas, B.J.; Yassour, M.; Levin, J.Z.; Thompson, D.A.; Amit, I.; Adiconis, X.; Fan, L.; Raychowdhury, R.; Zeng, Q.; et al. Full-length transcriptome assembly from RNA-Seq data without a reference genome. *Nat. Biotechnol.* **2011**, *29*, 644–652. [CrossRef]

55. Santibanez-Lopez, C.E.; Francke, O.F.; Ureta, C.; Possani, L.D. Scorpions from Mexico: From Species Diversity to Venom Complexity. *Toxins* **2016**, *8*, 2. [CrossRef]

56. Artimo, P.; Jonnalagedda, M.; Arnold, K.; Baratin, D.; Csardi, G.; de Castro, E.; Duvaud, S.; Flegel, V.; Fortier, A.; Gasteiger, E.; et al. ExPASy: SIB bioinformatics resource portal. *Nucleic Acids Res.* **2012**, *40*, W597–W603. [CrossRef]

57. Geer, L.Y.; Domrachev, M.; Lipman, D.J.; Bryant, S.H. CDART: Protein homology by domain architecture. *Genome Res.* **2002**, *12*, 1619–1623. [CrossRef]

58. Petersen, T.N.; Brunak, S.; von Heijne, G.; Nielsen, H. SignalP 4.0: Discriminating signal peptides from transmembrane regions. *Nat. Methods* **2011**, *8*, 785–786. [CrossRef]

59. Kall, L.; Krogh, A.; Sonnhammer, E.L. A combined transmembrane topology and signal peptide prediction method. *J. Mol. Biol.* **2004**, *338*, 1027–1036. [CrossRef]

60. Pineda, S.S.; Chaumeil, P.A.; Kunert, A.; Kaas, Q.; Thang, M.W.C.; Le, L.; Nuhn, M.; Herzig, V.; Saez, N.J.; Cristofori-Armstrong, B.; et al. ArachnoServer 3.0: An online resource for automated discovery, analysis and annotation of spider toxins. *Bioinformatics* **2018**, *34*, 1074–1076. [CrossRef]

61. Moller, S.; Croning, M.D.; Apweiler, R. Evaluation of methods for the prediction of membrane spanning regions. *Bioinformatics* **2001**, *17*, 646–653. [CrossRef]

62. Katoh, K.; Standley, D.M. MAFFT multiple sequence alignment software version 7: Improvements in performance and usability. *Mol. Biol. Evol.* **2013**, *30*, 772–780. [CrossRef]

63. Trifinopoulos, J.; Nguyen, L.T.; von Haeseler, A.; Minh, B.Q. W-IQ-TREE: A fast online phylogenetic tool for maximum likelihood analysis. *Nucleic Acids Res.* **2016**, *44*, W232–W235. [CrossRef]

64. Kalyaanamoorthy, S.; Minh, B.Q.; Wong, T.K.F.; von Haeseler, A.; Jermiin, L.S. ModelFinder: Fast model selection for accurate phylogenetic estimates. *Nat. Methods* **2017**, *14*, 587–589. [CrossRef]

65. Hoang, D.T.; Chernomor, O.; von Haeseler, A.; Minh, B.Q.; Vinh, L.S. UFBoot2: Improving the Ultrafast Bootstrap Approximation. *Mol. Biol. Evol.* **2018**, *35*, 518–522. [CrossRef]

66. Kapustin, Y.; Souvorov, A.; Tatusova, T.; Lipman, D. Splign: Algorithms for computing spliced alignments with identification of paralogs. *Biol. Direct.* **2008**, *3*, 20. [CrossRef]

Article

The Sequence and a Three-Dimensional Structural Analysis Reveal Substrate Specificity among Snake Venom Phosphodiesterases

Anwar Ullah [1],*, Kifayat Ullah [1], Hamid Ali [1], Christian Betzel [2],* and Shafiq ur Rehman [3]

[1] Department of Biosciences, COMSATS University Islamabad, Park Road, Tarlai Kalan, Islamabad 45550, Pakistan; kafy.biodiesel@gmail.com (K.U.); hamidpcmd@yahoo.com (H.A.)
[2] Institute of Biochemistry and Molecular Biology, University of Hamburg, Laboratory for Structural Biology of Infection and Inflammation, c/o DESY. Build. 22a, Notkestrasse 85, 22607 Hamburg, Germany
[3] Department of Botany, University of Okara, Okara 56300, Punjab, Pakistan; evergreenpk@gmail.com
* Correspondence: anwar.ms90@yahoo.com (A.U.); Christian.Betzel@uni-hamburg.de (C.B.)

Received: 25 July 2019; Accepted: 3 September 2019; Published: 28 October 2019

Abstract: (1) Background. Snake venom phosphodiesterases (SVPDEs) are among the least studied venom enzymes. In envenomation, they display various pathological effects, including induction of hypotension, inhibition of platelet aggregation, edema, and paralysis. Until now, there have been no 3D structural studies of these enzymes, thereby preventing structure–function analysis. To enable such investigations, the present work describes the model-based structural and functional characterization of a phosphodiesterase from *Crotalus adamanteus* venom, named PDE_*Ca*. (2) Methods. The PDE_*Ca* structure model was produced and validated using various software (model building: I-TESSER, MODELLER 9v19, Swiss-Model, and validation tools: PROCHECK, ERRAT, Molecular Dynamic Simulation, and Verif3D). (3) Results. The proposed model of the enzyme indicates that the 3D structure of PDE_*Ca* comprises four domains, a somatomedin B domain, a somatomedin B-like domain, an ectonucleotide pyrophosphatase domain, and a DNA/RNA non-specific domain. Sequence and structural analyses suggest that differences in length and composition among homologous snake venom sequences may account for their differences in substrate specificity. Other properties that may influence substrate specificity are the average volume and depth of the active site cavity. (4) Conclusion. Sequence comparisons indicate that SVPDEs exhibit high sequence identity but comparatively low identity with mammalian and bacterial PDEs.

Keywords: snake venom; phosphodiesterases; amino acid sequence and three-dimensional structural analysis; variable substrate specificity; PDE_*Ca* structure–function relationship

Key Contribution: This is the first report on snake venom phosphodiesterase (SVPDE) that describes the overall structural properties, makes structural comparisons, and examines the structural basis of substrate specificity.

---

1. Introduction

Snake venom is a crude mixture that contains enzymatic and non-enzymatic proteins, peptides, organic compounds of low molecular weight, and inorganic compounds [1,2]. Proteins constitute the major portion (about 90%) of the total dry mass of crude snake venom, with or without catalytic activity, including neurotoxins, cardiotoxins, C-type lectins, proteinases, metalloproteinases, serine proteinases, phospholipases, hyaluronidases, acetylcholinesterases, L-amino acid oxidases, three-finger toxins, phospholipase A_2s, and nucleases [3–9]. Metalloproteinases, serine proteinases, phospholipases, and neurotoxins are the most widely studied snake venom proteins, as they occur in high concentrations and are relatively easy to purify [1–5,10–12]. Other enzymes, such as nucleases, exist in small quantities and are the least studied. Nucleases are capable of cleaving phosphodiester bonds in nucleic acids, and, in snake venom, they have been classified as endonucleases and exonucleases [13,14]. Phosphodiesterases are generally considered exonucleases [14].

Phosphodiesterases (E.C. No. 3.1.4.1) belong to the Ectonucleotide pyrophosphatase/ phosphodiesterase (E-NPP) family of metalloenzymes [13]. Generally, viperid venoms contain more phosphodiesterases (PDEs) than crotalid or elapid venoms [15,16]. Phosphodiesterases cleave phosphodiester bonds in polynucleotides in a sequential manner, starting at the 3′-end, and release 5′-mononucleotides [16]. PDEs have been shown to hydrolyze a wide variety of nucleotides, such as ATP, ADP, NAD^+, $NADP^+$, and GDP [4,17]. Because this enzyme degrades oligonucleotide fragments, there is increasing demand for purified PDE for use in the structural analysis of nucleic acids [17,18].

PDEs have been isolated from various snake venoms, including Deinagkistrodon acutus [19], Bothrops atrox [20], Bothrops alternatus [21], Cerastes vipera [22], Crotalus atrox [23], Crotalus adamanteus [24,25] Crotalus durissus terrificus [25], Protobothrops flavoviridis [26], and Vipera aspis [27]. In addition to snake venoms, they also occur in spider venoms [28]. The structures of human, mouse, and bacterial PDEs have been well studied compared to snake and spider venom PDEs.

Snake venom PDE was first reported by Uzawa [29]. This is one of the least studied enzymes in snake venom due to the fact that earlier reports showed it to be non-toxic and involved only in digestion [13]. However, recent reports indicate that PDE has a major role in envenomation by hydrolyzing DNA and RNA, releasing adenosine and other purine nucleosides [30,31]. Adenosine induces a variety of pathological and pharmacological effects, such as increased vascular permeability, hypotension, inhibition of platelet aggregation, edema, and paralysis [32,33]. Snake venom phosphodiesterases (SVPDEs) have also been used as therapeutic agents in various diseases and conditions, such as cerebrovascular and cardiovascular diseases, hypertension, and atherosclerosis [34].

SVPDEs are monomeric proteins of high molecular weight (98–140 kDa), with basic pIs (8.4–9.2), and metal cofactors, usually zinc, which are essential for catalytic activity [35–38]. Some studies report a dimeric structure [13,39,40]. Sometimes, a single venom may contain multiple PDE isoforms [13,37,40].

Although the amino acid sequences of PDEs from various snake species are available in the literature, there is no information regarding their three-dimensional (3D) structures. Therefore, it is difficult to correlate function with structure. In order to enable structure-function studies, here we present a model-based 3D structural characterization of the phosphodiesterase from *Crotalus adamanteus* venom. The PDE_*Ca* structure model was produced and validated using various software (I-TESSER, MODELLER 9v19, Swiss-Model, PROCHECK, ERRAT, and Verif3D). The sequence alignment, structure-based substrate specificity, maturation, and comparison with PDEs from other organisms are also discussed.

2. Results and Discussion

2.1. Sequence Alignment Analysis

The PDE_*Ca* precursor contains 851 amino acids with 830 amino acid residues in the mature form. Sequence alignment indicates high sequence identity among SVPDEs and comparatively low sequence identity (<65%) with their mammalian counterparts (Table 1). The average sequence identities among SVPDEs and mammalian phosphodiesterases are 90.6% and 58.3%, respectively. The metal ion-binding/active site residues (Zn^{+2} 1 (D153, T191, D358, H359), Zn^{+2} 2 (D311, H315, and H476), and Ca^{+2} (N751, D753, H755, D757) (PDE_*Ca* precursor numbering scheme) are fully conserved among all phosphodiesterases examined, except N751 and H755 in SVPDEs, where these have mutated to D751 and R755, respectively, in mammalian homologs (Figure 1). Amino acid residues around the metal ion binding and active sites are highly conserved among all phosphodiesterases examined. The amino acid sequence of PDE_*Ca* contains 33 cysteine residues, of which 32 form 16 disulfide bridges (Table 2).

The generated model and DiANNA web server [38] also confirmed the presence of sixteen disulfide bridges in PDE_*Ca*.

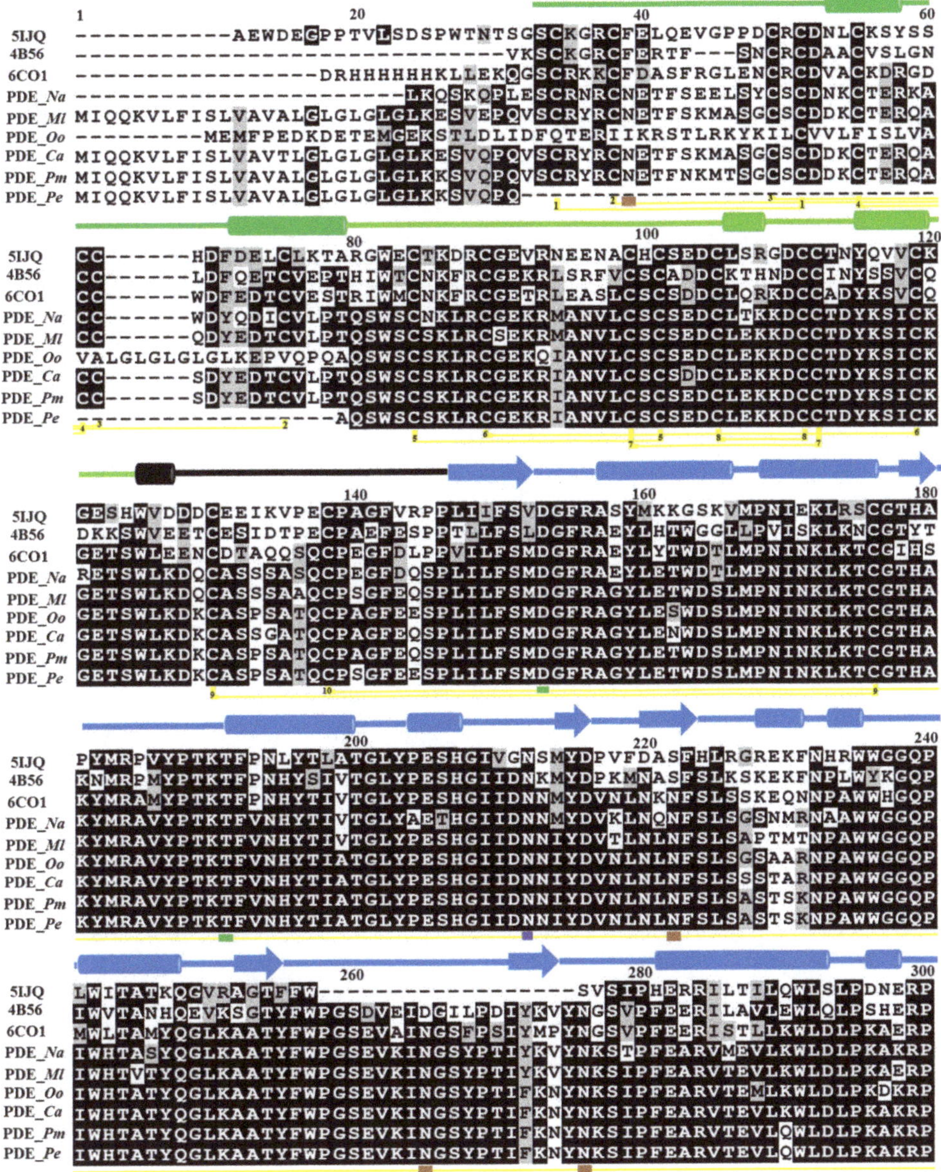

Figure 1. *Cont.*

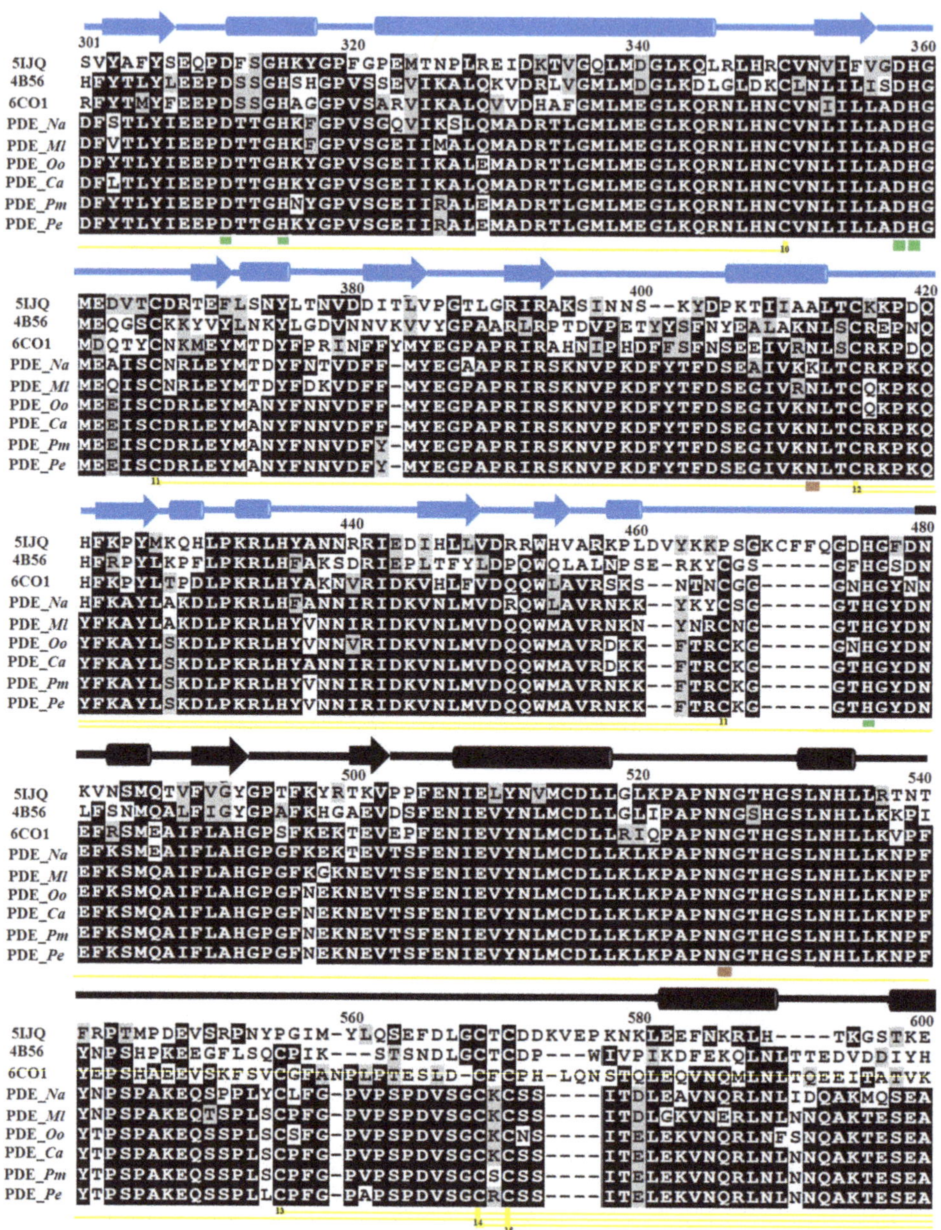

Figure 1. *Cont.*

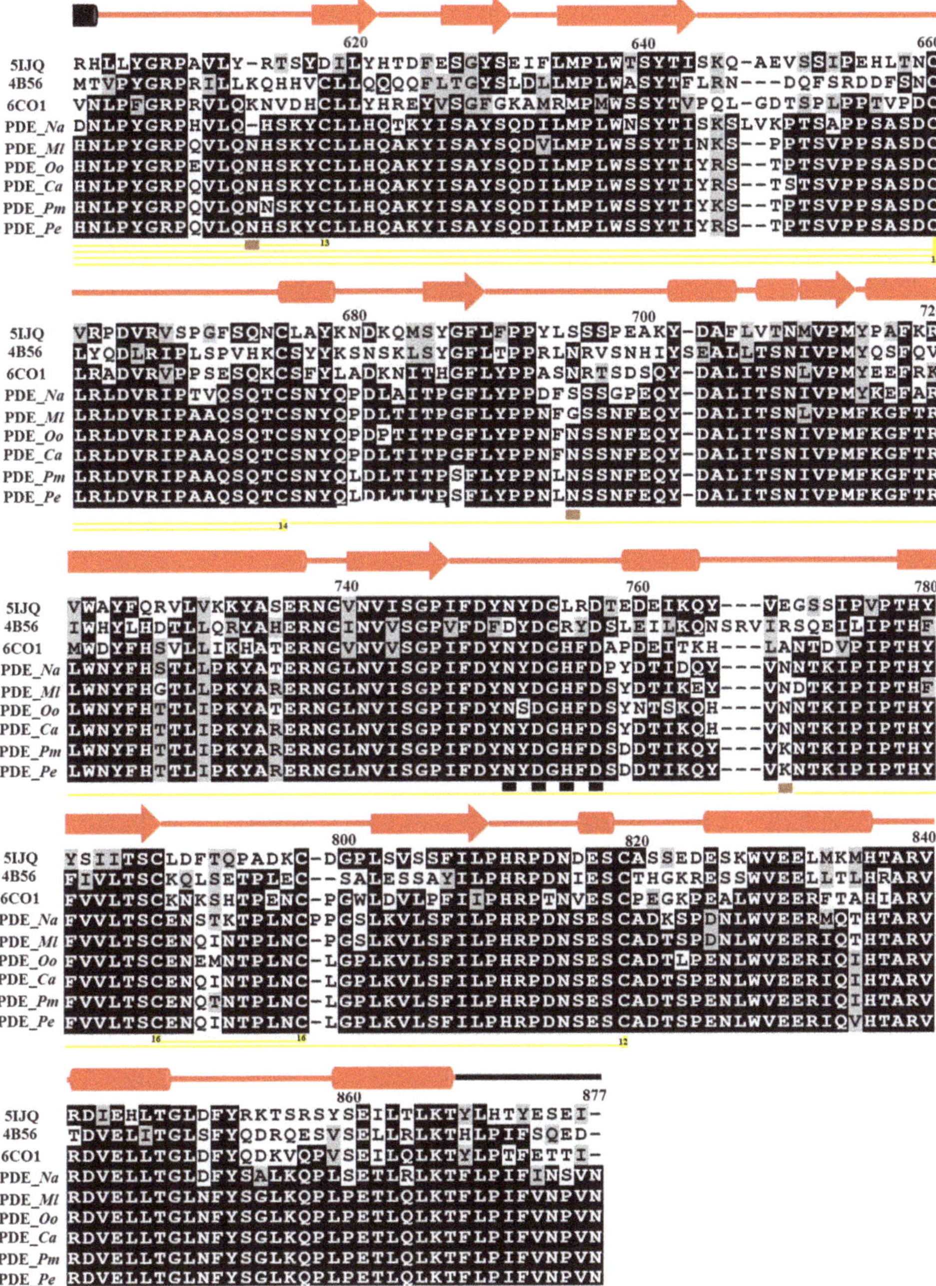

Figure 1. Sequence alignment of snake venom phosphodiesterases (SVPDEs) and mammalian homologs. 5IJQ; crystal of human Autotaxin (ENPP2); 4B56: Ectonucleotide pyrophosphatase phosphodiesterase-1 (NPP1) (*Mus musculus*); 6C01: human Ectonucleotide pyrophosphatase/phosphodiesterase 3 (ENPP3); 5GZ4, phosphodiesterase *Naja atra atra* (Gene Bank ID: A0A2D0TC04), PDE_*Ml*, phosphodiesterase *Macrovipera lebetina* (*Vl*) (Gene Bank ID: AHJ80885.1), PDE_*Oo*, phosphodiesterase *Ovophis okinavensis* (Gene Bank ID: BAN89426.1), PDE_*Ca*, phosphodiesterase *Crotalus adamanteus* (Gene Bank ID: JAS04699.1),

PDE_*Pm*, phosphodiesterase *Protobothrops mucrosquamatus* (Gene Bank ID: XP_015675293.1), PDE_*Pe*, phosphodiesterase *Protobothrops elegans* (Gene Bank ID: BAP39928.1). Residues involved in catalysis and metal ion binding are underlined with blue and black, respectively. Cysteine residues that form disulfide bridges are linked (yellow lines). Putatively N-glycosylated amino acid residues are underlined in brown. Amino acid residues in the somatomedin B domain, the somatomedin B-like domain, the ectonucleotide pyrophosphatase/phosphodiesterase domain (also called autotaxin), and the DNA/RNA non-specific domain, are colored green, light green, blue, and red, respectively. Secondary structural elements (alpha helices and beta strands) are shown above the sequence. Sequence numbering corresponds to the PDE_*Ca* precursor protein.

Table 1. Percent sequence identities among SVPDEs and their mammalian counterparts.

Proteins	PDE_*Ca*	PDE_*Pm*	PDE_*Ml*	PDE_*Pe*	PDE_*Oo*	5GZ4	6CO1	4B56	5IJQ
PDE_*Ca*	—	85%	93%	92%	96%	85%	64%	50%	47%
PDE_*Pm*	85%	—	92%	94%	95%	84%	63%	50%	48%
PDE_*Ml*	93%	92%	—	87%	90%	86%	63%	50%	46%
PDE_*Pe*	92%	94%	87%	—	95%	85%	64%	50%	49%
PDE_*Oo*	96%	95%	90%	95%	—	85%	64%	50%	48%
5GZ4	85%	84%	86%	85%	85%	—	64%	50%	47%
6CO1	64%	63%	63%	64%	64%	64%	—	53%	46%
4B56	50%	50%	50%	50%	50%	50%	53%	—	44%
5IJQ	47%	48%	46%	49%	48%	47%	46%	44%	—

PDE_Ca: Phosphodiesterase Crotalus adamanteus, PDE_Pm: Phosphodiesterase Protobothrops mucrosquamatus, PDE_Ml: Phosphodiesterase Macrovipera lebetina, PDE_Pe: Phosphodiesterase Protobothrops elegans, PDE_Oo: Phosphodiesterase Ovophis okinavensis, 5GZ4: Phosphodiesterase Naja atra atra, 6C01: human Ectonucleotide pyrophosphatase / phosphodiesterase 3 (ENPP3), 4B56: Ectonucleotide pyrophosphatase phosphodiesterase-1 (NPP1) (Mus musculus), 5IJQ: human Autotaxin (ENPP2).

Table 2. Cysteine residues participating in disulfide bridges.

1st Cysteine	2nd Cysteine
34	51
38	75
49	62
55	61
84	101
89	119
99	112
105	111
130	176
138	350
366	466
415	819
555	618
569	675
571	660
767	777

2.2. Domain Analysis

The primary structure of mature PDE_*Ca* contains four domains—the somatomedin B domain (residues 33–79), somatomedin B-like domain (residues 81–124), ectonucleotide pyrophosphatase/phosphodiesterase domain (also called autotaxin) (residues 147–479), and the DNA/RNA non-specific domain (residues 534–867) (Figure 1). The regions of amino acid residues from 124–146, 480–533, and 868–877 are connecting segments.

The ThreaDom (Threading-based Protein Domain Prediction) online web server [41] for domain conservation indicated that the domain architecture of PDE_*Ca* is fully conserved, with other proteins in the Protein Data Bank (PDB) containing a similar structure fold (Figure 2). The molecular weights of PDE_*Ca* in zymogen and the mature forms were 96.37 and 93.10 kDa, with corresponding pIs of 8.13 and 8.05, respectively [42]. The theoretically calculated molecular weights and pIs accord with the experimentally measured values for other SVPDEs [43–47].

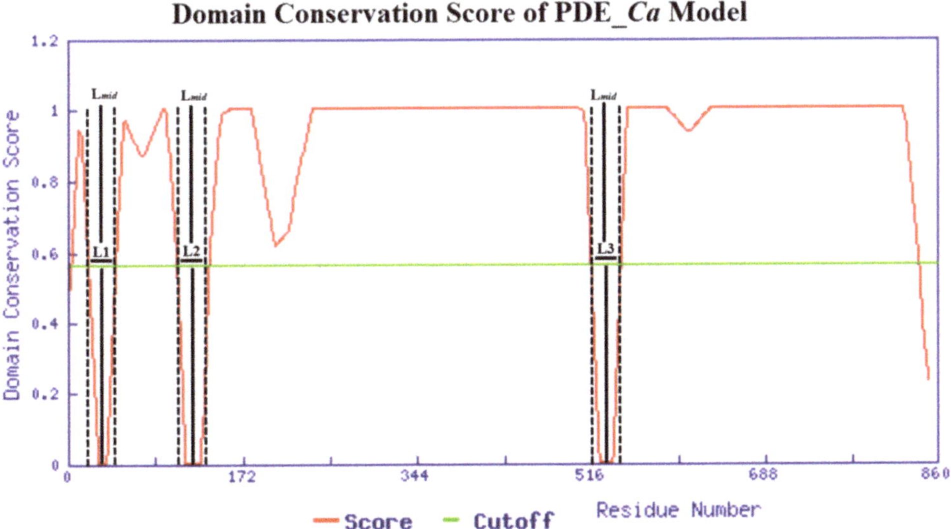

Figure 2. The ThreaDom-based domain prediction in the domain conservation score profile. Four domains are separated by three connecting segments. Vertical dotted lines indicate the start and end locations of each putative connecting segment. Vertical solid lines denote predicted boundaries at the middle of the connecting segments.

2.3. Glycosylation Sites

The primary structure of PDE_*Ca* contains 60 asparagine (N) residues (Protparam, [48]), of which nine were identified as potential glycosylation sites using the NetNGlyc 1.0 Server (N39, N222, N265, N276, N412, N526, N613, N695, and N771) (Figure 1). These asparagine residues are fully conserved among SVPDEs (Figure 1). Glycosylation sites were also confirmed with the Scan Prosite tool [49]. The primary structure of *Vipera lebetina* has also been shown to contain nine putative *N*-glycosylation sites [43]. PDEs of *B. jararaca* and *Walterinnesia aegyptia* contain 33% and 24% carbohydrates, respectively [48,49].

2.4. Homology Modeling and Model Evaluation

The homology model for three-dimensional (3D) structure characterization was produced using various online modeling servers (I-TESSER [50], MODELLER 9v19 program [51], and SWISS Model [52], using atomic coordinates of human ectonucleotide pyrophosphatase/phosphodiesterase 3 (PDB ID:

6C01, amino acid sequence identity 63.39%) as a template [53]. The best model was chosen based on analyses using the PROCHECK, ERRAT, and Verif3D software [54–57].

The PROCHECK analysis of the best model shows that >95% of the amino acid residues are in the most favored region of the graph (Figure 3). The remaining 5% are in the allowed region with no residues in the forbidden/disallowed region. ERRAT analysis shows an overall quality factor of 87.88 for the PDE_*Ca* model, which lies in the average quality range for 3D protein structures [55].

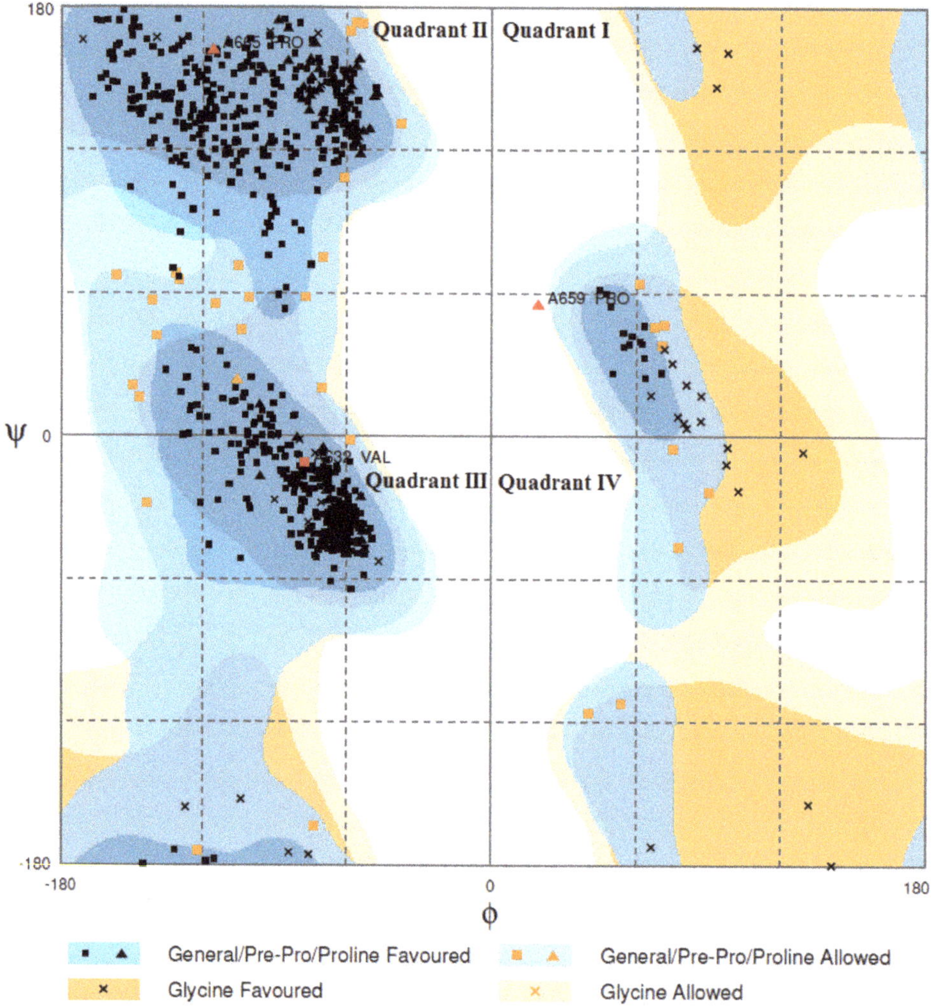

Figure 3. A Ramachandran plot of the modeled structure of PDE_*Ca*. In total, >95% of the amino acid residues are in the most favored region, while the remaining 5% are in the allowed region with no residues in the forbidden/disallowed region. Quadrant I displays a region where multiple conformations are allowed. Quadrant II shows the biggest region in the graph, with the most favorable conformations of atoms. Quadrant III shows the next biggest region in the graph, where the right-handed alpha helices lie. Quadrant IV has almost no outlined region. This conformation is disfavored due to steric clash.

2.5. Molecular Dynamics Simulation

GROMACS, AMBER16, MDWeb, and MDMobby [58–60] all produced the same results. MD simulation analysis indicated that all structural parameters, such as chirality, disulfide bonds, and the absence of steric clashes, were correct (Figure S1A). The two basic assessments (root mean square deviation (RMSD) and radius of gyration (RG)) used to validate the structures through MD simulation were analyzed. The RMSD deviation indicated that the PDE_*Ca* structure did not deviate more than 1 Å from the initial structure, and the RG was also maintained at around 31.7 Å (Figure S1B,C). Flexibility analysis (Bfactor and RMSD per residue), identified some flexible regions, located mostly in loop regions (Figure S1C,D). All these analyses indicate that the simulated structure does not exhibit critical structural deformations.

2.6. Overall Structure of PDE_Ca

The 3D structure of PDE_*Ca* is similar to that of the other members of the alkaline phosphatase-like superfamily (ALP-like superfamily) [25,61–63]. PDE_*Ca* has a complex structure. It is a multi-domain protein that consists of four domains—a somatomedin B domain, a somatomedin B-like domain, an ectonucleotide pyrophosphatase/phosphodiesterase domain (also called autotaxin), and a DNA/RNA non-specific domain (Figure 4). These domains are briefly described below.

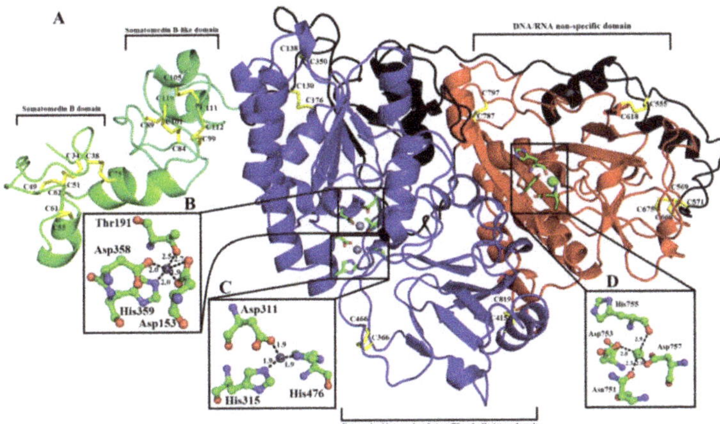

Figure 4. Overall structure of PDE_*Ca*: (**A**) cartoon representation. The active site and metal ion-binding residues are shown as green sticks. Zn^{+2} and Ca^{+2} ions are shown as gray and green spheres, respectively. Disulfide bridges are represented by yellow sticks. (**B–D**), residues involved in Zn^{+2} ion-binding, catalysis, and Ca^{+2} ion-binding are highlighted. Parts of the secondary structure belonging to the somatomedin B domain, somatomedin B-like domain, ectonucleotide pyrophosphatase/phosphodiesterase domain, DNA/RNA non-specific domain, and connecting are colored in green, light green, blue, red, and black, respectively.

2.6.1. Somatomedin B Domain

The somatomedin B domain (SMB) is located at the N-terminus of the protein and comprises amino acid residues 33–79 (Figures 1, 4 and 5). It has two alpha helices (Figures 4 and 5). The SMB domain is stabilized by four intrachain disulfide bridges, one salt bridge (between Asp52-Arg58) (Table 3), and extensive hydrogen bonding (14 intrachain H-bonds).

2.6.2. Somatomedin B-like Domain

The somatomedin B domain connects to another domain called the Somatomedin B-like domain (SMB-like). This domain consists of residues 81 to 124 (Figure 1). Like the somatomedin B domain, it also contains eight cysteine residues in four disulfide bridges (Figure 1). The secondary structure of this domain contains two alpha helices (Figures 4 and 5). Beside its disulfides, the SMB-like domain is stabilized by four salt bridges (Arg82-Glu85, Arg87-Asp104, Arg87-Asp98, Lys102-Asp98), and thirteen interchain H-bonds (four with the SMB domain and nine with the PDE domain), as well as 24 intrachain H-bonds.

Both the SMB and SMB-like domains are highly compact. Disulfide bridges are arranged in the centers of both domains, forming covalently bonded cores (Figure 4).

2.6.3. Ectonucleotide Pyrophosphatase/Phosphodiesterase Domain

The SMB-like domain connects to the catalytic domain, called the ectonucleotide pyrophosphatase/phosphodiesterase (ENPP/PDE) domain, through a short connecting segment of 22 amino acid residues (125–146) (Figure 1, 4 and 5).

The ENPP/PDE domain consists of amino acid residues 147–479 (Figure 1). It contains five cysteine residues, two of which make an intrachain disulfide bridge and three of which make interchain disulfide bridges (two with the connecting segment and one with the DNA/RNA non-specific domain) (Figures 1 and 4) (Table 2). This domain also contains the active site residues, together with the two zinc ions (Figure 4). One of the zinc ions ($Zn^{+2}1$) is coordinated by four amino acids (Asp153, Thr191, Asp358, and His59), and the other one ($Zn^{+2}2$) is coordinated by three residues (Asp311, His315, and His476). All these catalytic amino acid residues are fully conserved in mouse NPP1 [62], human NPP3 (61), human autotaxin ENPP2 [64], and bacterial PDE [29] (Figure 6).

The secondary structure of this domain contains 14 beta strands and 14 alpha helices (Figure 5). Of the fourteen alpha helices and beta strands, seven and five are short alpha helices and beta strands, respectively. This domain is stabilized by interchain disulfide bridges (one with the SMB-like domain and seven with the DNA/RNA non-specific domain), 13 salt bridges (Table 3), and numerous interchain H-bonds.

2.6.4. DNA/RNA Non-Specific Domain

This domain comprises residues 603 to 867 (Figure 1). It contains nine cysteines that form one intrachain and three interchain disulfides (one with the PDE domain and two with the connecting segment) (Figure 4). This domain also contains the Ca^{2+}-binding loop (Figure 4). The secondary structure of this domain contains seven beta strands and ten alpha helices (Figures 4 and 5). This domain is connected to the PDE domains through a long loop, called the lasso loop [65]. This domain is stabilized by four disulfides, eight salt bridges, and numerous H-bonds.

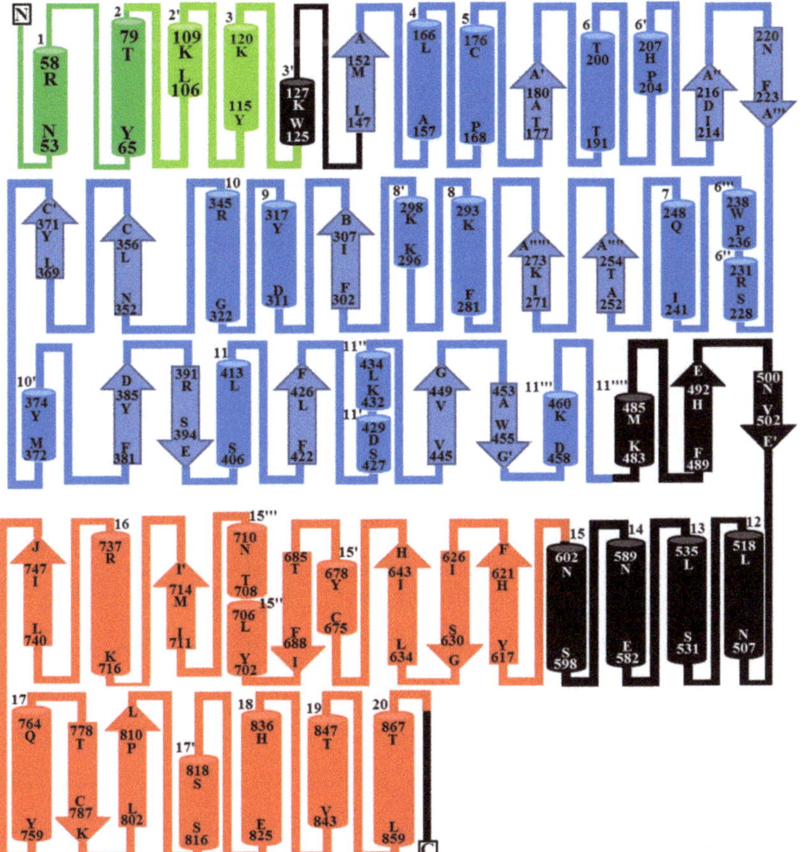

Figure 5. Topology diagram of PDE_*Ca*. The alpha helices (1–20) and beta strands (A–L) are represented as cylinders and arrows, respectively. Short alpha helices and beta strands are shown as primes (e.g., 17′). Secondary structures and amino acid residues in alpha helices and beta strands were assigned from the primary sequence using the program DSSP [66] and were confirmed with PyMOL from the tertiary structure. Parts of the secondary structure belonging to the somatomedin B, somatomedin B-like, ectonucleotide pyrophosphatase/phosphodiesterase, and DNA/RNA non-specific domains, as well as the connecting segments (UN), are colored in green, light green, blue, red, and black, respectively.

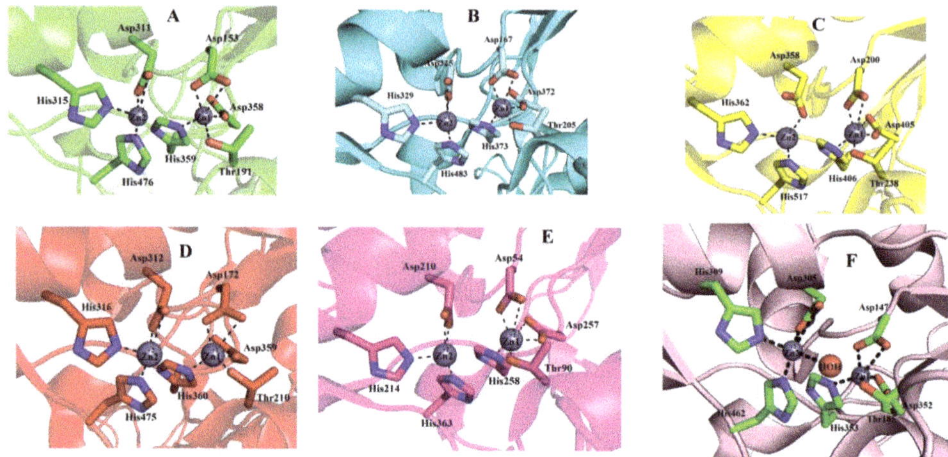

Figure 6. Active site comparison of (**A**) PDE_*Ca* with (**B**) human Ectonucleotide pyrophosphatase/ phosphodiesterase 3 (ENPP3), (**C**) Ectonucleotide pyrophosphatase-phosphodiesterase-1 (NPP1) (*Mus musculus*), (**D**) ENPP2 (human Autotaxin), (**E**) Xac Nucleotide Pyrophosphatase/Phosphodiesterase (*Xanthomonas axonopodis*), and (F) Phosphodiesterase *Naja atra*. Metal ion-binding amino acid residues are displayed as sticks and metal ions as grey spheres.

Table 3. Salt bridges in the PDE_*Ca* 3D structure.

Residue 1	Residue 2	Distance
NH1 ARG A 58	OD1 ASP A 52	3.59
NH1 ARG A 58	OD2 ASP A 52	2.60
NH2 ARG A 58	OD1 ASP A 52	2.69
NH2 ARG A 58	OD2 ASP A 52	3.30
NH2 ARG A 82	OE1 GLU A 85	3.39
NH1 ARG A 87	OD1 ASP A 104	2.58
NH1 ARG A 87	OD2 ASP A 104	3.54
NH2 ARG A 87	OD1 ASP A 98	2.84
NH2 ARG A 87	OD2 ASP A 98	3.93
NH2 ARG A 87	OD1 ASP A 104	3.31
NH2 ARG A 87	OD2 ASP A 104	2.62
NZ LYS A 102	OD2 ASP A 98	2.69
NZ LYS A 168	OD2 ASP A 158	2.87
ND1 HIS A 189	OD2 ASP A 352	3.42
NE2 HIS A 189	OD2 ASP A 352	3.82
NH1 ARG A 278	OE1 GLU A 302	2.85
NH2 ARG A 278	OE1 GLU A 302	2.88
NE2 HIS A 309	OD1 ASP A 305	2.89
NE2 HIS A 309	OD2 ASP A 305	2.95
NZ LYS A 337	OD2 ASP A 122	3.84
NH2 ARG A 339	OD1 ASP A 287	3.88
NH2 ARG A 339	OD2 ASP A 287	2.58
NE2 HIS A 353	OD1 ASP A 147	3.04
NE2 HIS A 353	OD1 ASP A 352	3.46
NE2 HIS A 353	OD2 ASP A 352	2.96

Table 3. *Cont.*

Residue 1	Residue 2	Distance
NH1 ARG A 384	OD2 ASP A 205	2.62
NH2 ARG A 384	OD2 ASP A 205	3.69
NH2 ARG A 384	OD2 ASP A 436	3.73
NZ LYS A 425	OD2 ASP A 727	2.56
NH1 ARG A 426	OD1 ASP A 465	3.74
NH1 ARG A 426	OE1 GLU A 467	3.98
NH2 ARG A 426	OD1 ASP A 465	2.70
NH2 ARG A 426	OD2 ASP A 465	3.11
NH2 ARG A 426	OE1 GLU A 467	3.26
NE2 HIS A 428	OD1 ASP A 816	2.79
NH2 ARG A 434	OD2 ASP A 210	2.94
NE2 HIS A 462	OD1 ASP A 305	3.83
NE2 HIS A 462	OD2 ASP A 305	2.68
NZ LYS A 469	OE1 GLU A 155	3.97
NE2 HIS A 515	OD1 ASP A 502	2.80
NE2 HIS A 515	OD2 ASP A 502	2.97
NH1 ARG A 588	OE1 GLU A 534	3.19
NH2 ARG A 588	OE1 GLU A 534	2.60
NH2 ARG A 645	OD1 ASP A 643	3.26
NH2 ARG A 645	OD2 ASP A 643	2.59
NZ LYS A 710	OE1 GLU A 804	3.97
NH1 ARG A 786	OE1 GLU A 791	3.76
NH2 ARG A 786	OD1 ASP A 788	3.97
NE2 HIS A 810	OE1 GLU A 791	3.37
NH1 ARG A 813	OE1 GLU A 492	3.24
NH2 ARG A 813	OD1 ASP A 816	2.69
NH1 ARG A 815	OE1 GLU A 818	2.75
NZ LYS A 840	OE1 GLU A 818	3.80

NH1 and NH2: Nitrogen atoms (amino groups) of the arginine side chain, OD1, and OD2: Oxygen atoms of aspartic acid side chains, OE: Oxygen atoms of glutamic acid side chains, NZ: Nitrogen atoms (amino groups) of lysine side chains.

2.6.5. Metal Ion-Binding Sites

SVPDEs are metalloenzymes that contain zinc and calcium ions [6,26,35,36,67]. The zinc ion participates in the active site and is important for catalytic activity [26,35–67]. In the modeled structure of PDE_*Ca*, two zinc ions and one calcium ion were found (Figure 4). Of the two zinc ions, one ($Zn^{+2}1$) is coordinated by amino acid residues Asp153, Thr191, His359, and Asp358 (Figure 4), and the other ($Zn^{+2}2$) is coordinated by amino acid residues Asp311, His315, and His476 (Figure 4). The Ca^{+2} is coordinated by amino acid residues Asn751, Asp753, His755, and Asp757 (Figure 4). The metal ion comparison with the human Ectonucleotide pyrophosphatase/phosphodiesterase 3, Ectonucleotide pyrophosphatase-phosphodiesterase-1 (*Mus musculus*), ENPP2 (Human Autotaxin), Xac Nucleotide Pyrophosphatase/Phosphodiesterase (*Xanthomonas axonopodis*), and Taiwan cobra (*Naja atra atra*) PDE (PDB ID: 5GZ4 and 5GZ5) shows that the amino acid residues coordinating these metal ions are fully conserved (Figures 1A–F and 6A–F).

2.7. Structural Basis for Substrate Specificity of Snake Venom Phosphodiesterases

For substrates other than oligonucleotides, SVPDEs display variable substrate specificity [43–47]. Among the SVPDEs for which substrate specificity has been studied, the PDEs from *Vipera lebetina* and *Daboia russelli russelli* hydrolyze ADP [43], while PDEs from *Crotalus adamanteus, Trimeresurus stejnegeri* and *Bothrops jararaca* hydrolyze ATP [45–47].

To explain these broad specificities of SVPDEs, the structures of two other PDEs from *Vipera lebetina* (PDE_Vl) and *Bothrops atrox* (PDE_Ba) were modeled, using the same modeling and validation programs used for the PDE_Ca structural model. The atomic coordinate of human Ectonucleotide pyrophosphatase/phosphodiesterase 3 (PDB ID: 6C01, amino acid sequence identity 64.09% and 62.91% with PDE_Vl and PDE_Ba, respectively) was used as a template. The Ramachandran plot analysis indicates that in both the modeled structures, 98% of amino acid residues were in the favored region of the plot, and 2% were in the allowed region (Figures S2 and S3). The ERRAT analysis shows an overall quality factor of 90 for the modeled structure, which lies in the average quality range for the protein 3D structures (Figures S4 and S5) [55]. The modeled structures of the PDE_Ca, PDE of *Vipera lebetina* (PDE_Vl), and *Bothrops atrox* (PDE_Ba) were compared, taking into account the active site residue composition, active site cavity volume and average depth, and the surface charge distribution (Figure 7, Table 4). The active site's amino acid residues are the same for these enzymes. However, the average active site's cavity volume, the average depth of the active site, and the surface charge distribution vary considerably (Figure 7, Table 4).

The average active site cavity volumes of PDE_Ca, PDE_Vl, and PDE_Ba are 6608.25, 3985.03, and 2243.11 Å^3, respectively, with corresponding average depths of 21.39, 12.54, and 9.86 Å, respectively (Table 4). These values indicate that the average active site cavity volume and depth of PDE_Ca is much larger than that of either PDE_Vl or PDE_Ba. These characteristics permit larger substrates (ATP) to access the active site of PDE_Ca, while preventing it for PDE_Vl.

Another factor that affects the substrate specificity of these enzymes is the surface charge distribution (Figure 7). The surface charge of PDE_Ca (overall and around the active site) is highly positive (Figure 7A), for PDE_Vl it is partially positive and negative (Figure 7B). Analysis of the active site cavity volume and its average depth indicate that SVPDEs with small active site cavity volumes and average depths (like *Vipera lebetina, Daboia russelli russelli,* and *Cerastes cerastes*) (Table 4) show a high preference for ADP, while other SVPDEs with large active site volumes and average depths (like PDEs from *Crotalus adamanteus, Trimeresurus stejnegeri,* and *Bothrops jararaca*) show a high preference for ATP.

Table 4. Average active site cavity volumes and average active site cavity depths of SVPDEs and their mammalian and bacterial counterparts.

Protein	Average Volume (Å3)	Average Depth (Å)
PD_Ca model	6608.25	21.39
PD_Vl model	3985.03	12.54
6C01	14,690.95	19.13
4B56	3651.33	16.30
2GSN	10,107.28	18.58
4ZG7	11,367.42	17.94

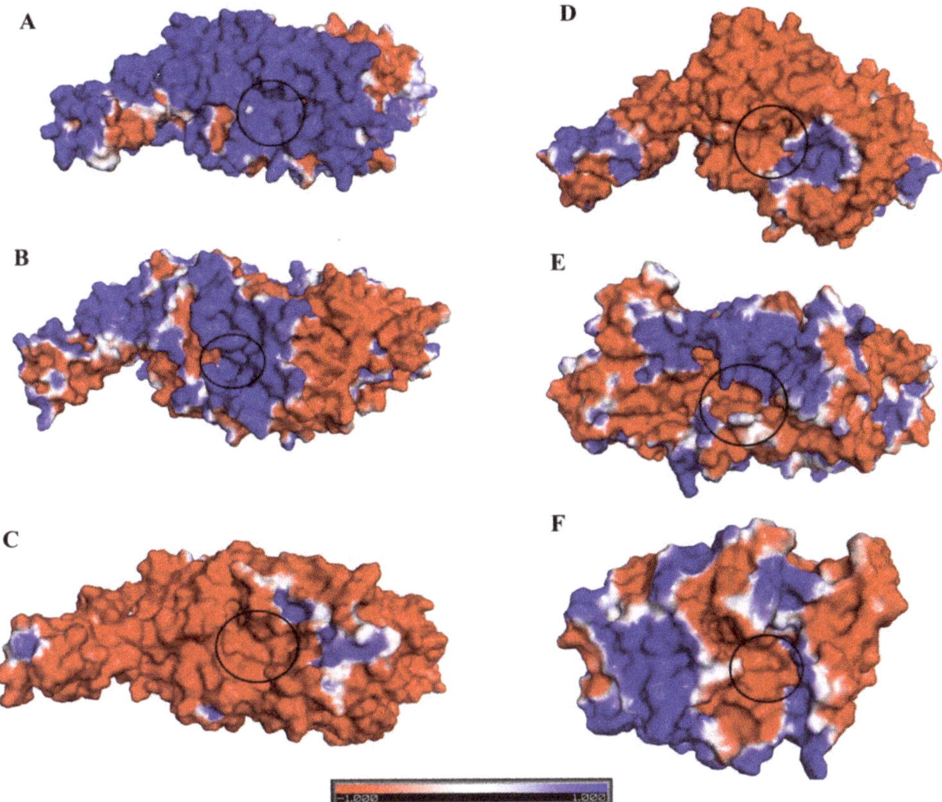

Figure 7. Surface charge distributions of (**A**) PDE_*Ca*, (**B**) PDE_*Vl*, (**C**) human Ectonucleotide pyrophosphatase/phosphodiesterase 3 (ENPP3), (**D**) Ectonucleotide pyrophosphatase-phosphodiesterase-1 (NPP1) (*Mus musculus*), (**E**) ENPP2 (human Autotaxin), and (**F**) Xac Nucleotide Pyrophosphatase/Phosphodiesterase (*Xanthomonas axonopodis*). Blue, red, and white represent the positive, negative, and neutral regions, respectively. Black circles indicate the location of the active-site pocket.

2.8. Structural Alignment between PDE_Ca, Human ENPP3, Mouse NPP1, Human Autotaxin, Xa NPP1, PDE_Vl, PDE_Ba and Naja atra atra PDE.

The structural alignment between PDE_*Ca*, human ENPP3 [53], mouse NPP1 [61], human autotaxin [63], Xa NPP1 [62], phosphodiesterase from *Vipera lebetina* (PDE_*Vl*), *Bothrops atrox* (PDE_*Ba*), and the PDE from Taiwan cobra (*Naja atra atra*; PDB ID: 5GZ4 and 5GZ5) shows that the three-dimensional structural folds of these enzymes are similar and that all of them align well, with an RMSD value range between 0.21 and 0.92 (average RMSD value of 0.61) (Table 5) (Figure 8). They have the same active site residues (Figures 1 and 6) and disulfide bridges (Figure 1). However, the amino acid residues in the loop regions vary considerably, both in composition and length (Figure 8). For this reason, the surface charge distribution also varies (Figure 7A–G), which may impart variable substrate specificity to these enzymes [45,68,69]. The overall surface charge for the SVPDEs is positive, while it is negative for human ENPP3, mouse NPP1, human autotaxin, and Xa NPP1 (Figure 7C–F). The average active site cavity volume and average depth also vary among these enzymes (Table 4).

The phylogenetic tree analysis indicates that PDE_*Ca* has a close evolutionary relationship with SVPDEs and PDEs from human beings and mice (Figure S6).

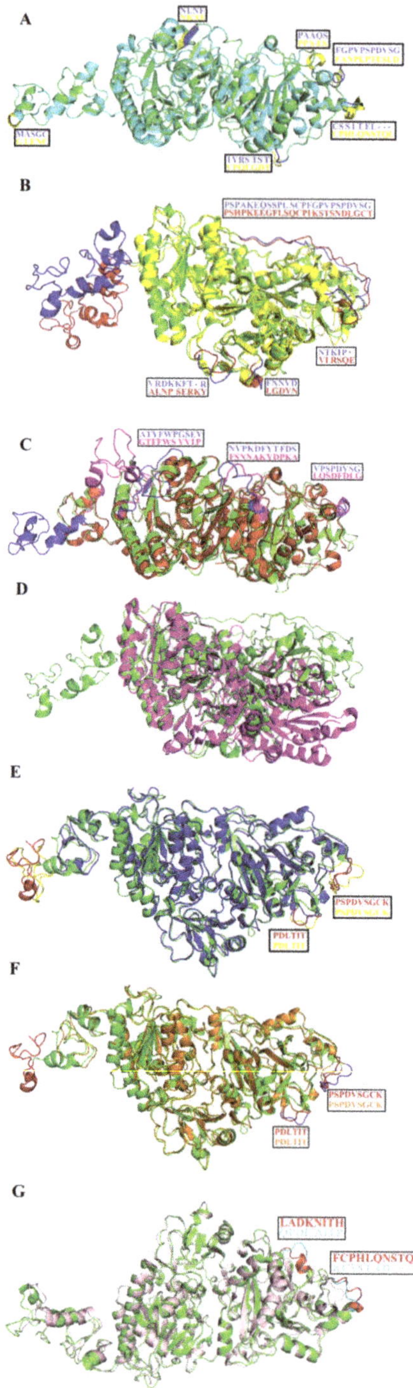

Figure 8. Structural alignment among SVPDEs (PDE_*Ca*, PDE_*Vl*, and PDE_*Ba*) and their mammalian and bacterial counterparts; (**A**) PDE_*Ca* (green) align human Ectonucleotide pyrophosphatase/ phosphodiesterase 3 (ENPP3) (Cyan), (**B**) PDE_*Ca* (green) align Ectonucleotide pyrophosphatase-

phosphodiesterase-1 (NPP1) (*Mus musculus*) (yellow), (**C**) PDE_*Ca* (green) align ENPP2 (human Autotaxin) (red), (**D**) PDE_*Ca* (green) align Xac Nucleotide Pyrophosphatase/Phosphodiesterase (*Xanthomonas axonopodis*) (magenta), (**E**) PDE_*Ca* (green) align *Vipera lebetina* phosphodiesterase (blue), (**F**) PDE_*Ca* (green) align *Bothrops atrox* phosphodiesterase (orange), (**G**) PDE_*Ca* (green) align 5ZG4 (light pink). Loops exhibiting differences and their corresponding amino acid residues are shown in boxes and colored in blue, red, magenta, orange, yellow, and cyan for PDE_*Ca*, ENPP3, NPP1, PDE_*Vl*, PDE_*Ba*, and 5GZ4, respectively.

Table 5. Root mean square deviation values of PDE_*Ca*, PDE_*Vl*, PDE_*Ba*, and their mammalian counterparts.

Protein	RMSD Value
PDE_Ca aligned 6C01	0.21
PDE_Ca aligned 4B56	0.72
PDE_Ca aligned 4ZG7	0.73
PDE_Ca aligned 2GSN	0.92
PDE_Ca aligned PDE_*Vl*	0.57
PDE_Ca aligned PDE_*Ba*	0.56
PDE_Ca aligned 5GZ4	0.60

2.9. Maturation Mechanism for SVPDEs

PDE_*Ca*, like other SVPDEs, is synthesized as a precursor protein (zymogen) [43,45]. The immature PDE_*Ca* contains 851 amino acid residues [64] (Figure 1), in which the first 23 amino acid residues belong to a signal peptide (confirmed with SignalP 3.0 [70], Figure 9A), eight amino acid residues to the activation peptide, and the remaining 820 amino acid residues to the mature protein [48]. The signal peptide prevents the protein from proper folding and is removed cotranslationally or by signal peptidases [71,72]. The Kyte and Doolittle hydropathy plot [42] indicates that this region is located in the hydrophilic part (Figure 9B). The function of the activation peptide is unknown. However, this is considered important for proper folding of the protein, as described for spider venom and plant proteins [2,73]. This part is removed by endopeptidases [74] (Figure 9D). It is also located in the hydrophilic region of the Kyte and Doolittle hydropathy plot [42], and it is exposed on the surface of the protein (and thereby accessible to peptidases). The remaining peptide does not undergo further processing.

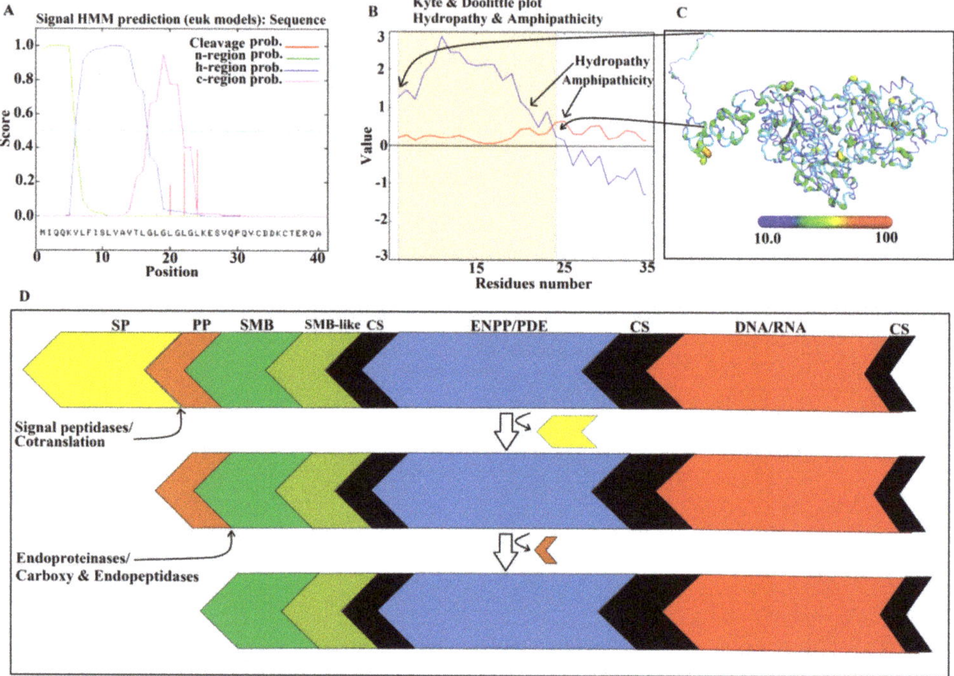

Figure 9. Maturation/processing mechanism for PDE_*Ca*. (**A**) A signalP-HMM prediction plot for PDE_*Ca*. (**B**) A Kyte and Doolittle plot for signal and activation peptides. (**C**) Ribbon representation of PDE_*Ca* colored by B-factor (temperature). (**D**) The prepropeptide of PDE_*Ca* with the signal peptide (colored in yellow), activation peptide (colored in brown), and the mature protein with four domains colored in green (somatomedin B domain), light green (somatomedin B-like domain), blue (Ectonucleotide pyrophosphatase/Phosphodiesterase domain), and red (DNA/RNA non-specific domain). The connecting segments are colored in black.

3. Materials and Methods

3.1. Sequence Retrieval and Multiple Sequence Alignment

The amino acid sequence of PDE_*Ca* (851 amino acid residues) (gene bank accession no. JAS04699.1; UniProt ID: A0A0F7Z2Q3) [64] was obtained from the NCBI (National Centre for Biotechnology Information) protein database (http://www.ncbi.nlm.nih.gov/protein). The signal peptide was identified using the SignalP 3.0 server [70] with default parameters. The amino acid sequence of PD_*Ca* was used as a query for searching homologous proteins from the non-redundant database by searching with the NCBI Protein BLAST using default parameters. The multiple sequence alignment of the selected homologous protein sequences, including the amino acid sequence of PD_*Ca*, was generated using MUSCLE [75]. The aligned sequences were edited and colored in Box-shade V3.21 [76].

3.2. Sequence Logo Generated from Multiple Sequence Alignment

The Weblogo 3.2 [77,78] was used to illustrate the conservation patterns of amino acids in the protein sequence, and graphically represent the multiple sequence alignment, using default parameters, except for the composition, which was done without adjustment.

3.3. In Silico Analysis of the Domain and Biochemical Properties of the PDE_Ca

The PDE_*Ca* primary sequence was analyzed for the presence of domains/motifs using the conserved domain search tool [79], available at http://www.ncbi.nlm.nih.gov/Structure/cdd/wrpsb.cgi. The protparam and Compute pI/MW tools from the ExPASy Proteomics server (http://web.expasy.org/compute_pi/) [42] were used to compute the isoelectric point (p*I*) and molecular weight of the protein.

3.4. Prediction of Ligand Binding and Glycosylation Sites

The 3DLigandSite-Ligand binding site prediction Server [80] was used for ligand binding amino acid residues in PDE_*Ca*, while putative glycosylation sites were predicted using the NetNGlyc 1.0 Server [48] and ScanProsite tool [81], with default parameters.

3.5. Disulfide Bond Prediction

The DiANNA web server [38] and Dinosolve [82–86] were used for prediction of disulfide bridges in PDE_*Ca*.

3.6. Homology Model Building of PDE_Ca

The 3D model of PDE_*Ca* was generated using various online proteins modeling programs, such as I-TESSER [87], the MODELLER 9v19 program [51], and SWISS Model [52], using the atomic coordinates of human Ectonucleotide pyrophosphatase/phosphodiesterase 3 (PDB ID: 6C01, amino acid sequence identity 63.39%) as a template [53]. The final model was selected based on the quality and validation reports generated by PROCHECK [50].

3.7. Molecular Dynamics Simulation

The modeled structure of PDE_*Ca* was validated through MD simulation using various programs, like AMBER16 [58], GROMACS [59], MDweb, and MDMoby [60]. The all-atom protein interaction was determined using the FF14SB force field [85]. The web-server H^{++} [84] was used to determine the protonation states of the amino acid side chain at pH 7.0. Chloride ions were used for system neutralization and were placed in a rectangular box of TIP3P water and extended to at least 15 Å from any protein atom. For the removal of bad contacts from the structure, the system was energy minimized for 500 conjugate gradients steps by applying a constant force constraint of 15 kcal/mol. Å^2. The system was then heated gradually from 0 to 300 K for 250 ps with a constant atom number, volume, and temperature (NVT) ensemble, at the same time that the protein was restrained with a constant force of 10 kcal/mol. Å^2. The equilibration step was carried out using the constant atom number, pressure, and temperature (NPT) ensemble for 500 ps, and the simulation was done for 100 ns with a 4 fs time step. The temperature and pressure were kept constant at 300 K and 1 atm, respectively, by Langevin coupling. The long-range electrostatic interactions were computed with the Particle–Mesh Ewald method (PME) [85], keeping the cut-off distance of 10 Å to Van der Waals interactions.

3.8. Model Validation

The build model of PDE_*Ca* was validated using the PROCHECK software [50,54], ERRAT version 2.0 [55], and Verify 3D [56,57].

3.9. Structure Superimposition

The build PDE_*Ca* protein model was aligned to homologous proteins using the PyMOL molecular graphics visualization program [86].

3.10. Surface Charge Analysis

Charge and radius calculations were carried out using the PDB2PQR server program [88]. The surface and charge were then visualized in ABPS Tools from the PyMOL molecular graphics visualization program [86].

4. Conclusions

In conclusion, a sequence and structural analysis of PDE_*Ca* was carried out using various computational biology programs. The amino acid sequence comparison analysis indicated that SVPDEs display high sequence identity (90.6%) with one another and comparatively low sequence identity (58.33%) with mammalian and bacterial PDEs. The three-dimensional model of PDE_*Ca*, produced using various modeling programs, was of good quality, as shown by the PROCHECK and ERRAT analysis. The modeled structure was further analyzed by molecular dynamic simulation, and the analysis indicated that all important structural parameters, such as chirality, disulfide bonds, and the absence of steric clashes, were correct. The root mean square deviation and radius of the gyration did not suffer significantly during model building and were maintained at 1 Å and 31.7 Å, respectively. The structural analysis indicated that the complex structure of PDE_*Ca* is folded into a multi-domain protein that comprises four domains—a somatomedin B domain, a somatomedin B-like domain, an Ectonucleotide pyrophosphatase/Phosphodiesterase domain (also called autotaxin), and a DNA/RNA non-specific domain. Structural comparisons with PDEs from other snake venoms, and mammalian and bacterial counterparts indicated that the surface charge distribution and the average active site cavity volume and depth vary considerably, which may contribute to their variable substrate specificity. Finally, during the maturation process, venom PDEs lose their signal and activation peptides to convert into the fully active mature forms. The structure of PDE_Cα presented in this paper is only a predicted structure. These conclusions need to be confirmed with experimental evidence [89].

Supplementary Materials: The following are available online at http://www.mdpi.com/2072-6651/11/11/625/s1. Figure S1. Molecular dynamic simulation analysis of PDE_*Ca*; Figure S2. Ramachandran plot of the modeled structures of *Vipera lebentia* phosphodiesterase model; Figure S3. Ramachandran plot of the modeled structures of *Bothrops atrox* phosphodiesterase model; Figure S4. Errors plot for the modeled structure of *Vipera lebentia*. The plot was generated by ERRAT2. The amino acid residues showing errors were shown by black lines; Figure S5. Errors plot for the modeled structure of *Bothrops atrox* model. The plot was generated by ERRAT2. The amino acid residues showing errors were shown by black lines; Figure S6. Phylogenetic relationships of PDEs based on protein sequences according to the neighbor-joining method without distance corrections.

Author Contributions: A.U. design of the work, conceptualization, supervision, and methodology; C.B. Funding Acquisition, Validation, analysis, and interpretation of data; K.U., writing—review; H.A., manuscript drafting and S.u.R. substantial revision and editing of the manuscript.

Funding: This research received no external funding.

Acknowledgments: The authors acknowledge financial support by the Cluster of Excellence "Advanced Imaging of Matter" of the Deutsche Forschungs-gemeinschaft (DFG) - EXC 2056 - project ID 390715994.

Conflicts of Interest: The authors declare no conflict of interest.

References

1. Kang, T.S.; Georgieva, D.; Genov, N.; Murakami, M.T.; Sinha, M.; Kumar, R.P.; Kaur, P.; Kumar, S.; Dey, S.; Sharma, S.; et al. Enzymatic toxins from snake venom: Structural characterization and mechanism of catalysis. *FEBS J.* **2011**, *278*, 4544–4576. [CrossRef] [PubMed]
2. Ullah, A.; Masood, R.; Ali, I.; Ullah, K.; Ali, H.; Akbar, H.; Betzel, C. Thrombin-like enzymes from snake venom: Structural characterization and mechanism of action. *Int. J. Biol. Macromol.* **2018**, *114*, 788–811. [CrossRef] [PubMed]
3. Ogawa, T.; Chijiwa, T.; Oda-Ueda, N.; Ohno, M. Molecular diversity and accelerated evolution of C-type lectin-like proteins from snake venom. *Toxicon* **2005**, *45*, 1–14. [CrossRef] [PubMed]
4. Kini, R.M.; Doley, R. Structure, function and evolution of three finger toxins: Mini proteins with multiple targets. *Toxicon* **2010**, *56*, 855–867. [CrossRef] [PubMed]

5. Fox, J.W. A brief review of the scientific history of several lesser-known snake venom proteins: L-amino acid oxidases, hyaluronidases and phosphodiesterases. *Toxicon* **2013**, *62*, 75–82. [CrossRef] [PubMed]

6. Ullah, A.; Masood, R.; Spencer, P.J.; Murakami, M.T.; Arni, R.K. Crystallization and preliminary X-ray diffraction studies of an L-amino-acid oxidase from *Lachesis muta* venom. *Acta Crystallogr. Sect. F Struct. Boil. Commun.* **2014**, *70*, 1556–1559. [CrossRef]

7. Ullah, A.; Magalhães, G.S.; Masood, R.; Mariutti, R.B.; Coronado, M.A.; Murakami, M.T.; Barbaro, K.C.; Arni, R.K. Crystallization and preliminary X-ray diffraction analysis of a novel sphingomyelinase D from *Loxosceles gaucho* venom. *Acta Crystallogr. Sect. F Struct. Boil. Commun.* **2014**, *70*, 1418–1420. [CrossRef]

8. De Oliveira, L.M.F.; Ullah, A.; Masood, R.; Zelanis, A.; Spencer, P.J.; Serrano, S.M.; Arni, R.K. Rapid purification of serine proteinases from Bothrops alternatus and Bothrops moojeni venoms. *Toxicon* **2013**, *76*, 282–290. [CrossRef]

9. Ullah, A.; Coronado, M.; Murakami, M.T.; Betzel, C.; Arni, R.K. Crystallization and preliminary X-ray diffraction analysis of an L-amino-acid oxidase from *Bothrops jararacussu* venom. *Acta Crystallogr. Sect. F Struct. Boil. Cryst. Commun.* **2012**, *68*, 211–213. [CrossRef]

10. Ullah, A.; Souza, T.D.A.C.B.D.; Masood, R.; Murakami, M.T.; Arni, R.K. Purification, crystallization and preliminary X-ray diffraction analysis of a class P-III metalloproteinase (BmMP-III) from the venom of *Bothrops moojeni*. *Acta Crystallogr. Sect. F Struct. Boil. Cryst. Commun.* **2012**, *68*, 1222–1225. [CrossRef]

11. Ullah, A.; Souza, T.A.; Betzel, C.; Murakami, M.T.; Arni, R.K. Crystallographic portrayal of different conformational states of a Lys49 phospholipase A_2 homologue: Insights into structural determinants for myotoxicity and dimeric configuration. *Int. J. Biol. Macromol.* **2012**, *51*, 209–214. [CrossRef] [PubMed]

12. Ullah, A.; Souza, T.A.; Abrego, J.R.; Betzel, C.; Murakami, M.T.; Arni, R.K. Structural insights into selectivity and cofactor binding in snake venom L-amino acid oxidases. *Biochem. Biophys. Res. Commun.* **2012**, *421*, 124–128. [CrossRef] [PubMed]

13. Mori, N.; Nikai, T.; Sugihara, H. Phosphodiesterase from the venom of crotalus ruber ruber. *Int. J. Biochem.* **1987**, *19*, 115–119. [CrossRef]

14. Stoynov, S.S.; Bakalova, A.T.; Dimov, S.I.; Mitkova, A.V.; Dolapchiev, L.B. Single-strand-specific DNase activity is an inherent property of the 140-kDa protein of the snake venom exonuclease. *FEBS Lett.* **1997**, *409*, 151–154. [CrossRef]

15. da Silva, N.J., Jr.; Aird, S.D. Prey specificity, comparative lethality and compositional differences of coral snake venoms. *Comp. Biochem. Physiol. Part C Toxicol. Pharmacol.* **2001**, *128*, 425–456. [CrossRef]

16. Sales, R.B.; Santoro, M.L. Nucleotidase and DNase activities in Brazilian snake venoms. *Comp. Biochem. Physiol. Part C Toxicol. Pharmacol.* **2008**, *147*, 85–95. [CrossRef]

17. Dhananjaya, B.L.; D'Souza, C.J.M. An overview on nucleases (DNase, RNase, and phosphodiesterase) in snake venoms. *Biochemistry* **2010**, *75*, 1–6.

18. Philipps, G.R. Purification and Characterization of Phosphodiesterase from *Crotalus* Venom. *Hoppe Seylers Z. Physiol. Chem.* **1975**, *356*, 1085–1096. [CrossRef]

19. Sugihara, H.; Nikai, T.; Komori, Y.; Katada, H.; Mori, N. Purification and characterization of the phosphodiesterase from the venom of *Akistrodon acutus* (China). *Jpn. J. Trop. Med. Hyg.* **1984**, *12*, 247–254. [CrossRef]

20. Björk, W. Purification of Phosphodiesterase from *Bothrops atrox* venoms, with special consideration of the elimination of monophosphatases. *J. Biol. Chem.* **1963**, *238*, 2487–2490.

21. Valério, A.A.; Corradini, A.C.; Panunto, P.C.; Mello, S.M.; Hyslop, S. Purification and Characterization of a Phosphodiesterase from *Bothrops alternatus* Snake Venom. *J. Protein Chem.* **2002**, *21*, 495–503. [CrossRef] [PubMed]

22. Saad, S.M.; Khan, S.; Ashraf, M. Purification of phosphodiesterase I from *Cerastes Vipera* venom, biochemical and biological properties of the purified enzyme. *Proc. Pak. Acad. Sci.* **2009**, *46*, 1–12.

23. Oka, J.; Ueda, K.; Hayaishi, O. Snake venom phosphodiesterase. Simple purification with blue sepharose and its application to poly (ADP-ribose) study. *Biochem. Biophys. Res. Commun.* **1978**, *80*, 841–848. [CrossRef]

24. Tatsuki, T.; Iwanaga, S.; Suzuki, T. A simple method for preparation of snake venom phosphodiesterase almost free from 5'-nucleotidase. *J. Biochem.* **1975**, *77*, 831–836. [CrossRef]

25. Laskowski, M., Sr. Purification and properties of venom phosphodiesterase. *Methods Enzymol.* **1980**, *65*, 276–284.

26. Kini, R.M.; Gowda, T.V. Rapid method for separation and purification of four isoenzymes of Phosphodiesrerase from *Trimeresurus flavorviridis* (Habu snake) venom. *J. Chromatogr.* **1984**, *291*, 299–305. [CrossRef]
27. Ballario, P.; Bergami, M.; Pedone, F. A simple method for the purification of phosphodiesterase from *Vipera aspis* venom. *Anal. Biochem.* **1977**, *80*, 646–651. [CrossRef]
28. Sannaningaiah, D.; Subbaiah, G.K.; Kempaiah, K. Pharmacology of spider venom toxins. *Toxin Rev.* **2014**, *33*, 206–220. [CrossRef]
29. Uzawa, S. Uber die phosphomonoesterase und die phosphodiesterase. *J. Biochem.* **1932**, *15*, 19–28. [CrossRef]
30. Aird, S.D. Ophidian envenomation strategies and the role of purines. *Toxicon* **2002**, *40*, 335–393. [CrossRef]
31. Aird, S.D. Taxonomic distribution and quantitative analysis of free purine and pyrimidine nucleosides in snake venoms. *Comp. Biochem. Physiol. Part B Biochem. Mol. Biol.* **2005**, *140*, 109–126. [CrossRef] [PubMed]
32. Hargreaves, M.B.; Stoggall, S.M.; Collis, M.G. Evidence that the adenosine receptor mediating relaxation in dog lateral saphenous vein and guinea-pig aorta is of the A2b subtype. *Br. J. Pharmacol.* **1991**, *102*, 198.
33. Sobrevia, L.; Yudilevich, D.L.; Mann, G.E. Activation of A2-purinoceptors by adenosine stimulates L-arginine transport (system y+) and nitric oxide synthesis in human fetal endothelial cells. *J. Physiol.* **1997**, *499*, 135–140. [CrossRef] [PubMed]
34. Uzair, B.; Khan, B.A.; Sharif, N.; Shabbir, F.; Menaa, F. Phosphodiesterases (PDEs) from Snake Venoms: Therapeutic Applications. *Protein Pept. Lett.* **2018**, *25*, 612–618. [CrossRef]
35. Iwanaga, S.; Suzuki, T. Enzymes in snake venom. In *Snake Venoms*; Lec, C.Y., Ed.; Springer: New York, NY, USA, 1979; Volume 19, pp. 61–158.
36. Pollack, S.E.; Tetsuo, U.; David, S.A. Snake venom phosphodiesterase: A zinc metalloenzyme. *J. Protein* **1983**, *2*, 1–12. [CrossRef]
37. Levy, Z.; Bdolah, A. Multiple molecular forms of snake venom phosphodiesterase from Vipera palastinae. *Toxicon* **1976**, *14*, 389–391. [CrossRef]
38. Ferrè, F.; Clote, P. DiANNA 1.1: An extension of the DiANNA web server for ternary cysteine classification. *Nucleic Acids Res.* **2006**, *34* (Suppl. 2), W182–W185. [CrossRef]
39. Perron, S.; Mackessy, S.P.; Hyslop, R.M. Purification and characterization of exonuclease from rattlesnake venom. *Acad. Sci.* **1993**, *25*, 21–22.
40. Mackessy, S.P. Venoms as Trophic Adaptations: An Ultrastructural Investigation of the Venom Apparatus and Biochemical Characterization of the Proteolytic Enzymes of the Northern Pacific Rattlesnake Crotalus Viridis Oreganus. Ph.D. Thesis, Washington State University, Pullman, WA, USA, 1989.
41. Xue, Z.; Xu, D.; Wang, Y.; Zhang, Y. ThreaDom: Extracting protein domain boundary information from multiple threading alignments. *Bioinformatics* **2013**, *29*, i247–i256. [CrossRef]
42. Gasteiger, E.; Hoogland, C.; Gattiker, A.; Duvaud, S.; Wilkins, M.R.; Appel, R.D.; Bairoch, A. Protein Identification and Analysis Tools on the ExPASy Server. In *The Proteomics Protocols Handbook*; Walker, J.M., Ed.; Humana Press: Totowa, NJ, USA, 2005; pp. 571–607.
43. Trummal, K.; Aaspõllu, A.; Tõnismägi, K.; Samel, M.; Subbi, J.; Siigur, J.; Siigur, E. Phosphodiesterase from Vipera lebetina venom—Structure and characterization. *Biochimie* **2014**, *106*, 48–55. [CrossRef]
44. Mitra, J.; Bhattacharyya, D. Phosphodiesterase from Daboia russelli russelli venom: Purification, partial characterization and inhibition of platelet aggregation. *Toxicon* **2014**, *88*, 1–10. [CrossRef] [PubMed]
45. Razzell, W.E.; Khorana, H.G. Studies on polynucleotides. III. Enzymic degradation; substrate specificity and properties of snake venom phosphodiesterase. *J. Biol. Chem.* **1959**, *234*, 2105–2113. [PubMed]
46. Peng, L.; Xu, X.; Shen, D.; Zhang, Y.; Song, J.; Yan, X.; Guo, M. Purification and partial characterization of a novel phosphodiesterase from the venom of Trimeresurus stejnegeri: Inhibition of platelet aggregation. *Biochimie* **2011**, *93*, 1601–1609. [CrossRef] [PubMed]
47. Santoro, M.L.; Vaquero, T.S.; Paes Leme, A.F.; Serrano, S.M. NPP-BJ, a nucleotide pyrophosphatase/phosphodiesterase from Bothrops jararaca snake venom, inhibits platelet aggregation. *Toxicon* **2009**, *54*, 499–512. [CrossRef]
48. Gupta, R.; Jung, E.; Brunak, S. Prediction of N-glycosylation Sites in Human Proteins. 2004. Available online: http://www.cbs.dtu.dk/services/NetNGlyc/ (accessed on 12 December 2018).
49. Sulkowski, E.; Bjork, W.; Laskowski, M. A specific and nonspecific alkaline monophosphatase in the venom of Bothrops atrox and their occurrence in the purified venom phosphodiesterase. *J. Biol. Chem.* **1963**, *238*, 2477–2486.

50. Laskowski, R.A.; MacArthur, M.W.; Thornton, J.M. PROCHECK: Validation of protein structure coordinates. In *International Tables of Crystallography: Volume F: Crystallography of Biological Macromolecules*; Rossmann, M.G., Arnold, E., Eds.; Kluwer Academic Publishers: Dordrecht, The Netherlands, 2001; pp. 722–725.

51. Webb, B.; Sali, A. Comparative Protein Structure Modeling Using MODELLER. *Curr. Protoc. Protein Sci.* **2016**, *86*, 3.

52. Waterhouse, A.; Bertoni, M.; Bienert, S.; Studer, G.; Tauriello, G.; Gumienny, R.; Heer, F.T.; de Beer, T.A.P.; Rempfer, C.; Bordoli, L.; et al. SWISS-MODEL: Homology modelling of protein structures and complexes. *Nucleic Acids Res.* **2018**, *46*, W296–W303. [CrossRef]

53. Gorelik, A.; Randriamihaja, A.; Illes, K.; Nagar, B. Structural basis for nucleotide recognition by the ectoenzyme CD203c. *FEBS J.* **2018**, *285*, 2481–2494. [CrossRef]

54. Laskowski, R.A.; MacArthur, M.W.; Moss, D.S.; Thornton, J.M. PROCHECK: A program to check the stereochemical quality of protein structures. *J. Appl. Crystallogr.* **1993**, *26*, 283–291. [CrossRef]

55. Colovos, C.; Yeates, T.O. Verification of protein structures: Patterns of nonbonded atomic interactions. *Protein Sci.* **1993**, *2*, 1511–1519. [CrossRef]

56. Bowie, J.; Luthy, R.; Eisenberg, D. A method to identify protein sequences that fold into a known three-dimensional structure. *Science* **1991**, *253*, 164–170. [CrossRef] [PubMed]

57. Lüthy, R.; Bowie, J.U.; Eisenberg, D. Assessment of protein models with three-dimensional profiles. *Nature* **1992**, *356*, 83–85. [CrossRef] [PubMed]

58. Maier, J.A.; Martinez, C.; Kasavajhala, K.; Wickstrom, L.; Hauser, K.E.; Simmerling, C. ff14SB: Improving the accuracy of protein side chain and backbone parameters from ff99SB. *J. Chem. Theory Comput.* **2015**, *11*, 3696–3713. [CrossRef] [PubMed]

59. Berendsen, H.J.C.; van der Spoel, D.; van Drunen, R. GROMACS: A message-passing parallel molecular dynamics implementation. *Comput. Phys. Commun.* **1995**, *91*, 43–56. [CrossRef]

60. Hospital, A.; Andrio, P.; Fenollosa, C.; Cicin-Sain, D.; Orozco, M.; Gelpí, J.L. MDWeb and MDMoby: An integrated web-based platform for molecular dynamics simulations. *Bioinformatics* **2012**, *28*, 1278–1279. [CrossRef]

61. Jansen, S.; Perrakis, A.; Ulens, C.; Winkler, C.; Andries, M.; Joosten, R.P.; Van Acker, M.; Luyten, F.P.; Moolenaar, W.H.; Bollen, M. Structure of NPP1, an Ectonucleotide pyrophosphatase/phosphodiesterase involved in tissue calcification. *Structure* **2012**, *20*, 1948–1959. [CrossRef]

62. Zalatan, J.G.; Fenn, T.D.; Brunger, A.T.; Herschlag, D. Structural and functional comparisons of nucleotide pyrophosphatase/phosphodiesterase and alkaline phosphatase: Implications for mechanism and evolution. *Biochemistry* **2006**, *45*, 9788–9803. [CrossRef]

63. Stein, A.J.; Bain, G.; Prodanovich, P.; Santini, A.M.; Darlington, J.; Stelzer, N.M.P.; Sidhu, R.S.; Schaub, J.; Goulet, L.; Lonergan, D.; et al. Structural Basis for Inhibition of Human Autotaxin by Four Potent Compounds with Distinct Modes of Binding. *Mol. Pharmacol.* **2015**, *88*, 982–992. [CrossRef]

64. Rokyta, D.R.; Margres, M.J.; Calvin, K. Post-transcriptional Mechanisms Contribute Little to Phenotypic Variation in Snake Venoms. *G3 Genes Genomes Genet.* **2015**, *5*, 2375–2382. [CrossRef]

65. Hausmann, J.; Kamtekar, S.; Christodoulou, E.; Day, J.E.; Wu, T.; Fulkerson, Z.; Albers, H.M.; Van Meeteren, L.A.; Houben, A.J.S.; Van Zeijl, L.; et al. Structural basis for substrate discrimination and integrin binding by autotaxin. *Nat. Struct. Mol. Boil.* **2011**, *18*, 198–204. [CrossRef]

66. Touw, W.G.; Baakman, C.; Black, J.; te Beek, T.A.; Krieger, E.; Joosten, R.P.; Vriend, G. A series of PDB-related databanks for everyday needs. *Nucleic Acids Res.* **2015**, *43*, D364–D368. [CrossRef] [PubMed]

67. Philipps, G.R. Purification and characterization of phosphodiesterase I from Bothrops atrox. *Biochim. Biophys. Acta* **1976**, *432*, 237–244.

68. Boman, H.G. On the specificity of the snake venom phosphodiesterase. *Ann. N. Y. Acad. Sci.* **1959**, *81*, 800–803. [CrossRef] [PubMed]

69. Ke, H.; Wang, H.; Ye, M. Structural Insight into the Substrate Specificity of Phosphodiesterases. *Pharmacol. Itch* **2011**, *204*, 121–134.

70. Bendtsen, J.D.; Nielsen, H.; von Heijne, G.; Brunak, S. Improved prediction of signal peptides: SignalP 3.0. *J. Mol. Biol.* **2004**, *340*, 783–795. [CrossRef]

71. Von Heijne, G. On the hydrophobic nature of signal sequences. *Eur. J. Biochem.* **1981**, *116*, 419–422. [CrossRef]

72. Lively, M.O.; Walsh, K.A. Hen oviduct signal peptidase is an integral membrane protein. *J. Biol. Chem.* **1983**, *258*, 9488–9495.

73. Ullah, A.; Mariutti, R.B.; Masood, R.; Caruso, I.P.; Costa, G.H.; de Freita, C.M.; Santos, C.R.; Zanphorlin, L.M.; Mutton, M.J.R.; Murakami, M.T.; et al. Crystal structure of mature 2S albumin from Moringa oleifera seeds. *Biochem. Biophys. Res. Commun.* **2015**, *468*, 365–371. [CrossRef]

74. Khan, A.R.; James, M.N. Molecular mechanisms for the conversion of zymogens to active proteolytic enzymes. *Protein Sci.* **1998**, *7*, 815–836. [CrossRef]

75. Edgar, R.C. MUSCLE: Multiple sequence alignment with high accuracy and high throughput. *Nucleic Acids Res.* **2004**, *32*, 1792–1797. [CrossRef]

76. Hofmann, K.; Baron, M.D. Institute for Animal Health Ash Road Pirbright, Surrey GU24 0, 2017 (NF U.K.).

77. Crooks, G.E.; Hon, G.; Chandonia, J.M.; Brenner, S.E. WebLogo: A sequence logo generator. *Genome Res.* **2004**, *14*, 1188–1190. [CrossRef] [PubMed]

78. Schneider, T.D.; Stephens, R.M. Sequence logos: A new way to display consensus sequences. *Nucleic Acids Res.* **1990**, *18*, 6097–6100. [CrossRef] [PubMed]

79. Marchler-Bauer, A.; Bryant, S.H. CD-Search: Protein domain annotations on the fly. *Nucleic Acids Res.* **2004**, *32*, W327–W331. [CrossRef] [PubMed]

80. Wass, M.N.; Kelley, L.A.; Sternberg, M.J. 3DLigandSite: Predicting ligand-binding sites using similar structures. *Nucleic Acids Res.* **2010**, *38*, W469–W473. [CrossRef] [PubMed]

81. De Castro, E.; Sigrist, C.J.A.; Gattiker, A.; Bulliard, V.; Langendijk-Genevaux, P.S.; Gasteiger, E.; Bairoch, N.; Hulo, A. ScanProsite: Detection of PROSITE signature matches and ProRule-associated functional and structural residues in proteins. *Nucleic Acids Res.* **2006**, *34*, W362–W365. [CrossRef]

82. Yaseen, A.; Li, Y. Dinosolve: A protein disulfide bonding prediction server using context-based features to enhance prediction accuracy. *BMC Bioinform.* **2013**, *14*, S9. [CrossRef]

83. Maier, J.A.; Martinez, C.; Kasavajhala, K.; Wickstrom, L.; Hauser, K.E.; Simmerling, C. ff14SB: Improving the accuracy of protein side chain and backbone parameters from ff99SB. *J. Chem. Theor. Comput.* **2015**, *8*, 3696–3713. [CrossRef]

84. Anandakrishnan, R.; Aguilar, B.; Onufriev, A.V. H++ 3.0: Automating pK prediction and the preparation of biomolecular structures for atomistic molecular modeling and simulation. *Nucleic Acids Res.* **2012**, *40*, W537–W541. [CrossRef]

85. Darden, T.; York, D.; Pedersen, L. Particle mesh Ewald: An N log (N) method for Ewald sums in large systems. *J. Chem. Phys.* **1993**, *98*, 10089–10092. [CrossRef]

86. DeLano, W.L. *The PyMOL Molecular Graphics System*; DeLano Scientific: San Carlos, CA, USA, 2002.

87. Roy, A.; Kucukural, A.; Zhang, Y. I-TASSER: A unified platform for automated protein structure and function prediction. *Nat. Protoc.* **2010**, *5*, 725–738. [CrossRef]

88. Dolinsky, T.J.; Nielsen, J.E.; McCammon, J.A.; Baker, N.A. PDB2PQR: Expanding and upgrading automated preparation of biomolecular structures for molecular simulations. *Nucleic Acids Res.* **2004**, *32*, W665–W667. [CrossRef] [PubMed]

89. Ullah, A.; Masood, R.; Hayat, Z.; Hafeez, A. Determining the Structures of the Snake and Spider Toxins by X-Rays. *Methods Mol. Biol.* **2020**, *2068*, 163–172.

Article

Discovery of the Gene Encoding a Novel Small Serum Protein (SSP) of *Protobothrops flavoviridis* and the Evolution of SSPs

Kento Inamaru [1], Ami Takeuchi [1], Marie Maeda [1], Hiroki Shibata [2], Yasuyuki Fukumaki [2], Naoko Oda-Ueda [3], Shosaku Hattori [4], Motonori Ohno [1] and Takahito Chijiwa [1,*]

[1] Department of Applied Life Science, Faculty of Bioscience and Biotechnology, Sojo University, Kumamoto 860–0082, Japan; inashimarushi@gmail.com (K.I.); amitakeuchi921@gmail.com (A.M.); yuis2.0603@gmail.com (M.M.); mohno218@ybb.ne.jp (M.O.)

[2] Medical Institute of Bioregulation, Research Center of Genetic Information, Kyushu University, Fukuoka 812–8582, Japan; hshibata@gen.kyushu-u.ac.jp (H.S.); yfukumak@gen.kyushu-u.ac.jp (Y.F.)

[3] Department of Biochemistry, Faculty of Pharmaceutical Sciences, Sojo University, Kumamoto 860–0082, Japan; naoko@ph.sojo-u.ac.jp

[4] Institute of Medical Science, University of Tokyo, Oshima-gun, Kagoshima 894-1531, Japan; shattori@ims.u-tokyo.ac.jp

* Correspondence: chijiwa@life.sojo-u.ac.jp; Tel.: +81-96-326-3984

Received: 27 January 2020; Accepted: 10 March 2020; Published: 12 March 2020

Abstract: Small serum proteins (SSPs) are low-molecular-weight proteins in snake serum with affinities for various venom proteins. Five SSPs, *Pf*SSP-1 through *Pf*SSP-5, have been reported in *Protobothrops flavoviridis* ("habu", *Pf*) serum so far. Recently, we reported that the five genes encoding these *Pf*SSPs are arranged in tandem on a single chromosome. However, the physiological functions and evolutionary origins of the five SSPs remain poorly understood. In a detailed analysis of the habu draft genome, we found a gene encoding a novel SSP, SSP-6. Structural analysis of the genes encoding SSPs and their genomic arrangement revealed the following: (1) SSP-6 forms a third SSP subgroup; (2) SSP-5 and SSP-6 were present in all snake genomes before the divergence of non-venomous and venomous snakes, while SSP-4 was acquired only by venomous snakes; (3) the composition of paralogous SSP genes in snake genomes seems to reflect snake habitat differences; and (4) the evolutionary emergence of SSP genes is probably related to the physiological functions of SSPs, with an initial snake repertoire of SSP-6 and SSP-5. SSP-4 and its derivative, SSP-3, as well as SSP-1 and SSP-2, appear to be venom-related and were acquired later.

Keywords: small serum proteins; *Protobothrops flavoviridis*; evolution; gene array; comparative genomics

Key Contribution: Discovery of the gene encoding a novel small serum protein (SSP), SSP-6. The proposal of a relationship between the composition of SSP genes and snake habitat conditions. The proposal of a relationship between the evolutionary emergence of SSP genes and the physiological functions of SSPs.

---

1. Introduction

The bites of viperid snakes, including *Protobothrops flavoviridis* (*Pf*), cause a variety of symptoms, including bleeding, necrosis, edema, and neurotoxicity, and can be fatal in severe cases. Recent transcriptomic and proteomic studies have identified multiple components of viperid venoms [1–3], including phospholipases A_2 [4–7], metalloproteases (snake venom metalloproteases, SVMPs) [8–11], and serine proteases [12,13]. Many of these venom proteins have isoforms. In contrast to neurotoxic group IA PLA$_2$s of elapid (Elapinae and Hydrophiinae) venoms, group IIA-PLA$_2$s of viperid (Viperinae and Crotalinae) venoms [14], such as hemolytic neutral [Asp^{49}]PLA$_2$s [15,16], edema-inducing basic [Asp^{49}]PLA$_2$s [17,18], neurotoxic highly basic [Asp^{49}]PLA$_2$s [19], and myotoxic [Lys^{49}]PLA$_2$s [15,20–22], diversified via accelerated evolution, in which nucleotide substitutions occurred predominantly at non-synonymous sites.

In contrast, snakes bitten by themselves or other snakes do not show severe symptoms as humans do. Snake serum is able to neutralize or inhibit snake venom activities. Phospholipase A$_2$ inhibitors (PLIs) [23–27] and habu serum factor (HSF) [28,29], which inhibit the hemorrhage induced by SVMPs, are able to neutralize these venom activities. Recently, we found a low-molecular-weight serum protein that specifically binds to myotoxic [Lys^{49}]PLA$_2$ isozymes and revealed that this is a homolog of Small Serum Protein-2 (SSP-2), a human prostatic secretory protein superfamily of 94 amino acids (PSP94) [30]. From *P. flavoviridis* serum, five SSPs, *Pf*SSP-1, *Pf*SSP-2, *Pf*SSP-3, *Pf*SSP-4, and *Pf*SSP-5, have been identified to date [30,31]. However, in terms of blood content, *Pf*SSP-4 and *Pf*SSP-5 are significantly less abundant than *Pf*SSP-1, *Pf*SSP-2, and *Pf*SSP-3 [31]. SSPs are two-domain proteins [31]. The variable N-terminal domains are thought to be involved in binding diverse target molecules, whereas the C-terminal domain, which is largely conserved among the five SSPs, is assumed to be involved in forming oligomers with HSF [32]. *Pf*SSP-2 and *Pf*SSP-5 show high affinity for triflin, a neurotoxin-like protein that blocks muscle contraction [33,34]. *Pf*SSP-1 and *Pf*SSP-4 show affinity for HV1, a low-molecular-weight SVMP that induces apoptosis of vascular endothelial cells [11,35]. *Pf*SSP-3 binds to flavorase, a non-hemorrhagic SVMP [36]. *Pf*SSP-2 binds to [Lys^{49}]PLA$_2$s [37]. cDNAs encoding five *Pf*SSPs have been isolated and sequenced [30,31]. Interestingly, the cDNAs encoding *Pf*SSP-3 and *Pf*SSP-4 are interrupted by nonsense mutations at the same site on the fourth exon, so as to express truncated mature proteins. The genome fragment containing the genes encoding *Pf*SSP-1 and *Pf*SSP-2 has also been isolated and sequenced [38].

Recently, we revealed that genes for the five PfSSPs, *PfSSP-4*, *PfSSP-5*, *PfSSP-1*, *PfSSP-2*, and *PfSSP-3*, are arranged in tandem in this order on one chromosome of *P. flavoviridis* [39]. According to the configuration of nucleotide sequences in the introns, such as long interspersed nuclear elements (LINEs), DNA transposons, and repetitive sequences, the five *PfSSPs* can be divided into two subgroups: the Long SSP subgroup consists of *PfSSP-1*, *PfSSP-2*, and *PfSSP-5*, and the Short SSP subgroup consists of *PfSSP-3* and *PfSSP-4*. Mathematical analysis of the nucleotide sequences of *PfSSPs* showed that *PfSSPs* in the Short SSP subgroup have evolved in an accelerated manner, whereas those in the Long SSP subgroup have evolved alternately in accelerated and neutral manners. Ortholog analysis of *SSP* genes from five snakes, including non-venomous snakes, suggested that these genes emerged in the order of their configuration on the chromosome. Moreover, a comparison of the arrays of *SSP* genes of five snakes showed that the genome segment encompassing *SSP-1* to *SSP-2* of *Protobothrops* has been inverted. Chromosome inversion appears to have preserved non-synonymous nucleotide substitutions, providing evidence of accelerated evolution [39].

In the present study, we discovered a gene encoding a novel *P. flavoviridis* small serum protein, named *PfSSP-6*, in the 5′ region upstream of the array of five *PfSSPs* [39]. From a detailed structural analysis, we propose: (1) a novel classification of SSPs, (2) an evolutionary scenario to explain SSP paralogs, (3) a relationship between arrays of *SSP* paralogs in the snake genome and environmental conditions, (4) relationships between SSP evolution and physiological functions of their products, suggesting an initial repertoire of *SSP-6*, *SSP-5*, and *SSP-4*, and the subsequent appearance of venom-related *SSP-1*, *SSP-2*, and *SSP-3*.

2. Results and Discussion

2.1. Discovery of the Gene Encoding a Novel SSP, PfSSP-6, Far Upstream of the Array of Five PfSSP Genes

blastn and tblastx analyses of the habu (HabAm1) database [40] using the nucleotide sequence of *PfSSP-5* as a query revealed that Scaffold 2858 contains an approximately 3.7 kbp sequence similar to the nucleotide sequence of *PfSSP-5*. The amino acid sequence of its N-terminal domain differs from those of the five known *PfSSPs*, whereas its C-terminal domain is very similar. Ten cysteines are conserved among *PfSSPs*. Therefore, this nucleotide sequence was determined to encode a novel type of SSP which was named *PfSSP-6*. To sequence *PfSSP-6*, genomic PCR was performed using the draft nucleotide sequence of the corresponding region in Scaffold 2858 of the Amami Island *P. flavoviridis* genome as a reference. A 1453 bp genome segment that encompassed the 5′ terminus of the putative first exon to the 3′ terminus of the putative third exon of *PfSSP-6*, and another 2375 bp segment that encompassed the 5′ terminal of the putative third exon to 85 bp downstream from the putative fourth exon of *PfSSP-6* were then acquired. Finally, the 3642 bp sequence of *PfSSP-6* was determined. Referring to the construction of *PfSSP-5*, definitive exon–intron boundaries of *PfSSP-6* were identified. *PfSSP-6* consists of four exons and three introns and encodes a 111 amino acid protein, including a 19 amino acid signal peptide. The deduced amino acid sequence of the mature protein encoded by *PfSSP-6* shows 33%–61% identity with the other five PfSSPs, and the positions of its 10 cysteine residues are conserved (Figure 1). Referring to the draft nucleotide sequence of Scaffold 2858 encompassing *PfSSP-6* to *PfSSP-4*, which is the 5′ terminal gene of the array of five *PfSSPs* [39], further genomic PCR with the Amami–Oshima *P. flavoviridis* genome was performed and the 12,406 bp sequence of the intergenic region between *PfSSP-6* and *PfSSP-4*, named *Pf*I-Reg64, was determined (Figure 2).

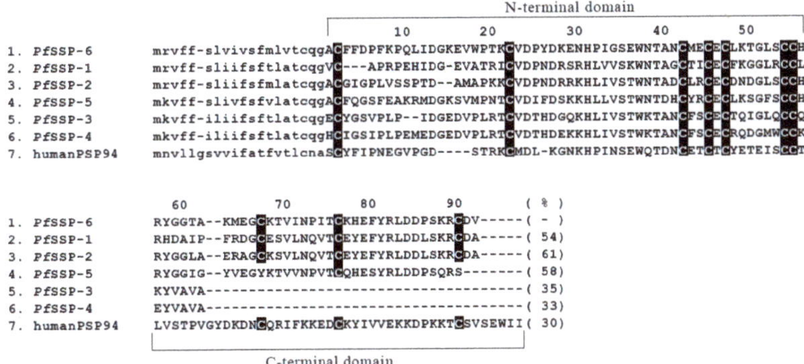

Figure 1. Alignment of the deduced amino acid sequences encoded by the open reading frames (ORFs) of six *PfSSPs* and human *PSP94*. Position numbers refer to amino acid residues of the mature proteins. Signal peptide sequences are in lower case letters. The cysteines are shaded. Abbreviations: *Pf*, *P. flavoviridis*. References: *PfSSP-6* (this study); *PfSSP-1* (AB360906.1); *PfSSP-2* (AB360907.1); *PfSSP-5* (AB360910.1); *PfSSP-3* (AB360908.1); *PfSSP-4* (AB360909.1) [31]; human *PSP94* (NP_002434.1) [41]. Numerals in parentheses show percent identities of *Pf*SSPs and human PSP94 with *Pf*SSP-6.

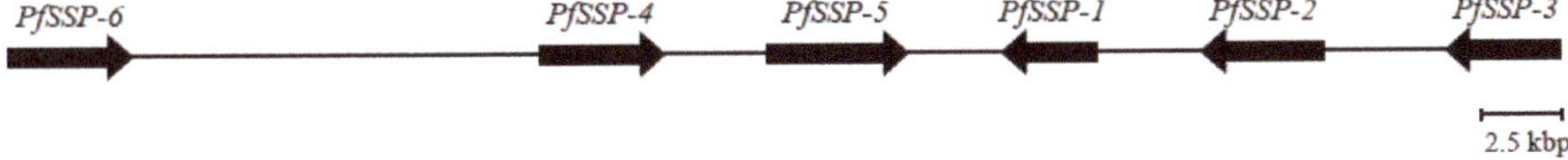

Figure 2. Schematic representation of the 16,048 bp genome segment containing *PfSSP-6* in the 5′ region upstream of the array of five *PfSSPs*. Bold arrows indicate the areas and transcription directions of the genes in the segment.

2.2. Sequence Configurations Classify the Six PfSSPs Into Three Subgroups

Introns of *PfSSP-6* contained insertions of specific nucleotide sequences, fragments of L1, chicken repeat-1 (CR1), and Gypsy LINEs, fragments of a reverse transcriptase (RT) domain of L2 LINE, fragments of Mariner and hobo-Ac-Tam3 (hAT) DNA transposons, and repetitive sequences, as in the other five *PfSSPs* (Figure 3). The five inserted fragments, L1 and CR1 LINEs in the first intron and Mariner-iii, Gypsy-i, and Gypsy-ii in the third intron, are conserved in all *PfSSPs*. These insertions must therefore have occurred before the divergence of the six *PfSSPs*. Second, configurations of the nucleotide sequences inserted into the second or third intron classified the six *PfSSPs* into two subgroups, Long SSPs and Short SSPs [39]. Long *PfSSPs* are characterized by the fragment of the RT domain of L2 LINE in the third intron. However, the nucleotide sequence of the fragment of the RT domain of L2 LINE in the third intron of *PfSSP-6* differs from those inserted into the other three genes of conventional Long SSPs, *PfSSP-1*, *PfSSP-2*, and *PfSSP-5*. The fragment of L2 LINE in the third intron of three *PfSSPs*, *PfSSP-1*, *PfSSP-2*, and *PfSSP-5*, is truncated in the 3′ terminal region (Figure 1). On the other hand, the fragment of L2 LINE in the third intron of *PfSSP-6* is truncated from the 5′ terminal region, as in typical LINEs [42]. L2 LINE is composed of two open reading frames, ORF1 and ORF2, in which ORF1 encodes an RNA-binding protein and ORF2 encodes a two-domain protein consisting of an endonuclease (EN) and an RT domain [42]. The RT domain of L2 LINE consists of 10 subdomains numbered from zero to IX and a carboxy-terminal conserved region (CTCR) which is thought to serve as the scaffold of reverse-transcription of L2 LINE [43]. A 320 bp section of the L2 LINE fragment in *PfSSP-1* encodes three subdomains, zero to II, of the RT domain. A 431 bp section of that in *PfSSP-2* encodes four subdomains, zero to III, of the RT domain, and 1011 bp of the L2 LINE fragment in *PfSSP-5* encodes nine subdomains, zero to VIII, of the RT domain [39]. However, 1240 bp of the L2 LINE fragment in *PfSSP-6* encode eight subdomains, III to X, and CTCR of the RT domain. This indicates that this L2 LINE is truncated in the 5′ terminal region. It is highly likely that the nucleotide sequence from the 3′ downstream region of the third exon of *PfSSP-6* to the 5′ terminal of the inserted L2 LINE fragment of *PfSSP-6* has disappeared, accompanied by 5′ truncation of L2 LINE. These characteristics indicate that *PfSSP-6* should be classified as a novel Long SSP. Interestingly, body map analysis using semi-quantitative RT-PCR showed that *PfSSP-6* is strongly expressed in the stomach and weakly in the liver (data not shown). It seems that the product of *PfSSP-6* is irrelevant to its role in blood. Thus, the three configurations of inserted nucleotide sequences classify the six *PfSSPs* into three subgroups, conventional and novel Long SSPs, and Short SSPs.

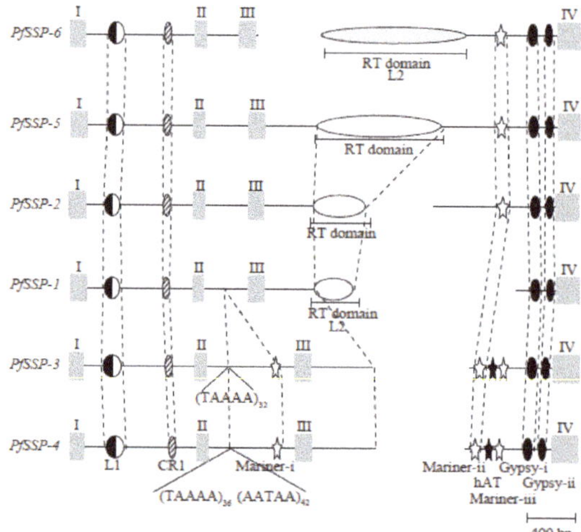

Figure 3. Schematic configurations of nucleotide sequences in the introns of six *PfSSPs*. Gray bars represent exons. Half-closed, hatched, open, and closed ellipses represent fragments of L1, CR1, L2, and Gypsy LINEs, respectively. Since the fragment of L2 LINE in the third intron of *PfSSP-6* differs from those of L2 LINE in the third introns of *PfSSP-1, 2,* and *5*, the ellipse of *PfSSP-6* is shown in gray. Open and closed stars represent the fragments of Mariner and hAT DNA transposons. Positions of the corresponding fragments are linked with dashed lines. Positions of repetitive sequences (TAAAA and AATAA) are indicated with carets and numbers of repetitions are indicated as subscripts.

2.3. Configurations of SSP Paralogs Relevant to Snake Habitat Conditions

Following Chijiwa et al. [39], blastn and tblastx analysis of the draft genomes of seven snakes *Crotalus viridis* (*Cv*, venomous), *Deinagkistrodon acutus* (*Da*, venomous), *Ophiophagus hannah* (*Oh*, venomous), *Python bivittatus* (*Pb*, non-venomous), *Protobothrops mucrosquamatus* (*Pm*, venomous), *Thamnophis sirtalis* (*Ts*, non-venomous), and *Vipera berus* (*Vb*, venomous), in addition to the habu, *P. flavoviridis*, revealed orthologous relationships and configurations of *SSP*s (Figure 4). The genome of the non-venomous Burmese python, *P. bivittatus* (India), contains an ortholog of *PfSSP-6*, named *PbSSP-6*, and three paralogs of *PfSSP-5*, named *PbSSP-5α*, *PbSSP-5β*, and *PbSSP-5γ*(Ψ). The genome of the garter snake, *T. sirtalis*, a North American colubrid, contains an ortholog of *PfSSP-6*, named *TsSSP-6*, an ortholog of *PfSSP-4*, called *TsSSP-4*, and two paralogs of *PfSSP-5*, *TsSSP-5α*, and *TsSSP-5β*. The genome of the European adder, *V. berus*, a viperid, contains orthologs of *PfSSP-6* and *PfSSP-4* in one scaffold (2247), called *VbSSP-6* and *VbSSP-4* in this study. However, the nucleotide sequences of the second and fourth exons of *VbSSP-4* remain unknown. In addition, the ortholog of *PfSSP-5*, named *VbSSP-5*, was also found in another *V. berus* scaffold (15,659). The genome of the prairie rattlesnake, *C. viridis*, (North America) possesses orthologs of *PfSSP-6*, *PfSSP-4*, and *PfSSP-5* on Chromosome 9, named *CvSSP-6*, *CvSSP-4*, and *CvSSP-5* in this study. The genome of the king cobra, *O. hannah*, (India) has orthologs of *PfSSP-6* and *PfSSP-4*, named *OhSSP-6* and *OhSSP-4*(Ψ), and three paralogs of *PfSSP-5*, named *OhSSP-5α*, *OhSSP-5β*, and *OhSSP-5γ* in one scaffold (4527). One ortholog and two paralogs of *OhSSP*s were renamed in this study. Two nucleotide segments, previously annotated as *OhSSP-1* and *OhSSP-2* [39], were acquired via genomic PCR to determine their nucleotide sequences. Their nucleotide and deduced amino acid sequences revealed that they are paralogs of *PfSSP-5* and should be renamed *OhSSP-5β* and *OhSSP-5γ*. Therefore, the nucleotide sequence, already annotated as *OhSSP-5*(Ψ) [39], was also renamed *OhSSP-5α*. Structural analysis of the L2 LINE fragment in the third intron showed that *SSP-1*, *SSP-2*, and *SSP-5* belong to the conventional Long SSP subgroup. In addition, the locations of those alleles also suggested that *OhSSP-5β* and *OhSSP-5γ* are evolutionarily related to *SSP-1* and *SSP-2* in *D. acutus*, *P. mucrosquamatus*, and *P. flavoviridis*. The genome of the hundred-pace viper, *D. acutus*, (Southeast Asia) contained orthologs *PfSSP-6*, *PfSSP-4*, *PfSSP-1*, *PfSSP-2*, and *PfSSP-3* in one scaffold (405), named *DaSSP-6*, *DaSSP-4*, *DaSSP-1*, *DaSSP-2*, and *DaSSP-3* in this study. Only the nucleotide segment corresponding to the second exon of the ortholog of *DaSSP-5* was found in the intergenic region between *DaSSP-4* and *DaSSP-1*. Therefore, this fragmented *DaSSP-5* is described as *DaSSP-5δ*(Ψ) in Figure 4. The genome of the Taiwan habu, *P. mucrosquamatus*, (Taiwan) contained an ortholog of *PfSSP-6*, named *PmSSP-6*, and orthologs of *PfSSP-5*, *PfSSP-1*, *PfSSP-2*, and *PfSSP-3* in one scaffold (462), named *PmSSP-5*, *PmSSP-1*, *PmSSP-2*, and *PmSSP-3*. In addition, an ortholog of *PfSSP-4*, named *PmSSP-4*, was also found in another scaffold (21,362). Orthologs of *PfSSP-6* are conserved in the genomes of all eight snakes, whether venomous or non-venomous. Chijiwa et al. showed that *SSP-5* and *SSP-4* were the initial genes in the conventional Long and Short SSP subgroups, respectively [39]. Moreover, the current study revealed that the initial repertoire of SSP genes in the genomes of all snakes should be two genes, encoding SSP-6 for the novel Long SSP subgroup and SSP-5 for the conventional Long SSP subgroup, and that the gene encoding SSP-4 was acquired specifically in the genomes of venomous snakes.

Configurations of *SSP* paralogs in each snake genome were used to classify the eight snakes into three groups. Non-venomous *P. bivittatus* formed the first group, in which two genes encoding SSP-6 and SSP-5 were present in the genome. *T. sirtalis*, *V. berus*, *C. viridis*, and *O. hannah* formed a second group in which three genes encoding SSP-6, SSP-5, and SSP-4 were present. *D. actus*, *P. mucrosquamatus*, and *P. flavoviridis* formed a third group with two genes encoding SSP-1 and SSP-2, in addition to the initial three genes, *SSP-6*, *SSP-4*, and *SSP-5*. This result suggests that the configuration of *SSP* paralogs is relevant to habitat characteristics of each snake. Snakes in the third group, *D. actus*, *P. mucrosquamatus*, and *P. flavoviridis*, inhabit the Orient, where the warm and humid climate might provide richer and more diversified prey than in Europe and America. It is likely that their venom proteins have become varied, and that the serum proteins that neutralize those venoms then also diversified. *O. hannah* did not need to develop novel varieties of IIA-PLA$_2$ isozymes; it had another type of venom PLA$_2$, the neurotoxic IA-PLA$_2$, a lethal component. Therefore, *OhSSP-5β* and *OhSSP-5γ*, corresponding to *SSP-1* and/or *SSP-2*, may have had no need to become derivatives as the counterpart of variable IIA-PLA$_2$s.

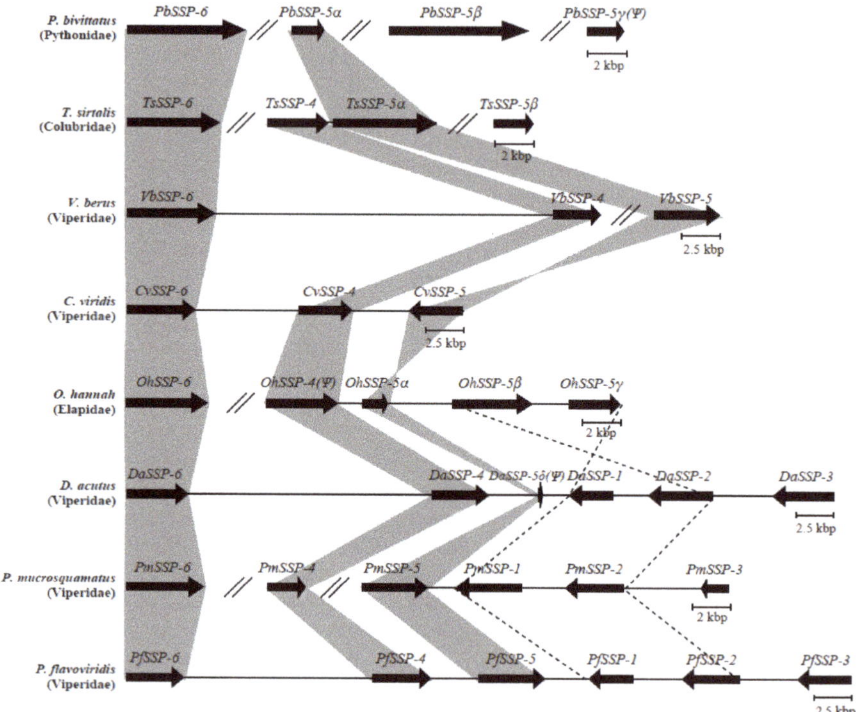

Figure 4. Schematic comparison of the arrays of *SSPs* of eight snake taxa. Abbreviations: *Cv*: *C. viridis*; *Da*: *D. acutus*; *Oh*: *O. hannah*; *Pb*: *P. bivittatus*; *Pf*: *P. flavoviridis*; *Pm*: *P. mucrosquamatus*; *Ts*: *T. sirtalis*; *Vb*: *V. berus*. Bold arrows indicate the areas and transcription directions of the genes. Orthologs of *SSP-6*, *SSP-5*, and *SSP-4* in each snake genome are linked with gray. Inverted genome segments of *O. hannah*, *D. acutus*, *P. mucrosquamatus*, and *P. flavoviridis* are linked with dashed lines. Double slashes indicate interruptions of the nucleotide sequences.

2.4. Diversified SSPs Acquired by Advanced Snakes Have More Complex Venom Compositions

Genes encoding SSP paralogs of each snake were analyzed mathematically. The K_A/K_S ratio, which is the relative ratio of synonymous to nonsynonymous substitutions between the ORFs (Tables 1–8), or K_N, which is the rate of substituted nucleotides between the introns, were calculated (Tables 9–13). However, for genes for which full-length nucleotide sequences of exons or introns remained unknown, the rate of K_A/K_S for *DaSSP-5δ(Ψ)*, *PbSSP-5γ*, and *VbSSP-4*, or K_Ns for *PbSSPs*, *TsSSPs*, and *VbSSPs* were not calculated. In the previous section, we proposed that the initial repertoire of *SSPs* in the genomes of venomous snakes comprised *SSP-6*, *SSP-5*, and *SSP-4*. Our mathematical analysis suggested that these three differ in their characteristics. For any snake, the K_A/K_S ratio estimated between the ORFs of *SSP-6* and *SSP-5* was the lowest, or considerably lower than the K_A/K_S ratios estimated between other paralogs. On the other hand, the K_A/K_S ratios estimated between *SSP-6* and *SSP-4* and between *SSP-5* and *SSP-4* were close to one. Our interpretation of these results is as follows. SSP-6 or SSP-5 are irrelevant for neutralizing venom proteins and have constitutive or essential roles, such as digestion or blood homeostasis. Therefore, nucleotide sequences of *SSP-6* and *SSP-5* have been conserved. On the other hand, *SSP-4*, acquired in the genomes of venomous snakes, may have encoded the first SSP with a role specific to venom neutralization in the event of accidental bites. Therefore, *SSP-4* may have had to be more plastic than *SSP-5* and *SSP-6*.

Table 1. K_A/K_S ratios estimated between the ORFs of *C. viridis* SSPs.

	CvSSP-4	*CvSSP-5*	*CvSSP-6*
CvSSP-4		0.749	0.879
CvSSP-5			0.306
CvSSP-6			

Table 2. K_A/K_S ratios estimated between the ORFs of *D. acutus* SSPs.

	DaSSP-1	*DaSSP-2*	*DaSSP-3*	*DaSSP-4*	*DaSSP-6*
DaSSP-1		1.61	0.934	0.823	0.832
DaSSP-2			0.849	0.705	1.03
DaSSP-3				1.77	0.878
DaSSP-4					0.830
DaSSP-6					

Table 3. K_A/K_S ratios estimated between the ORFs of *O. hannah* SSPs.

	OhSSP-4(Ψ)	*OhSSP-5α*	*OhSSP-5β*	*OhSSP-5γ*	*OhSSP-6*
OhSSP-4(Ψ)		0.931	0.875	0.700	0.919
OhSSP-5α			0.836	0.719	0.666
OhSSP-5β				0.832	0.685
OhSSP-5γ					0.509
OhSSP-6					

Table 4. K_A/K_S ratios estimated between the ORFs of *P. bivittatus* SSPs.

	PfSSP-5α	*PfSSP-5β*	*PbSSP-6*
PbSSP-5α		1.594	0.343
PbSSP-5β			0.296
PbSSP-6			

Table 5. K_A/K_S ratios estimated between the ORFs of *P. mucrosquamatus* SSPs.

	PmSSP-1	*PmSSP-2*	*PmSSP-3*	*PmSSP-4*	*PmSSP-5*	*PmSSP-6*
PmSSP-1		1.49	0.891	0.821	0.639	0.694
PmSSP-2			0.825	0.662	0.476	0.620
PmSSP-3				1.35	0.479	0.780
PmSSP-4					0.586	0.889
PmSSP-5						0.370
PmSSP-6						

Table 6. K_A/K_S ratios estimated between the ORFs of *P. flavoviridis* SSPs.

	PfSSP-1	*PfSSP-2*	*PfSSP-3*	*PfSSP-4*	*PfSSP-5*	*PfSSP-6*
PfSSP-1		1.80	0.660	0.790	0.597	0.792
PfSSP-2			0.808	0.781	0.504	0.891
PfSSP-3				1.40	0.599	0.990
PfSSP-4					0.670	1.07
PfSSP-5						0.626
PfSSP-6						

Table 7. K_A/K_S ratios estimated between the ORFs of *T. sirtalis* SSPs.

	TsSSP-4	*TsSSP-5α*	*TsSSP-5β*	*TsSSP-6*
TsSSP-4		1.06	0.603	0.651
TsSSP-5α			0.675	0.447
TsSSP-5β				0.659
TsSSP-6				

Table 8. K_A/K_S ratios estimated between the ORFs of *V. berus* SSPs.

	VbSSP-5	*VbSSP-6*
VbSSP-5		0.604
VbSSP-6		

Table 9. K_N values estimated between the introns of *C. viridis* SSPs.

	CvSSP-4	*CvSSP-5*	*CvSSP-6*
CvSSP-4		0.319	0.358
CvSSP-5			0.372
CvSSP-6			

Table 10. K_N values estimated between the introns of *D. acutus* SSPs.

	DaSSP-1	*DaSSP-2*	*DaSSP-3*	*DaSSP-4*	*DaSSP-6*
DaSSP-1		0.0227	0.248	0.253	0.249
DaSSP-2			0.247	0.251	0.246
DaSSP-3				0.0050	0.285
DaSSP-4					0.288
DaSSP-6					

Table 11. K_N values estimated between the introns of *O. hannah* SSPs.

	OhSSP-4(Ψ)	*OhSSP-5α*	*OhSSP-5β*	*OhSSP-5γ*	*OhSSP-6*
OhSSP-4(Ψ)		0.331	0.340	0.324	0.530
OhSSP-5α			0.0615	0.0801	0.277
OhSSP-5β				0.0857	0.226
OhSSP-5γ					0.275
OhSSP-6					

Table 12. K_N values estimated between the introns of *P. mucrosquamatus* SSPs.

	PmSSP-1	*PmSSP-2*	*PmSSP-3*	*PmSSP-4*	*PmSSP-5*	*PmSSP-6*
PmSSP-1		0.154	0.338	0.342	0.339	0.374
PmSSP-2			0.281	0.284	0.262	0.316
PmSSP-3				0.0397	0.293	0.346
PmSSP-4					0.296	0.349
PmSSP-5						0.295
PmSSP-6						

Table 13. K_N values estimated between the introns of *P. flavoviridis* SSPs.

	PfSSP-1	*PfSSP-2*	*PfSSP-3*	*PfSSP-4*	*PfSSP-5*	*PfSSP-6*
PfSSP-1		0.0317	0.251	0.258	0.248	0.231
PfSSP-2			0.254	0.261	0.253	0.229
PfSSP-3				0.0283	0.261	0.279
PfSSP-4					0.270	0.287
PfSSP-5						0.267
PfSSP-6						

The K_A/K_S ratios estimated between *DaSSP-1* and *DaSSP-2*, *DaSSP-3* and *DaSSP-4*, *PmSSP-1* and *PmSSP-2*, *PmSSP-3* and *PmSSP-4*, *PfSSP-1* and *PfSSP-2*, and *PfSSP-3* and *PfSSP-4* were 1.61, 1.77, 1.49, 1.35, 1.80, and 1.42, respectively (Table 2, Table 5, and Table 6), and the rates of K_N were 0.0227, 0.005, 0.154, 0.0397, 0.0317, and 0.0283, respectively (Table 10, Table 12, and Table 13). These results showed that the branching of these genes, especially late genes such as *SSP-1*, *SSP-2*, or *SSP-3*, occurred in an accelerated manner, and that the time that passed after their divergence was very short. In addition, the K_A/K_S ratios estimated between *SSP-1* and *SSP-5* or *SSP-2* and *SSP-5* of *D. acutus*, *P. mucrosquamatus*, and *P. flavoviridis* were around 0.7. This result also supports the idea that *SSP-1* or *SSP-2* and *SSP-5* are evolutionarily related, as suggested above. That is, *SSP-1* and *SSP-2* were recently derived from *SSP-5* and then diversified in an accelerated manner to accommodate venom proteins. SSP-3, the truncated SSP acquired as the successor to SSP-4, is also thought to bind more venom proteins than SSP-4, as do SSP-1 and SSP-2 relative to SSP-5. Therefore, the K_A/K_S ratios estimated between *SSP-3* and *SSP-4* also show considerably higher values. The many reports that venom proteins bind to SSP-1 [35], SSP-2 [33,37], or SSP-3 [36] also support the above idea. Since animal venoms work as defense mechanisms, as tools to catch prey, or simply to enhance digestion, they should be sensitive to the surrounding environment. Because venom proteins have become more diversified in environments where there are more diverse prey, serum proteins required to neutralize venom activities, such as *SSP-1*, *SSP-2*, and *SSP-3*, have also diversified in an accelerated manner. Even among conventional Long SSPs, *SSP-1*, *SSP-2*, and *SSP-5* have evolved in an accelerated or neutral manner, depending on whether they deal with venom components. On the other hand, *SSP-3* and *SSP-4*, which specifically arose as anti-venom proteins, have evolved in an accelerated manner.

Chijiwa et al. proposed that most nucleotide substitutions at non-synonymous sites occur only immediately after gene duplication. Then random mutations accumulate over time, and selective pressure that leaves "neutral" mutations at synonymous sites erases the traces of accelerated evolution [39]. However, in the genomes of the viperids *P. flavoviridis* and *P. mucrosquamatus*, inversion of the genome segment encompassing *SSP-1* to *SSP-2* occurred, and subsequent accumulation of random mutations was suppressed [39]. These findings are also applicable to *D. acutus DaSSP-1* and *DaSSP-2*. Since the *SSP-3* allele is located in the 3′ region downstream of the inverted genome segment containing *SSP-1* and *SSP-2*, the inversion may also have suppressed accumulation of random mutations in *PfSSP-3*.

3. Materials and Methods

3.1. Materials

P. flavoviridis specimens were provided by the Institute of Medical Sciences of the University of Tokyo. The tail of an *O. hannah* specimen was provided by the Japan Snake Center. That of a *P. mucrosquamatus* was provided by the Medical Institute of Bioregulation, at the Research Center of Genetic Information, Kyushu University. High-molecular-weight genomic DNA was prepared from livers or tails of the snakes according to the method of Blin and Stafford [44]. Total RNA was prepared from various snake organs, according to the ISOGEN protocol (Nippon Gene, Toyama, Japan). Restriction endonucleases and KOD plus DNA polymerase were purchased from Nippon Gene and TOYOBO (Osaka, Japan), respectively. Other reagents and antibiotics were from Nacalai Tesque (Kyoto, Japan) and TAKARA BIO (Shiga, Japan). Specific oligonucleotide primers were synthesized by GENNET (Fukuoka, Japan).

3.2. Cloning and Sequencing of the Genome Segment Containing PfSSP-6

A dedicated database, HabAm1, [40] was constructed to carry out blastn and tblastx analysis with the nucleotide sequences of *PfSSPs* (*PfSSP-1–PfSSP-5*) as queries. Exon–intron boundaries were then determined based on the five *PfSSPs*. Referring to the nucleotide sequence of Scaffold 2858, the sense primer SSP6-5UTR-1, 5′-ggC gTC CCT CCT TCT CCT Tg-3′, which anneals specifically to the first exon of *PfSSP-6*, and the antisense primer SSP6ex3-2, 5′-CTC gCA TTC CAT ACA ATT ggC Tg-3′, which anneals specifically to the third exon of *PfSSP-6*, were used to amplify the 1453 bp genome fragment (Table 14). The sense primer, SSP6ex3-1, 5′-TgT ggC CAA CCA AAT gCg Tgg-3′, which anneals specifically to the third exon of *PfSSP-6*, and the antisense primer SSP6-3flank-1, 5′-CAg CTA TgC ATg CCT TAT ATC AC-3′, which anneals specifically to 85 bp 3′ downstream of the fourth exon of *PfSSP-6*, were then used to amplify the 2363 bp genome fragment (Table 14). Amplified genome fragments were ligated to the pCR™-Blunt II-TOPO® vector (Life Technologies, Carlsbad, CA, USA) and transformed with DH5α-competent cells (TAKARA BIO, Shiga, Japan). Nucleotide sequences were determined using an ABI 3130xl capillary sequencer. The 1453 bp PCR fragment overlapped with the 2363 bp PCR fragment by 89 bp. The physical structure of the 3727 bp segment encompassing the first exon of *PfSSP-6* to 85 bp 3′ downstream of the fourth exon of *PfSSP-6* was determined. This 3727 bp DNA fragment contained four exons encoding *Pf*SSP-6. Moreover, to acquire the nucleotide sequence of the intergenic region between *PfSSP-6* and *PfSSP-4*, named *Pf*I-Reg64, genomic PCR was carried out against the Amami–Oshima *P. flavoviridis* genome and the sense primer Ireg64-1, 5′-CTC CAT gCA AAg gAg gAT TTC C-3′, which anneals to the 3′ terminus of the third intron of *PfSSP-6*, and the antisense primer Ireg64-6, 5′-TAg gCC TTg ACA CAT gAT ggC-3′, which anneals to the middle portion of *Pf*I-Reg64, were used to amplify the 7717 bp genome fragment, named *Pf*IREG64-I (Table 14). The *Pf*IREG64-I fragment was also cloned and sequenced. The 7717 bp *Pf*IREG64-I overlapped with the 3727 bp *PfSSP-6* by 474 bp. The sense primer Ireg64-5, 5′-CAT TgT TgA gCA ACC CTT ggC-3′, which anneals 2501 bp 5′ upstream of Ireg64-6, and the antisense primer Ireg64-8 5′-ggA CTA TTA AgC AgT ggA ATg gC-3′, which anneals 2340 bp 5′ upstream of the first exon of *PfSSP-4* (3′ terminal of *Pf*IReg-64), were then used to amplify the 5283 bp genome fragment, named *Pf*IREG64-II (Table 14). The *Pf*IREG64-II fragment was also cloned and sequenced. The 5283 bp *Pf*IREG64-II overlapped with the 7717 bp *Pf*IREG64-I by 2523 bp. The sense primer Ireg64-9, 5′-ggC CCT CTT CCA Agg ACA AgC-3′, which anneals 455 bp 5′ upstream of Ireg64-8, and the antisense primer Ireg64-10, 5′-ACC TCg TTC CTC CAg CCA CT-3′, which anneals to the 5′ terminus of the first intron of *PfSSP-4*, were then used the 2971 bp genome fragment, named *Pf*IREG64-III (Table 14). The *Pf*IREG64-III fragment was also cloned and sequenced. The 2971 bp *Pf*IREG64-III overlapped with the 5267 bp *Pf*IREG64-II by 455 bp. Finally, the physical structure of the 16,248 bp segment encompassing the third intron of *PfSSP-6* to the first intron of *PfSSP-4* was completely established. The nucleotide sequences of *PfSSP-6* and the genome segment from *PfSSP-6* to *PfSSP-4* are available from the Genbank/EMBL/DDBJ databases under Accession No. LC518073.

Table 14. Primers used to acquire the nucleotide sequences from the genome domain encompassing *PfSSP-6* to *PfSSP-4*. The symbols (f) or (r) after the position numbers indicate the directions of the primers. Forward or reverse denote whether the direction of elongation was the same or opposite to that of transcription. Nucleotide positions refer to nucleotide sequences reported in this study (LC518073).

Name	Positions	Nucleotide Sequence (GC Content: %, Tm: °C)
SSP6-5UTR-1	1–20 (f)	5′- ggC gTC CCT CCT TCT CCT Tg -3′ (65, 66)
SSP6ex3-2	1431–1453 (r)	5′- CTC gCA TTC CAT ACA ATT ggC Tg-3′ (48, 68)
SSP6ex3-1	1365–1385 (f)	5′- TgT ggC CAA CCA AAT gCg Tgg -3′ (57, 66)
SSP6-3UTR-2	3581–3604 (r)	5′- ACA TgA gAg ATT TAT TCC AgT gTg-3′(38, 66)
SSP6-3flank-1	3705–3727 (r)	5′- CAg CTA TgC ATg CCT TAT ATC AC -3′ (43, 66)
Ireg64-1	3254–3275 (f)	5′- CTC CAT gCA AAg gAg gAT TTC C -3′ (50, 66)
Ireg64-6	10,950–10,970 (r)	5′- TAg gCC TTg ACA CAT gAT ggC -3′ (52, 64)
Ireg64-5	8449–8469 (f)	5′- CAT TgT TgA gCA ACC CTT ggC -3′ (52, 64)
Ireg64-8	13,709–13,731 (r)	5′- ggA CTA TTA AgC AgT ggA ATg gC -3′ (48, 68)
Ireg64-9	13,277–13,297 (f)	5′- ggC CCT CTT CCA Agg ACA AgC -3′ (62, 68)
Ireg64-10	16,228–16,247 (r)	5′- ACC TCg TTC CTC CAg CCA CT -3′ (60, 64)

3.3. RepeatMasker Analysis of the Nucleotide Sequence of PfSSP-6

A dedicated database was constructed with repetitive sequences of the genomes of various organisms collected from Repbase of the Genetic Information Research Institute [45]. RepeatMasker utilized the nucleotide sequences of *PfSSP-6* against the database via BLAST+, RMBlast (NCBI), and Tandem Repeats Finder (Boston University) [46].

3.4. Determination of Nucleotide Sequences and Chromosomal Configurations of Genes Encoding Orthologs of PfSSPs from Seven Snake Taxa

Draft nucleotide sequences of the genomes of seven taxa, *C. viridis* (*Cv*, venomous) [47], *D. acutus* (*Da*, venomous) [48], *O. hannah* (*Oh*, venomous) [49], *P. bivittatus* (*Pb*, non-venomous) [50], *P. mucrosquamatus* (*Pm*, venomous) [51], *T. sirtalis* (*Ts*, non-venomous) [52], and *V. berus* (*Vb*, venomous) [53], were downloaded to create a dedicated genome database. Referring to the nucleotide and amino acid sequences of *PfSSPs* deduced using tblastn or blastn, the nucleotide sequences encoding orthologs of *PfSSPs* and their flanking regions in each snake genome were ascertained. The *T. sirtalis* genome segment containing *TsSSP-4* and *TsSSP-5α* and the three *P. bivittatus* genes, *PbSSP-5α*, *PbSSP-5β*, and *PbSSP-5γ* (Ψ), were identified in separate scaffolds; therefore, their locations and arrangements are tentative.

3.5. Determining the Nucleotide Sequences of SSP Paralogs of P. mucrosquamatus and O. hannah

To acquire complete nucleotide sequences of *PmSSP-3*, *PmSSP-4*, *OhSSP-1*, *OhSSP-2*, *OhSSP-5*, and *OhSSP-6*, genomic PCR was performed on the *O. hannah* and *P. mucrosquamatus* genomes to amplify two overlapping nucleotide segments separately. These included the 5′ segment of the gene encompassing the first exon to the second exon, and the 3′ segment of the gene encompassing the second exon to the fourth exon of each gene.

For *PmSSP-3* (*Pm* Scaffold 462), the sense primer, PmSSP34-5UTR, 5′-CAA ggg TTg gTC TTg gTT TTT g-3′, which anneals to the 5′ terminus of the first exon of *PmSSP-3* and *PmSSP-4*, and the antisense primer, PmSSP3ex2-R, 5′-ggT AgA gAA AAg CCC CCA AAg-3′, which anneals to the second exon of *PmSSP-3*, were used to amplify the 1169 bp 5′ segment of *PmSSP-3* (Table 15). The sense primer, PmSSP3-F, 5′-TgC TTT ggg ggC TTT TCT C-3′, which anneals to the middle portion of the second exon of *PmSSP-3*, and the antisense primer, PmSSP34-R, 5′-CTT gAC TgA GAC TgA AgT TCC-3′, which anneals to the 311 bp 3′ region downstream of the fourth exon of *PmSSP-3* and *PmSSP-4*, were then used to amplify the 2722 bp 3′ segment of *PmSSP-3* (Table 15). The 5′ segment of *PmSSP-3* overlapped with the 3′ segment of *PmSSP-3* by 31 bp. The physical structure of the 3860 bp segment encompassing the first exon of *PmSSP-3* to the 311 bp at the 3′ region downstream of the fourth exon of *PmSSP-3* was completed.

Table 15. Primers utilized to determine nucleotide sequences of genome fragments containing *SSPs* of *P. mucrosquamatus* and *O. hannah*. The symbols (f) or (r) after the position numbers indicate the directions of the primers. Forward or reverse indicate whether the direction of elongation was the same or opposite to that of transcription. Nucleotide positions refer to the nucleotide sequences reported in this study. Abbreviations: *Oh*: *O. hannah*; *Pm*: *P. mucrosquamatus*.

Name	Scaffold	Nucleotide Sequence (GC Content: %, Tm: °C)
PmSSP34-5UTR (f)	*Pm* Scaffold 462	5′- CAA ggg TTg gTC TTg gTT TTT g -3′ (45, 64)
PmSSP3ex2-R (r)	*Pm* Scaffold 462	5′- ggT AgA gAA AAg CCC CCA AAg -3′ (52, 64)
PmSSP3-F (f)	*Pm* Scaffold 462	5′- TgC TTT ggg ggC TTT TCT C -3′ (47, 56)
PmSSP34-R (r)	*Pm* Scaffold 462	5′- CTT gAC TgA GAC TgA AgT TCC -3′ (45, 62)
PmSSP4ex2-R (r)	*Pm* Scaffold 462	5′- CgT TTC Agg TAA Agg AAT ACT C -3′ (41, 62)
PmSSP4-F (f)	*Pm* Scaffold 21,362	5′- gAg TAT TCC TTT ACC TgA AAC g -3′ (41, 62)
OhSSPs-5UTR (f)	*Oh* Scaffold 4527	5′- ATA AAT Tgg Agg AgC RgA TTC CT -3′ (43, 66)
OhSSP5-ex2-R (r)	*Oh* Scaffold 4527	5′- CTC AgC TTC AAA gCC CCA gg -3′ (60, 64)
OhSSP5-F (f)	*Oh* Scaffold 4527	5′- gAg CAT gCT TTA CCT ggg gC -3′ (60, 64)
OhSSP5-R (r)	*Oh* Scaffold 47,978	5′- TCC ATg TgT AgA gAT CAA ACA Cg -3′ (43, 66)
OhSSP2-ex2-R (r)	*Oh* Scaffold 4527	5′- CTC AgC TTC AAA gAg CCC TCT -3′ (52, 64)
OhSSP2-F (f)	*Oh* Scaffold 4527	5′- gAg CAT gCT ATA gAg ggC TCT -3′ (52, 64)
OhSSP2-R (r)	*Oh* Scaffold 4527	5′- gAT CAA ACA TCA CAg CgC TgC -3′ (52, 64)
OhSSP1-ex2-R (r)	*Oh* Scaffold 4527	5′- TTA Agg AAC ACT CCA AAg CAC C -3′ (52, 64)
OhSSP1-F (f)	*Oh* Scaffold 4527	5′- gAg ggT gCT TTg gAg TgT TCC -3′ (45, 64)
OhSSP1-R (r)	*Oh* Scaffold 4527	5′- gAT CAg ACA CCA CAg CTg Tgg -3′ (57, 66)
OhSSP6-ex2-R (r)	*Oh* Scaffold 10,541	5′- TAA ACT gAg gTT TAA AgA gAT CCA -3′ (33, 64)
OhSSP6-F (f)	*Oh* Scaffold 10,541	5′- gCA gCA TgC TTC ATg gAT CTC -3′ (52, 64)
OhSSP6-R (r)	*Oh* Scaffold 12,359	5′- CCg TgT gAA AAg NTC AgA CAT C -3′ (50, 66)

With regard to *PmSSP-4*, the sense primer, PmSSP34-5UTR, described above, and the antisense primer PmSSP4ex2-R, 5′-CgT TTC Agg TAA Agg AAT ACT C-3′, which anneals to the second exon of *PmSSP-4* based on the nucleotide sequence of *Pm* Scaffold 21,362, were used to amplify the 1139 bp 5′ portion of *PmSSP-4* (Table 15). Using *Pm* Scaffold 21,362, the sense primer, PmSSP4-F, 5′-gAg TAT TCC TTT ACC TgA AAC g-3′, which anneals to the middle portion of the second exon of *PmSSP-4*, and the antisense primer, PmSSP34-R, described above, were then used to amplify the 2999 bp 3′ segment of *PmSSP-4*. The 5′ segment of *PmSSP-4* overlapped with the 3′ segment of *PmSSP-4* by 22 bp (Table 15). The physical structure of the 4118 bp segment encompassing the first exon of *PmSSP-4* to the 311 bp at the 3′ region downstream of the fourth exon of *PmSSP-4* was sequenced.

For *OhSSP-5α* (*Oh* Scaffold 4527), the sense primer, OhSSPs-5UTR, 5′-ATA AAT Tgg Agg AgC RgA TTC CT-3′, which anneals to the common nucleotide sequence of the 5′ UTR of *OhSSPs*, and the antisense primer, OhSSP5-ex2-R, 5′-CTC AgC TTC AAA gCC CCA gg-3′, which anneals to the second exon of *OhSSP-5*, were used to amplify the 1107 bp 5′ segment of *OhSSP-5α* (Table 15). The sense primer OhSSP5-F, 5′-gAg CAT gCT TTA CCT ggg gC-3′, which anneals to the middle portion of the second exon of *OhSSP-5α* (*Oh* Scaffold 4527), and the antisense primer, OhSSP5-R 5′-TCC ATg TgT AgA gAT CAA ACA Cg-3′, which anneals to the middle portion of the fourth exon of *OhSSP-5α* (*Oh* Scaffold 47,978), were the used to amplify the 2130 bp 3′ part of *OhSSP-5α* (Table 15). The 5′ segment of *OhSSP-5α* overlapped with the 3′ segment of *OhSSP-5α* by 32 bp. The sequence of the 3205 bp segment encompassing the first exon of *OhSSP-5α* to the fourth exon of *OhSPP-5α* was determined.

In regard to *OhSSP-5β*, the sense primer, OhSSPs-5UTR, described above, and the antisense primer, OhSSP2-ex2-R, 5′-CTC AgC TTC AAA gAg CCC TCT-3′, which anneals to the second exon of *OhSSP-5β* (*Oh* Scaffold 4527), were used to amplify the 2456 bp 5′ section of *OhSSP-5β* (Table 15). The sense primer, OhSSP2-F, 5′-gAg CAT gCT ATA gAg ggC TCT-3′, which anneals to the middle portion of the second exon of *OhSSP-5β*, and the antisense primer, OhSSP2-R, 5′-gAT CAA ACA TCA

CAg CgC TgC-3′, which anneals to the fourth exon of *OhSSP-5β*, were then used to amplify the 2180 bp 3′ section of *OhSSP-5β* (Table 15). The 5′ segment of *OhSSP-5β* overlapped with 3′ segment of *OhSSP-5β* by 42 bp. The 4594 bp segment encompassing the first exon of *OhSSP-5β* to the fourth exon of *OhSPP-5β* was sequenced.

For *OhSSP-5γ*, the sense primer, OhSSPs-5UTR, described above, and the antisense primer, OhSSP1-ex2-R, 5′-TTA Agg AAC ACT CCA AAg CAC C-3′, which anneals to the second exon of *OhSSP-5γ* (*Oh* Scaffold 4527), were used to amplify the 1534 bp 5′ segment of *OhSSP-5γ* (Table 15). The sense primer, OhSSP1-F, 5′-gAg ggT gCT TTg gAg TgT TCC-3′, which anneals to the middle portion of the second exon of *OhSSP-5γ* (*Oh* Scaffold 4527), and the antisense primer, OhSSP1-R, 5′-gAT CAg ACA CCA CAg CTg Tgg-3′, which anneals to the fourth exon of *OhSSP-5γ*, were then used to amplify the 1999 bp 3′ segment of *OhSSP-5γ*. The 5′ half segment of *OhSSP-5γ* overlapped with the 3′ half segment of *OhSSP-5γ* by 26 bp (Table 15). The structure of the 3508 bp segment encompassing the first exon of *OhSSP-5γ* to the fourth exon of *OhSPP-5γ* was deciphered.

For *OhSSP-6*, the sense primer, OhSSPs-5UTR, described above, and the antisense primer, OhSSP6-ex2-R, 5′-TAA ACT gAg gTT TAA AgA gAT CCA-3′, which anneals to the second exon of *OhSSP-6* (*Oh* Scaffold 10,541), were used to amplify the 1734 bp 5′ segment of *OhSSP-6* (Table 15). The sense primer, OhSSP6-F, 5′-gCA gCA TgC TTC ATg gAT CTC-3′, which anneals to the middle portion of the second exon of *OhSSP-6* (*Oh* Scaffold 10,541), and the antisense primer, OhSSP6-R, 5′-CCg TgT gAA AAg NTC AgA CAT C-3′, which anneals to the fourth exon of *OhSSP-6* (*Oh* Scaffold 12,359), were then used to amplify the 3461 bp 3′ segment of *OhSSP-6* (Table 15). The 5′ section of *OhSSP-6* overlapped with the 3′ segment of *OhSSP-6* by 43 bp. The sequence of the 5152 bp segment encompassing the first exon of *OhSSP-6* to the fourth exon of *OhSPP-6* was determined. Nucleotide sequences of *OhSSP-6*, *OhSSP-5α*, *OhSSP-5β*, *OhSSP-5γ*, *PmSSP-4*, and *PmSSP-3* are available in the Genbank/EMBL/DDBJ databases under Accession Nos. LC518074–LC518078 and LC519888.

3.6. Expression Analysis of PfSSP-6 mRNA Using Semi-Quantitative RT-PCR

First-strand cDNAs from snake organs were synthesized by reverse transcription and primer extension with a SMART cDNA Library Construction Kit (Clontech, California, USA). Based on nucleotide sequences of the genes encoding *PfSSP-6*, the sense primer SSP6-5UTR-1, described above, and the antisense primer SSP6-3UTR-2, 5′- ACA TgA gAg ATT TAT TCC AgT gTg -3′, which anneals to the 3′ terminal of the fourth exon of *PfSSP-6*, were designed (Table 14). cDNA of β-actin, designated as ACTB, was amplified as an internal standard with the sense primer, SHU7, 5′-CAg AgC AAg AgA ggT ATC CN-3′ (N = G, A, T, C), and the antisense primer, SHU8, 5′-TAg ATg ggC ACA gTg Tgg gN-3′, as described previously [54].

3.7. Mathematical Analysis

Alignment of the amino acid sequences of snake SSPs was performed using ClustalX software. Nucleotide sequences of ORFs encoding the mature SSPs were rearranged and gaps in the aligned amino acid sequences were removed using PAL2NAL. The rates of synonymous (K_S) and nonsynonymous (K_A) substitutions per site between the ORFs of the genes were calculated using the Nei–Gojobori method, as implemented in PAML [55]. After removing LINEs, DNA transposons, and indels (insertion/deletion) from the introns, alignment of introns was performed using ClustalX. Values of K_N that estimated rates of substituted nucleotides between the introns of *SSPs* were calculated from the aligned sequence data.

Author Contributions: K.I. and T.C. designed experiments and performed analyses. K.I., A.T., M.M., H.S., Y.F., S.H., N.O.-U., M.O., and T.C. prepared materials from live specimen. K.I. performed sequencing and manual assembly from *PfSSP-6* to *PfSSP-4*. T.C. supervised the project. K.I. and T.C. wrote the manuscript with contributions from all other authors. All authors have read and agreed to the published version of the manuscript.

Funding: This work was partially supported by the Ministry of Education, Science, Sports and Culture, Grant-in Aid for Scientific Research (C), 2014-2016 (26340095, Takahito Chijiwa).

Conflicts of Interest: The authors declare no conflicts of interest.

References

1. Aird, S.D.; Watanabe, Y.; Villar-Briones, A.; Roy, M.C.; Terada, K.; Mikheyev, A.S. Quantitative high-throughput profiling of snake venom gland transcriptomes and proteomes (*Ovophis okinavensis* and *Protobothrops flavoviridis*). *BMC Genomics* **2013**, *14*, 790. [CrossRef] [PubMed]
2. Rodrigues, R.S.; Boldrini-França, J.; Fonseca, F.P.P.; de la Torre, P.; Henrique-Silva, F.; Sanz, L.; Calvete, J.J.; Rodrigues, V.M. Combined snake venomics and venom gland transcriptomic analysis of *Bothropoides pauloensis*. *J. Proteomics* **2012**, *75*, 2707–2720. [CrossRef] [PubMed]
3. Rokyta, D.R.; Wray, K.P.; Margres, M.J. The genesis of an exceptionally lethal venom in the timber rattlesnake (*Crotalus horridus*) revealed through comparative venom-gland transcriptomics. *BMC Genomics* **2013**, *14*, 394. [CrossRef] [PubMed]
4. Brunie, S.; Bolin, J.; Gewirth, D.; Sigler, P.B. The refined crystal structure of dimeric phospholipase A_2 at 2.5 Å Access to a shielded catalytic center. *J. Biol. Chem.* **1985**, *260*, 9742–9749. [PubMed]
5. Holland, D.R.; Clancy, L.L.; Muchmore, S.W.; Ryde, T.J.; Einspahr, H.M.; Finzel, R.L.; Heinrikson, R.L.; Watenpaugh, K.D. The crystal structure of a lysine 49 phospholipase A_2 from the venom of the cottonmouth snake at 2.0-Å resolution. *J. Biol. Chem.* **1990**, *265*, 17649–17656.
6. Renetseder, R.; Brunie, S.; Dijkstra, B.W.; Drenth, J.; Sigler, P.B. A comparison of the crystal structures of phospholipase A_2 from bovine pancreas and Crotalus atrox venom. *J. Biol. Chem.* **1985**, *260*, 11627–11634.
7. Suzuki, A.; Matsueda, E.; Yamane, T.; Ashida, T.; Kihara, H.; Ohno, M. Crystal Structure Analysis of Phospholijase A_2 from *Trimeresurus flavoviridis* (Habu Snake) Venom at 1.5 Å Resolution. *J. Biochem.* **1995**, *117*, 730–740. [CrossRef]
8. Fox, J.W.; Serrano, S.M. Structural considerations of the snake venom metalloproteinases, key members of the M12 reprolysin family of metalloproteinases. *Toxicon* **2005**, *45*, 969–985. [CrossRef]
9. Omori-Satoh, T.; Sadahiro, S. Resolution of the major hemorrhagic component of *Trimeresurus flavoviridis* venom into two parts. *Biochim. Biophys. Acta* **1979**, *580*, 392–404. [CrossRef]
10. Kini, R.M. Excitement ahead: Structure, function and mechanism of snake venom phospholipase A_2 enzymes. *Toxicon* **2003**, *42*, 827–840. [CrossRef]
11. Masuda, S.; Hayashi, H.; Atobe, H.; Morita, T.; Araki, S. Purification, cDNA cloning and characterization of the vascular apoptosis-inducing protein, HV1, from *Trimeresurus flavoviridis*. *J. Biochem.* **2001**, *268*, 3339–3345. [CrossRef]
12. Matsui, T.; Fujimura, Y.; Titani, K. Snake venom proteases affecting hemostasis and thrombosis. *Biochim. Biophys. Acta* **2000**, *1447*, 146–156. [CrossRef]
13. Carone, S.E.I.; Menaldo, D.L.; Sartim, M.A.; Bernardes, C.P.; Caetano, R.C.; da Silva, R.R.; Cabral, H.; Barraviera, B.; Ferreira Junior, R.S.; Sampaio, S.V. BjSP, a novel serine protease from *Bothrops jararaca* snake venom that degrades fibrinogen without forming fibrin clots. *Toxicol. Appl. Pharmacol.* **2018**, *357*, 50–61. [CrossRef] [PubMed]
14. Kini, R.M. *Venom Phospholipase A_2 Enzymes: Structure, Function and Mechanism*; Kini, R.M., Ed.; John Wiley & Sons: Hoboken, NJ, USA, 1997.
15. Kihara, H.; Uchikawa, R.; Hattori, S.; Ohno, M. Myotoxicity and physiological effects of three *Trimeresurus flavoviridis* phospholipases A_2. *Biochem. Int.* **1992**, *28*, 895–903. [PubMed]
16. Oda, N.; Ogawa, T.; Ohno, M.; Sasaki, H.; Sakaki, Y.; Kihara, H. Cloning and Sequence Analysis of cDNA for *Trimeresurus flavoviridis* Phospholipase A_2, and Consequent Revision of the Amino Acid Sequence. *J. Biochem.* **1990**, *108*, 816–821. [CrossRef]
17. Yamaguchi, Y.; Shimohigashi, Y.; Chijiwa, T.; Nakai, M.; Ogawa, T.; Hattori, S.; Ohno, M. Characterization, amino acid sequence and evolution of edema-inducing, basic phospholipase A_2 from *Trimeresurus flavoviridis* venom. *Toxicon* **2001**, *39*, 1069–1076. [CrossRef]
18. Posada Arias, S.; Rey-Suárez, P.; Pereáñez, J.A.; Acosta, C.; Rojas, M.; Delazari dos Santos, L.; Ferreira, R.S., Jr.; Núñez, V. Isolation and Functional Characterization of an Acidic Myotoxic Phospholipase A2 from Colombian *Bothrops asper* Venom. *Toxins* **2017**, *9*, 342. [CrossRef] [PubMed]
19. Chijiwa, T.; Hamai, S.; Tsubouchi, S.; Ogawa, T.; Deshimaru, M.; Oda-Ueda, N.; Hattori, S.; Kihara, H.; Tsunasawa, S.; Ohno, M. Interisland Mutation of a Novel Phospholipase A_2 from *Trimeresurus flavoviridis* Venom and Evolution of Crotalinae Group II Phospholipases A_2. *J. Mol. Evol.* **2003**, *57*, 546–554. [CrossRef]

20. Liu, S.Y.; Yoshizumi, K.; Oda, N.; Ohno, M.; Tokunaga, F.; Iwanaga, S.; Kihara, H. Purification and Amino Acid Sequence of Basic Protein II, a Lysine-49-Phospholipase A_2 with Low Activity, from *Trimeresurus flavoviridis* Venom. *J. Biochem.* **1990**, *107*, 400–408. [CrossRef]

21. Yoshizumi, K.; Liu, S.Y.; Miyata, T.; Saita, S.; Ohno, M.; Iwanaga, S.; Kihara, H. Purification and amino acid sequence of basic protein I, a lysine-49-phospholipase A_2 with low activity, from the venom of *Trimeresurus flavoviridis* (Habu snake). *Toxicon* **1990**, *28*, 43–54. [CrossRef]

22. Teixera, L.F.; de Carvalho, L.H.; de Castro, O.B.; Bastos, J.F.S.; Néry, N.M.; Oliveira, G.A.; Kayano, A.M.; Soares, A.M.; Zuliani, J.P. Local and systemic effects of BdipTX-I, a Lys-49 phospholipase A2 isolated from *Bothrops diporus* snake venom. *Toxicon* **2018**, *141*, 55–64. [CrossRef] [PubMed]

23. Nobuhisa, I.; Deshimaru, M.; Chijiwa, T.; Nakashima, K.; Ogawa, T.; Shimohigashi, Y.; Fukumaki, Y.; Hattori, S.; Kihara, H.; Ohno, M. Structures of genes encoding phospholipase A_2 inhibitors from the serum of *Trimeresurus flavoviridis* snake. *Gene* **1997**, *191*, 31–37. [CrossRef]

24. So, S.; Chijiwa, T.; Ikeda, N.; Nobuhisa, I.; Oda-Ueda, N.; Hattori, S.; Ohno, M. Identification of the B Subtype of γ-Phospholipase A_2 Inhibitor from *Protobothrops flavoviridis* Serum and Molecular Evolution of Snake Serum Phospholipase A_2 Inhibitors. *J. Mol. Evol.* **2008**, *66*, 298–307. [CrossRef] [PubMed]

25. Gimenes, S.N.C.; Ferreira, F.B.; Silveira, A.N.P.; Rodrigues, R.S.; Yoneyama, K.A.G.; Izabel dos Santos, J.; Fontes, M.R.d.M.; de Campos Brites, V.L.; Santos, A.L.Q.; Borges, M.H.; et al. Isolation and biochemical characterization of a γ-type phospholipase A2 inhibitor from *Crotalus durissus collilineatus* snake serum. *Toxicon* **2014**, *81*, 58–66. [CrossRef] [PubMed]

26. Oliveira, C.Z.; Santos-Filho, N.A.; Menaldo, D.L.; Boldrini-Franca, J.; Giglio, J.R.; Calderon, L.A.; Stabeli, R.G.; Rodrigues, F.H.S.; Tasic, L.; da Silva, S.L.; et al. Structural and Functional Characterization of a γ-Type Phospholipase A2 Inhibitor from *Bothrops jararacussu* Snake Plasma. *Curr. Top. Med. Chem.* **2011**, *11*, 2509–2519. [CrossRef]

27. Serino-Silva, C.; Morais-Zani, K.; Toyama, M.H.; Toyama, D.d.O.; Gaeta, H.H.; Rodrigues, C.F.B.; Aguiar, W.d.S.; Tashima, A.K.; Grego, K.F.; Tanaka-Azevedo, A.M. Purification and characterization of the first γ-phospholipase inhibitor (γPLI) from *Bothrops jararaca* snake serum. *PLoS ONE* **2018**, *13*, e0193105. [CrossRef]

28. Yamakawa, Y.; Omori-Satoh, T. Primary Structure of the Antihemorrhagic Factor in Serum of the Japanese Habu: A Snake Venom Metalloproteinase Inhibitor with a Double-Headed Cystatin Domain. *J. Biochem.* **1992**, *112*, 583–589. [CrossRef]

29. Deshimaru, M.; Tanaka, C.; Tokunaga, A.; Goto, M.; Terada, S. Efficient Purification of an Antihemorrhagic Factor (HSF) in Serum of Japanese Habu (*Trimeresurus flavoviridis*). *Fukuoka Univ. Sci. Rep.* **2003**, *33*, 45–53.

30. Aoki, N.; Sakiyama, A.; Deshimaru, M.; Terada, S. Identification of novel serum proteins in a Japanese viper: Homologs of mammalian PSP94. *Biochem. Biophys. Res. Commun.* **2007**, *359*, 330–334. [CrossRef]

31. Aoki, N.; Matsuo, H.; Deshimaru, M.; Terada, S. Accelerated evolution of small serum proteins (SSPs)-The PSP94 family proteins in a Japanese viper. *Gene* **2008**, *426*, 7–14. [CrossRef]

32. Shioi, N.; Narazaki, M.; Terada, S. Novel function of antihemorrhagic factor HSF as an SSP-binding protein in Habu (*Trimeresurus flavoviridis*) serum. *Fukuoka Univ. Sci. Rep.* **2011**, *41*, 177–184.

33. Aoki, N.; Sakiyama, A.; Kuroki, K.; Maenaka, K.; Kohda, D.; Deshimaru, M.; Terada, S. Serotriflin, a CRISP family protein with binding affinity for small serum protein-2 in snake serum. *Biochim. Biophys. Acta* **2008**, *1784*, 621–628. [CrossRef] [PubMed]

34. Yamazaki, Y.; Koike, H.; Sugiyama, Y.; Motoyoshi, K.; Wada, T.; Hishinuma, S.; Mita, M.; Morita, T. Cloning and characterization of novel snake venom proteins that block smooth muscle contraction. *Eur. J. Biochem.* **2002**, *269*, 2708–2715. [CrossRef] [PubMed]

35. Shioi, N.; Ogawa, E.; Mizukami, Y.; Abe, S.; Hayashi, R.; Terada, S. Small serum protein-1 changes the susceptibility of an apoptosis-inducing metalloproteinase HV1 to a metalloproteinase inhibitor in habu snake (*Trimeresurus flavoviridis*). *J. Biochem.* **2013**, *153*, 121–129. [CrossRef]

36. Shioi, N.; Nishijima, A.; Terada, S. Flavorase, a novel non-haemorrhagic metalloproteinase in *Protobothrops flavoviridis* venom, is a target molecule of small serum protein-3. *J. Biochem.* **2015**, *158*, 37–48. [CrossRef] [PubMed]

37. Chijiwa, T.; So, S.; Hattori, S.; Yoshida, A.; Oda-Ueda, N.; Ohno, M. Suppression of severe lesions, myonecrosis and hemorrhage, caused by *Protobothrops flavoviridis* venom with its serum proteins. *Toxicon* **2013**, *76*, 197–205. [CrossRef]

38. Tanaka, Y.; Shioi, N.; Terada, S.; Deshimaru, M. Structural Organization and Evolution of a Cluster of Small Serum Protein Genes of *Protobothrops Flavoviridis* Snake. *Fukuoka Univ. Sci. Rep.* **2013**, *43*, 59–66.

39. Chijiwa, T.; Inamaru, K.; Takeuchi, A.; Maeda, M.; Yamaguchi, K.; Shibata, H.; Hattori, S.; Oda-Ueda, N.; Ohno, M. Unique structure (construction and configuration) and evolution of the array of small serum protein genes of *Protobothrops flavoviridis* snake. *Biosci. Rep.* **2019**, *39*, BSR20190560. [CrossRef]

40. Shibata, H.; Chijiwa, T.; Oda-Ueda, N.; Nakamura, H.; Yamaguchi, K.; Hattori, S.; Matsubara, K.; Matsuda, Y.; Yamashita, A.; Isomoto, A.; et al. The habu genome reveals accelerated evolution of venom protein genes. *Sci. Rep.* **2018**, *8*, 1–11. [CrossRef]

41. Ito, Y.; Tsuda, R.; Kimura, H. Ultrastructural localizations of beta-microseminoprotein, a prostate-specific antigen, in human prostate and sperm: Comparison with gamma-seminoprotein, another prostate-specific antigen. *J. Lab. Clin. Med.* **1989**, *114*, 272–277.

42. Haas, N.B.; Grabowski, J.M.; Sivitz, A.B.; Burch, J.B. Chicken repeat 1 (CR1) elements, which define an ancient family of vertebrate non-LTR retrotransposons, contain two closely spaced open reading frames. *Gene* **1997**, *197*, 305–309. [CrossRef]

43. Kajikawa, M.; Ohshima, K.; Okada, N. Determination of the entire sequence of turtle CR1: The first open reading frame of the turtle CR1 element encodes a protein with a novel zinc finger motif. *Mol. Biol. Evol.* **1997**, *14*, 1206–1217. [CrossRef] [PubMed]

44. Blin, N.; Stafford, D.W. A general method for isolation of high molecular weight DNA from eukaryotes. *Nucleic Acids Res.* **1976**, *3*, 2023–2308. [CrossRef] [PubMed]

45. Smit, A.; Hubley, R.; Green, P. RepeatMasker-Open-3.0 [WWW Document] RepeatMasker-Open3.0. 1996. Available online: http://repeatmasker.org (accessed on 12 March 2018).

46. Benon, G. Tandem repeats finder: A program to analyze DNA sequences. *Nucleic Acids Res.* **1999**, *27*, 573–580. [CrossRef]

47. Pasquesi, G.I.M.; Adams, R.H.; Card, D.C.; Schield, D.R.; Corbin, A.B.; Perry, B.W.; Reyes-Velasco, J.; Ruggiero, R.P.; Vandewege, M.W.; Shortt, J.A.; et al. Squamate reptiles challenge paradigms of genomic repeat element evolution set by birds and mammals. *Nat. Commun.* **2018**, *9*, 1–11. [CrossRef]

48. Yin, W.; Wang, Z.; Li, Q.; Lian, J.; Zhou, Y.; Lu, B.; Jin, L.; Qiu, P.; Zhang, P.; Zhu, W.; et al. Evolutionary trajectories of snake genes and genomes revealed by comparative analyses of five-pacer viper. *Nat. Commun.* **2016**, *7*, 1–11. [CrossRef]

49. Vonk, F.J.; Casewell, N.R.; Henkel, C.V.; Heimberg, A.M.; Jansen, H.J.; McCleary, R.J.; Kerkkamp, H.M.; Vos, R.A.; Guerreiro, I.; Calvete, J.J.; et al. The king cobra genome reveals dynamic gene evolution and adaptation in the snake venom system. *Proc. Natl. Acad. Sci. USA* **2013**, *110*, 20651–20656. [CrossRef]

50. Castoe, T.A.; de Koning, A.P.; Hall, K.T.; Card, D.C.; Schield, D.R.; Fujita, M.K.; Ruggiero, R.P.; Degner, J.F.; Daza, J.M.; Gu, W.; et al. The Burmese python genome reveals the molecular basis for extreme adaptation in snakes. *Proc. Natl. Acad. Sci. USA* **2013**, *110*, 20645–20650. [CrossRef]

51. Aird, S.D.; Arora, J.; Barua, A.; Qiu, L.; Terada, K.; Mikheyev, A.S. Population Genomic Analysis of a Pitviper Reveals Microevolutionary Forces Underlying Venom Chemistry. *Genome Biol. Evol.* **2017**, *9*, 2640–2649. [CrossRef]

52. Warren, W.C.; Wilson, R.K. *Direct Submission*; Genome Institute; Washington University School of Medicine: St. Louis, MO, USA, 2015.

53. Liu, Y.; Hughes, D.; Dinh, H.; Dugan, S.; Jhangiani, S.; Lee, S.; Okwuonu, G.; Santibanez, J.; Bandaranaike, D.; Chao, H.; et al. *Direct Submission*; Baylor College of Medicine: Houston, TX, USA, 2014.

54. Chijiwa, T.; Nakasone, H.; Irie, S.; Ikeda, N.; Tomoda, K.; Oda-Ueda, N.; Hattori, S.; Ohno, M. Structural characteristics and evolution of the *Protobothrops elegans* pancreatic phospholipase A_2 gene in contrast with those of *Protobothrops* genus venom phospholipase A_2 genes. *Biosci. Biotechnol. Biochem.* **2013**, *77*, 97–102. [CrossRef]

55. Yang, Z. PAML 4: Phylogenetic Analysis by Maximum Likelihood. *Mol. Biol. Evol.* **2007**, *24*, 1586–1591. [CrossRef] [PubMed]

Article

A Novel $\alpha_{IIb}\beta_3$ Antagonist from Snake Venom Prevents Thrombosis without Causing Bleeding

Yu-Ju Kuo [1], Ching-Hu Chung [1], Tzu-Yu Pan [2], Woei-Jer Chuang [3,*] and Tur-Fu Huang [1,2,*]

[1] Department of Medicine, Mackay Medical College, New Taipei City 25245, Taiwan; d01443001@ntu.edu.tw (Y.-J.K.); chchung@mmc.edu.tw (C.-H.C.)
[2] Graduate Institute of Pharmacology, College of Medicine, National Taiwan University, Taipei 10051, Taiwan; r02443003@ntu.edu.tw
[3] Department of Biochemistry, National Cheng Kung University Medical College, Tainan 70101, Taiwan
* Correspondence: wjcnmr@mail.ncku.edu.tw (W.-J.C.); turfu@ntu.edu.tw (T.-F.H.)

Received: 7 November 2019; Accepted: 19 December 2019; Published: 21 December 2019

Abstract: Life-threatening thrombocytopenia and bleeding, common side effects of clinically available $\alpha_{IIb}\beta_3$ antagonists, are associated with the induction of ligand-induced integrin conformational changes and exposure of ligand-induced binding sites (LIBSs). To address this issue, we examined intrinsic mechanisms and structure–activity relationships of purified disintegrins, from *Protobothrops flavoviridis* venom (i.e., *Trimeresurus flavoviridis*), TFV-1 and TFV-3 with distinctly different pro-hemorrhagic tendencies. TFV-1 with a different $\alpha_{IIb}\beta_3$ binding epitope from that of TFV-3 and chimeric 7E3 Fab, i.e., Abciximab, decelerates $\alpha_{IIb}\beta_3$ ligation without causing a conformational change in $\alpha_{IIb}\beta_3$, as determined with the LIBS antibody, AP5, and the mimetic, drug-dependent antibody (DDAb), AP2, an inhibitory monoclonal antibody raised against $\alpha_{IIb}\beta_3$. Consistent with their different binding epitopes, a combination of TFV-1 and AP2 did not induce FcγRIIa-mediated activation of the ITAM–Syk–PLCγ2 pathway and platelet aggregation, in contrast to the clinical antithrombotics, abciximab, eptifibatide, and disintegrin TFV-3. Furthermore, TFV-1 selectively inhibits Gα_{13}-mediated platelet aggregation without affecting talin-driven clot firmness, which is responsible for physiological hemostatic processes. At equally efficacious antithrombotic dosages, TFV-1 caused neither severe thrombocytopenia nor bleeding in FcγRIIa-transgenic mice. Likewise, it did not induce hypocoagulation in human whole blood in the rotational thromboelastometry (ROTEM) assay used in perioperative situations. In contrast, TFV-3 and eptifibatide exhibited all of these hemostatic effects. Thus, the $\alpha_{IIb}\beta_3$ antagonist, TFV-1, efficaciously prevents arterial thrombosis without adversely affecting hemostasis.

Keywords: arterial thrombosis; antiplatelet agent; integrin $\alpha_{IIb}\beta_3$; bleeding side effect; snake venom proteins; disintegrins

Key Contribution: Current $\alpha_{IIb}\beta_3$ antagonists are efficacious anti-thrombotics, but have significant adverse bleeding effects. This study clarifies a pathologically intrinsic mechanism in drug-induced thrombocytopenia and identifies a new candidate that may lead to development of safer anti-thrombotics with significantly reduced bleeding risk.

---

1. Introduction

Integrin $\alpha_{IIb}\beta_3$ is a member of the integrin family of adhesion receptors and remains in an inactive state in normal circulation, preventing undesirable thrombus formation [1]. Upon endothelial injury and agonist-stimulated platelet activation, the $\alpha_{IIb}\beta_3$ complex undergoes a dramatic conformational change and assumes an active intermediate affinity state, which is recognized by the sequence,

HHLGGAKQAGV, at the C terminus of the fibrinogen γ chain and Arg-Gly-Asp (RGD) sequences in the α chains of ligands [2].

Disintegrins are Arg-Gly-Asp (RGD)/Lys-Gly-Asp (KGD)-containing, cysteine-rich proteins in many snake venoms [3]. Based on the RGD/KGD mimetic sequence, disintegrins acts as $\alpha_{IIb}\beta_3$ antagonists and potential antithrombotic agents [4–6]. Currently, three clinically available $\alpha_{IIb}\beta_3$ antagonists, represented by abciximab, eptifibatide, and tirofiban, are used as potent antithrombotics for their rapid action and high efficacy. However, their use is primarily limited to patients undergoing percutaneous coronary intervention because of significant bleeding risk [7,8].

Life-threatening thrombocytopenia and bleeding, common severe side effects of these RDG-mimetic drugs and $\alpha_{IIb}\beta_3$ antagonists [9,10], are associated with induction of drug-induced integrin conformational changes and ligand-induced binding site (LIBSs) exposure of integrin $\alpha_{IIb}\beta_3$ [11,12]. Various reports indicate that thrombocytopenia after exposure to $\alpha_{IIb}\beta_3$ antagonists is linked to drug-dependent antibody binding (DDAb) to platelets [10]. These antibodies are exquisitely drug-dependent and detect conformational changes in $\alpha_{IIb}\beta_3$ elicited by antagonist binding. Drug-induced platelet activation in the circulation is associated with symptoms of disseminated intravascular coagulation due to platelet-activating antibodies [13,14]. These antibodies are believed to form complexes with antagonists/platelet factor on the platelets, on endothelial cells, or in plasma, and the resulting ternary complex binds to platelet FcγRIIa (CD32) in an Fc-dependent fashion [15]. The complex induces FcγRIIa-mediated downstream activation of signaling components, resulting in platelet consumption and leading to thrombocytopenia and bleeding [16,17]. Therefore, at doses where current $\alpha_{IIb}\beta_3$ antagonists exhibit efficacious antithrombotic activity, they also increase bleeding risk [18]. Thus, it is crucial to develop new antagonists that do not cause bleeding.

To address this issue, two disintegrins with different hemorrhagic properties, TFV-1 and TFV-3, were purified from *Protobothrops flavoviridis* venom. Previously, we found that eptifibatide or disintegrin TMV-2, in combination with 10E5 or AP2, the inhibitory monoclonal antibodies (mAbs) raised against $\alpha_{IIb}\beta_3$, specifically triggered platelet activation [19]. Because of the unique properties of these mAbs, we used them as probes to clarify binding to $\alpha_{IIb}\beta_3$ and the pro-hemorrhagic tendencies of disintegrins. TFV1, exhibiting a binding motif distinct from those of TFV-3 and abciximab, decelerated $\alpha_{IIb}\beta_3$ ligation without causing a conformational change of integrin $\alpha_{IIb}\beta_3$. At efficacious antithrombotic doses, TFV-1 prevents thrombus formation without increasing bleeding risk in the FcγRIIa transgenic mouse model, in contrast to TFV-3 and abciximab. Taken together, the pathological mechanism in $\alpha_{IIb}\beta_3$ antagonist-induced thrombocytopenia and the structure–activity relationship of TFV-1 and TFV-3 may help to advance development of new, safer $\alpha_{IIb}\beta_3$ antagonists with minimal effects on normal physiological hemostasis.

2. Results

2.1. Purification and Characterization of TFV1 and TFV3

Venom of *Protobothrops flavoviridis*, the Okinawa habu, was fractionated into three subfractions using an FPLC Superdex 75 column chromatography (Figure 1A). Fraction III (*) exhibiting potent platelet inhibitory activity was collected and further refractionated on a C_{18} reverse-phase HPLC column (Figure 1B). Two antiplatelet fractions eluted at approximately 10 and 20 min, and the dried purified proteins were lyophilized and named TFV-1 and TFV-3, respectively. Purified TFV-1 and TFV-3 behave as single-chain peptides by SDS-PAGE, since their mobilities remained the same in the presence or absence of 2% β-mercaptoethanol (Figure 1C). With MALDI-TOF, their molecular masses were 7310 and 7646 Da, respectively (Figure 1D,E).

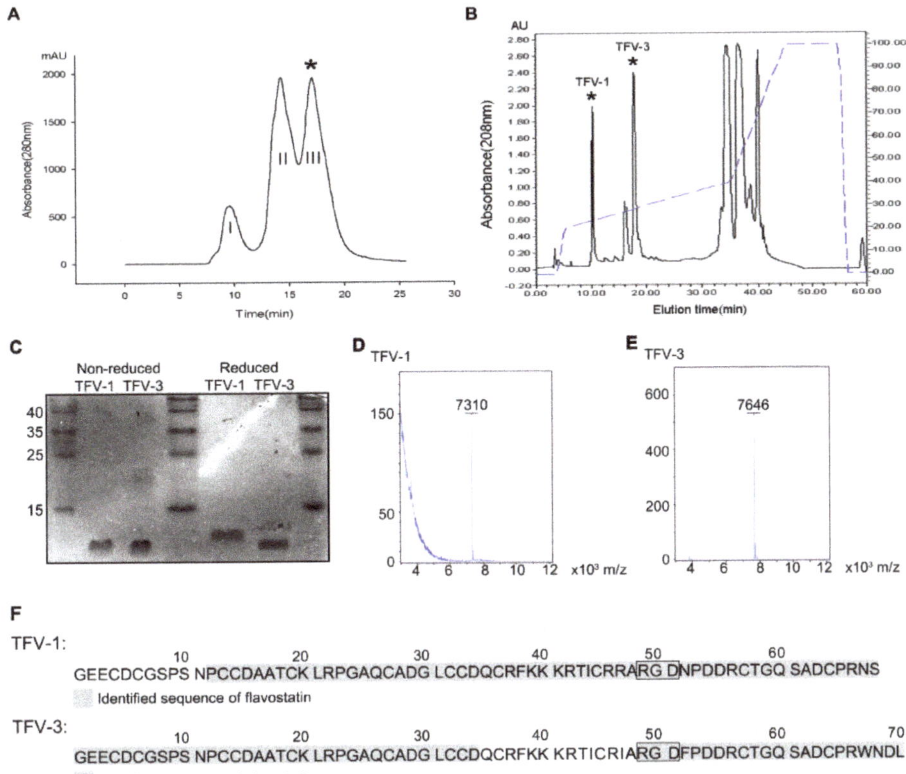

Figure 1. Purification and characterization of TFV1 and TFV3 from *Protobothrops flavoviridis* venom. (**A**) Purification of TFV1 and TFV3. 500 mg of crude venom was applied to a Superdex G-75 column. 0.01 N Ammonium bicarbonate in 0.15 N NaCl was used as the eluent at a flow rate of 0.75 mL/min. Fraction III (*, elution time ~15–17 min) exhibited potent inhibitory activity on collagen (10 µg/mL) and induced platelet aggregation. Therefore, this fraction was collected and further purified by reverse-phase HPLC. (**B**) Purification of TFV-1 and TFV-3 using reverse-phase HPLC. The antiplatelet fraction III (*) from the Superdex 75 column was applied to a C_{18} reverse-phase HPLC column equilibrated in 0.1% TFA at a flow rate of 0.8 mL/min. Chromatography was carried out with a two-solvent gradient (buffer A, 0.1% TFA in distilled water; buffer B, 80% acetonitrile with 0.1% TFA). Fractions were eluted over 60 min with a gradient of 0–80% acetonitrile (dashed line). TFV-1 eluted in approximately 24% acetonitrile at about 10 min. TFV-3 eluted in approximately 28% acetonitrile and an elution time of ~20 min. (**C**) TFV-1 and TFV-3 were run on 15% SDS-PAGE in the presence and absence of 2% β-mercaptoethanol. Gels were stained with Coomassie brilliant blue. Molecular masses of TFV-1 and TFV-3 were estimated at ~7 kDa. (**D,E**) MALDI-TOF mass spectra of TFV-1 and TFV-3 showed peaks with molecular masses of 7310 and 7646 Da, respectively. (**F**) Sequence determination of TFV-1 and TFV-3 using mass spectrometry. TFV-1 and TFV-3 sequences are marked in gray. Based on the MS/MS results, flavostatin was identified in sample TFV-1 (upper), while trimestatin was identified in sample TFV-3 (lower), which possesses a WNDL tetrapeptide at the C-terminus. The Arg-Gly-Asp (RGD) sequence common to both is indicated in a box.

To determine their sequences, high-energy collisional dissociation fragmentation was employed with liquid chromatography (LC)–tandem mass spectrometry (MS/MS). The results derived from top-down (Figure S1) and bottom-up approaches provided information on the sequences near the protein C- and N-termini, respectively. The partial sequence of TFV-1 exhibited 84% sequence identity with the flavostatin [20] (Figure 1F), a disintegrin purified from the venom of *Protobothrops flavoviridis*; and the partial sequence of TFV-3 had 80% sequence identity with trimestatin [21] (Figure 1F), another disintegrin from the same venom.

2.2. Aggregation Studies of the RGD-Containing Disintegrins, TFV-1 and TFV-3

In human platelet-rich plasma (PRP), both TFV-1 (Figure S2A) and TFV-3 (Figure S2B) caused concentration-dependent inhibition of platelet aggregation induced by collagen (10 µg/mL) or adenosine 5′-diphosphate (ADP, 20 µM). In human platelet suspensions (PS), TFV-1 (Figure S2C) and TFV-3 (Figure S2D) blocked the platelet aggregation caused by collagen (10 µg/mL) or thrombin (0.1 U/mL) in a concentration-dependent manner (Table 1). These results indicated that TFV-1 and TFV-3 inhibit platelet aggregation by blocking an essential step of platelet aggregation with a similar IC_{50} irrespective of agonists used.

Table 1. IC_{50} of TFV-1 and TFV-3 in human PRP and human PS. PRP, platelet-rich plasma; PS, washed platelet suspension; ADP, adenosine 5′-diphosphate. The data are presented as means ± SEM ($n = 5$).

Antithrombotic Agents	TFV-1 (µg/mL)		TFV-3 (µg/mL)	
Inducer	PRP	PS	PRP	PS
ADP (20 µM)	0.720 ± 0.025		0.133 ± 0.003	
Collagen (10 µg/mL)	1.090 ± 0.148	0.530 ± 0.03	0.323 ± 0.124	0.217 ± 0.007
Thrombin (0.1 U/mL)		0.467 ± 0.02		0.301 ± 0.014

2.3. TFV-1 Binds to an Epitope of Integrin $\alpha_{IIb}\beta_3$, Different from Those Bound by TFV-3 and Abciximab

To examine the binding motif of TFV-1 and TFV-3 toward the $\alpha_{IIb}\beta_3$ receptor, we used FITC-TFV-1 (Figure 2A) or FITC-TFV-3 (Figure 2B) as probes to examine the effects of a monoclonal antibody (mAb) raised against the $\alpha_{IIb}\beta_3$ domain, on binding of disintegrins to platelets. It is well established that the binding site of mAb 7E3 is near the RGD binding site of the βA-domains [22], whereas mAb 10E5 interacts with the cap subdomain of the α_{IIb} β-propeller [12]. FITC-TFV-1 binding to $\alpha_{IIb}\beta_3$ was competitively inhibited by mAb 10E5, but not mAb 7E3, while FITC-TFV-3 binding to $\alpha_{IIb}\beta_3$ was competitively blocked by mAb 7E3 but not mAb 10E5 (Figure 2A,B).

We previously reported that mAb 7E3 shares the same binding site with RGD-containing $\alpha_{IIb}\beta_3$ antagonists rhodostomin and trigramin [5,23], which cause thrombocytopenia and bleeding owing to their effects on a conformational change of integrin $\alpha_{IIb}\beta_3$. Since the humanized version of a function-blocking mAb, c7E3 (i.e., abciximab) has been reported to bind to the βA domains and subsequently induces exposure of ligand-induced binding sites and consequent thrombocytopenia [9,24], we used abciximab as a positive control (Figure 2C). Interestingly, we found that TFV-3 competitively inhibited mAb 7E3 binding to platelet $\alpha_{IIb}\beta_3$, while TFV-1 did not affect binding of mAb 7E3. Furthermore, TFV-1 competitively reduced binding of mAb 10E5 to platelets, while abciximab and TFV-3 did not (Figure 2D). Together, these data demonstrated that the RGD-bearing disintegrins TFV-1 and TFV-3 inhibit agonist-induced platelet aggregation via $\alpha_{IIb}\beta_3$ receptor blockade. Furthermore, the binding site of TFV-3 is close to the βA domains and similar to that of abciximab, while the binding site of TFV-1 is near the $\alpha_{IIb}\beta_3$-propeller domain.

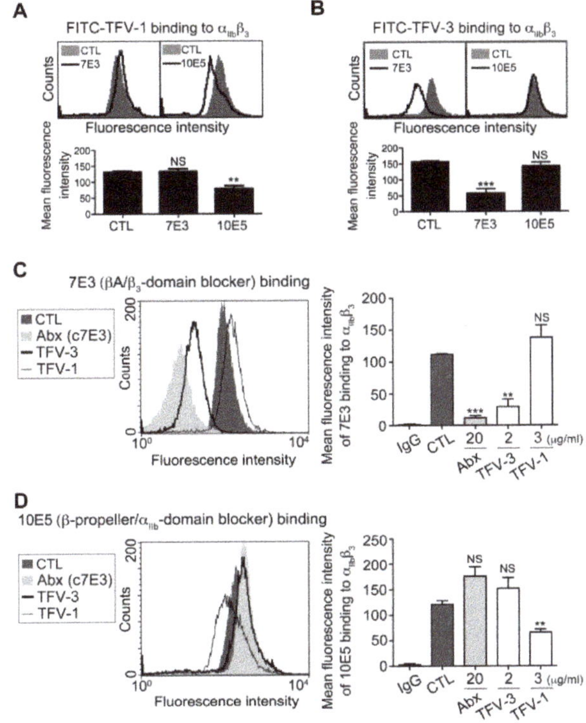

Figure 2. Effect of TFV-1 and TFV-3 on expression of integrin $\alpha_{IIb}\beta_3$, probed by mAb 7E3 and 10E5 on unstimulated platelet. (**A,B**) Human PS was incubated with PBS (CTL), 7E3 ((**A**), 20 µg/mL), or 10E5 ((**B**), 20 µg/mL), probed with 5 µg/mL FITC-TFV-1 or FITC-TFV-3, as indicated, and subsequently analyzed by flow cytometry (mean ± SEM, n ≥ 10, * $p < 0.05$, ** $p < 0.01$, *** $p < 0.001$ compared with control group by Dunnett's test; NS, non-significance). (**C,D**) Human PS was incubated with PBS (CTL), abciximab, TFV-3, or TFV-1, and then probed with 20 µg/mL mAb 7E3 (**C**) and 10E5 (**D**) raised against $\alpha_{IIb}\beta_3$. Finally, the expression of mAb binding to $\alpha_{IIb}\beta_3$ was analyzed by flow cytometry using FITC-conjugated anti-IgG mAb as a secondary antibody (mean ± SEM, error bars, n ≥ 8, ** $p < 0.01$, *** $p < 0.001$ compared with control group by Dunnett's test; n.s, non-significance).

2.4. TFV-1 Binding to Integrin $\alpha_{IIb}\beta_3$ Does Not Prime the Resting $\alpha_{IIb}\beta_3$ to Bind Ligand

Immune thrombocytopenia occurs on first exposure to RGD-mimetic agents. That is, platelet count usually declines sharply within hours of the commencement of drug administration, demonstrating the presence of a naturally occurring antiplatelet antibody in patients who took these kinds of drugs [11]. Previous reports have revealed that upon binding of RGD-mimetic drugs to integrin $\alpha_{IIb}\beta_3$, the ligand-binding capacity increased in the activated integrin and intrinsic antibodies recognized conformational changes in $\alpha_{IIb}\beta_3$ induced by drugs [12]. Thus, we tested the 'priming' effect of these $\alpha_{IIb}\beta_3$ antagonists. In this assay, the ability of agents to induce resting integrin $\alpha_{IIb}\beta_3$ to adopt a high-affinity ligand binding conformation was judged by measuring the intensity of fluorescence-conjugated fibrinogen (Figure 3A) or PAC-1 (Figure 3B) binding to platelets [25], and compared with platelets without agent administration. TFV-3, RGD-mimetic agent eptifibatide, and abciximab increased fibrinogen (Figure 3A) or PAC-1 (Figure 3B) binding to platelets. In contrast, TFV-1 did not prime integrin $\alpha_{IIb}\beta_3$ to high-affinity state, demonstrating that TFV-1 does not induce major conformational changes in the integrin β_3 subunit or prime the receptor.

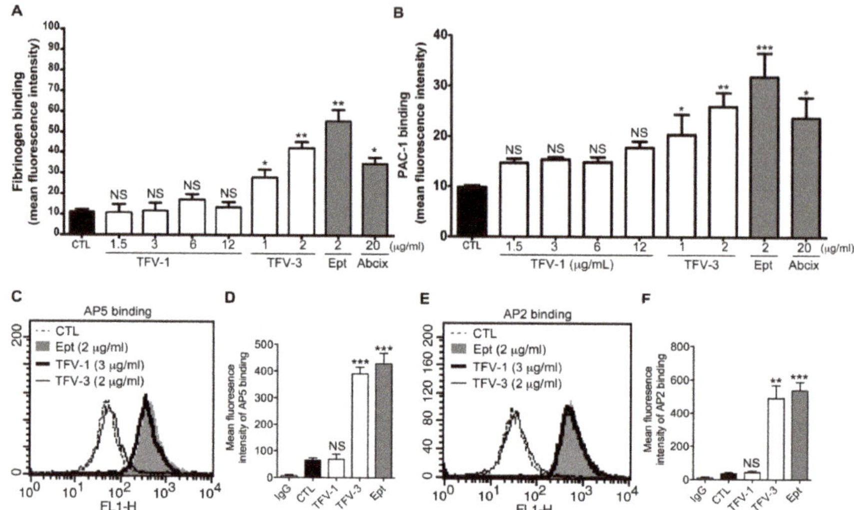

Figure 3. Effect of TFV-1 and TFV-3 on the conformational change of $\alpha_{IIb}\beta_3$, ligand-induced binding sites (LIBS) exposure, and intrinsic monoclonal antibody recruitment. (**A,B**) Eptifibatide (Ept), abciximab, or various concentrations of TFV-1 or TFV-3 were added to human PS and then the platelets were fixed with 1% paraformaldehyde. After quenching the paraformaldehyde with glycine and washing, fluorescent fibrinogen ((**A**), 200 µg/mL) or PAC-1 ((**B**), 10 µg/mL) was added for 30 min at 37 °C. After washing twice, bound fluorescent fibrinogen or PAC-1 was detected by flow cytometry. The data shown are the mean fluorescence intensity of platelets in the presence of each compound. (mean ± SEM, error bars, n ≥ 5, *$p < 0.05$, **$p < 0.01$ and ***$p < 0.001$ compared with the control group by paired Newman–Keuls test; NS, non-significance) (**C–F**) Washed human platelets were incubated with PBS (Ctl), Eptifibatide (Ept, 2 µg/mL), TFV-1 (3 µg/mL) or TFV-3 (2 µg/mL), and then probed with 20 µg/mL mAb AP5 (**C,D**) or AP2 (**E,F**) raised against the ligand-induced binding site and the intrinsic antibody binding site of $\alpha_{IIb}\beta_3$, respectively. Levels of AP5 and AP2 binding to $\alpha_{IIb}\beta_3$ were analyzed by flow cytometry with fluorescein isothiocyanate-conjugated anti-IgG mAb as a secondary antibody. (mean ± SEM, error bars, n ≥ 6, **$p < 0.01$ and *** $p < 0.001$ compared with the control group by paired Newman–Keuls test; NS, non-significance).

2.5. TFV-1 Does Not Cause a Conformational Change of $\alpha_{IIb}\beta_3$ Identified by LIBS Antibody AP5 or mAb AP2

Binding of RGD mimetic to integrin $\alpha_{IIb}\beta_3$ causes conformational changes of $\alpha_{IIb}\beta_3$ and exposure of cryptic epitopes, termed LIBS (ligand-induced binding sites), which are recognized by mAbs PMI-1, AP5, and D3 [26,27]. Therefore, we investigated the effect of TFV1 and TFV3 on the conformational change of $\alpha_{IIb}\beta_3$ using the LIBS antibody, AP5, raised against conformation-dependent epitopes (Figure 3C,D) or mAb AP2 raised against $\alpha_{IIb}\beta_3$ [28] (Figure 3E,F) as a platform. After treatment with TFV-3 and eptifibatide, the level of AP5 or AP2 binding to $\alpha_{IIb}\beta_3$ was increased to 4–4.5× higher than in resting platelets, whereas TFV-1 did not enhance AP5 and AP2 binding to $\alpha_{IIb}\beta_3$, compared with resting platelets.

2.6. Combination of TFV-1 with AP2 Does Not Induce FcγRIIa-Mediated Activation of the Downstream ITAM/Syk/PLCγ2 Pathway and Platelet Aggregation

We previously reported that upon drug-induced LIBS exposure, the mimetic drug-dependent antibody (DDAb) mAb, AP2, which was raised against the epitopes of $\alpha_{IIb}\beta_3$, recruited FcγRIIa [19] and elicited FcγRIIa-mediated platelet aggregation [29] and platelet consumption [11]. The recent report also indicated that eptifibatide-induced thrombocytopenia is associated with an increase in circulating procoagulant platelet-derived microparticles [14]. Here, we found that rhodostomin, eptifibatide

(1 μg/mL; Figure 4A), and TFV-3 (1 μg/mL; right panel of Figure 4B) showed a similar activating effect as when combined with AP2, while the combination of TFV-1 (1.5 μg/mL; left panel of Figure 4B) with AP2 did not induce platelet aggregation. Consistent with this finding, TFV-3 (2 μg/mL)/AP2 induced time-dependent phosphorylation of signal molecules, including focal adhesion kinase (FAK), Src, Syk, PI3K, and PLCγ2, while TFV-1 (2 μg/mL)/AP2 did not (Figure 4D). Pretreatment of FcγRIIa mAb (i.e., CD32), an ITAM-bearing transmembrane receptor responsible for $\alpha_{IIb}\beta_3$ outside-in signaling [16], completely inhibited platelet aggregation caused by TFV-3/AP2 (Figure 4C), suggesting that FcγRIIa is essential for activation.

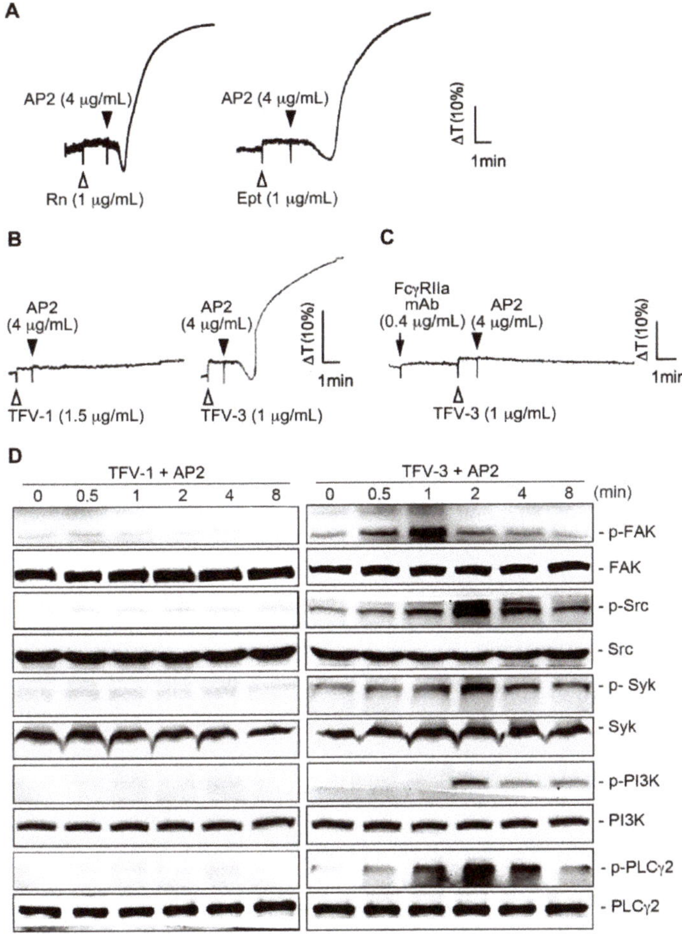

Figure 4. Effect of the combination of TFV-1 or TFV-3 with mAb AP2 on FcγRIIa-mediated downstream signaling and platelet aggregation. (**A,B**) Platelet suspension was incubated with rhodostomin ((**A**), left panel, Rn), eptifibatide ((**A**), right panel, Ept), TFV-1 ((**B**), left panel) or TFV-3 ((**B**), right panel) 1 min before addition of mAb AP2. (**C**) Human platelets were preincubated with anti-FcγRIIa mAb at 37 °C for 3 min, and TFV-3 and mAb AP2 were then added. These tracings in (**A–C**) are the representative ones that were reproducible at least three times. (**D**) Platelets were treated with TFV-1 (1.5 μg/mL)/AP2 (4 μg/mL) or TFV-3 (1 μg/mL)/AP2 (4 μg/mL) and aliquots were removed at indicated time periods. The extent of FAK, Src, Syk, PI3K, PLCγ2 phosphorylation was then determined by immunoblotting. Typical traces shown represent at least three independent experiments with similar results.

2.7. Combination of TFV-1 with AP2 Does Not Increase Intracellular Ca2+ Mobilization and P-selectin Expression in Human Platelets

FcγRIIa-mediated phosphorylation of ITAM tyrosine residues recruits and activates the tyrosine kinase, Syk, leading to intracellular Ca^{2+} mobilization, activation of MAPK/NF-κB pathways, and cell activation [30]. We found that pretreatment of Fura-2-loaded platelets with TFV-3 (1.5 μg/mL)/AP2 (4 μg/mL) increased cytosolic Ca2+ mobilization, while TFV-1 (1.5 μg/mL)/AP2 (4 μg/mL) did not (Figure 5A,B).

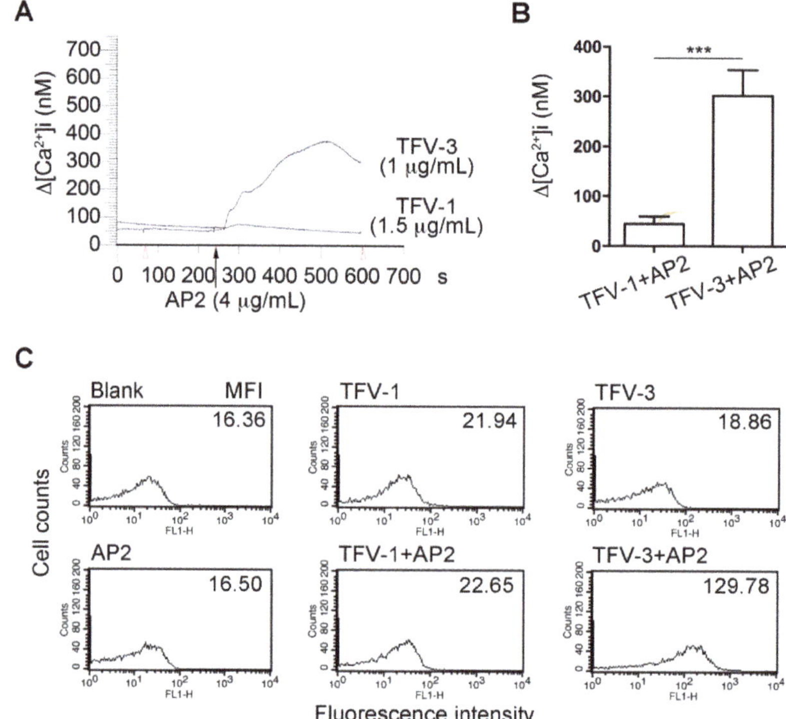

Figure 5. Effects of TFV-1/AP2 and TFV-3/AP2 on intracellular calcium mobilization and P-selectin expression in platelet suspension. (**A,B**) Fura-2 loaded platelets were resuspended in Tyrode's buffer containing 1mM $CaCl_2$, and the mAb, AP2 (4 μg/mL), was added to Fura-2-loaded platelets in the presence of TFV-1 (1.5 μg/mL) or TFV-3 (1 μg/mL), which were preincubated with platelets for 3 min. The level of $[Ca^{2+}]i$ was continuously monitored. (mean ± SEM, error bars, $n = 5$, *** $p < 0.001$ compared with the TFV-3/AP2 group by Dunnett's test). (**C**) Flow cytometric analysis of P-selectin expression of platelet in the presence of TFV-1 (1 μg/mL), TFV-3 (1.5 μg/mL), AP2 (4 μg/mL), or the combination of TFV-1/AP2 or TFV-3/AP2. These experiments were repeated at least three times and only a representative tracing was shown.

Upon platelet activation, P-selectin (CD62), a member of the C-type lectin family, is rapidly translocated from α-granules of platelets to the cell surface [31,32]. Therefore, P-selectin expression on the platelet surface has been widely used to characterize platelet activation [33]. Neither AP2 nor disintegrin alone could cause P-selectin expression (Figure 5C). However, the presence of TFV-3/AP2 (1/4 μg/mL) markedly increased expression of P-selectin, in contrast to TFV-1/AP2.

Shrinking and consolidation of fibrin clots are responsible for hemostatic processes and wound healing from within [34–36]; thus, we assessed the effect of TFV-1 and TFV-3 on thrombin-induced clot retraction in human PRP (Figure 6A,B). Like abciximab, TVF-3 also inhibited thrombin-induced clot retraction while TFV-1 did not, suggesting that TFV-1 has minimal effect on clot firmness during the hemostatic process.

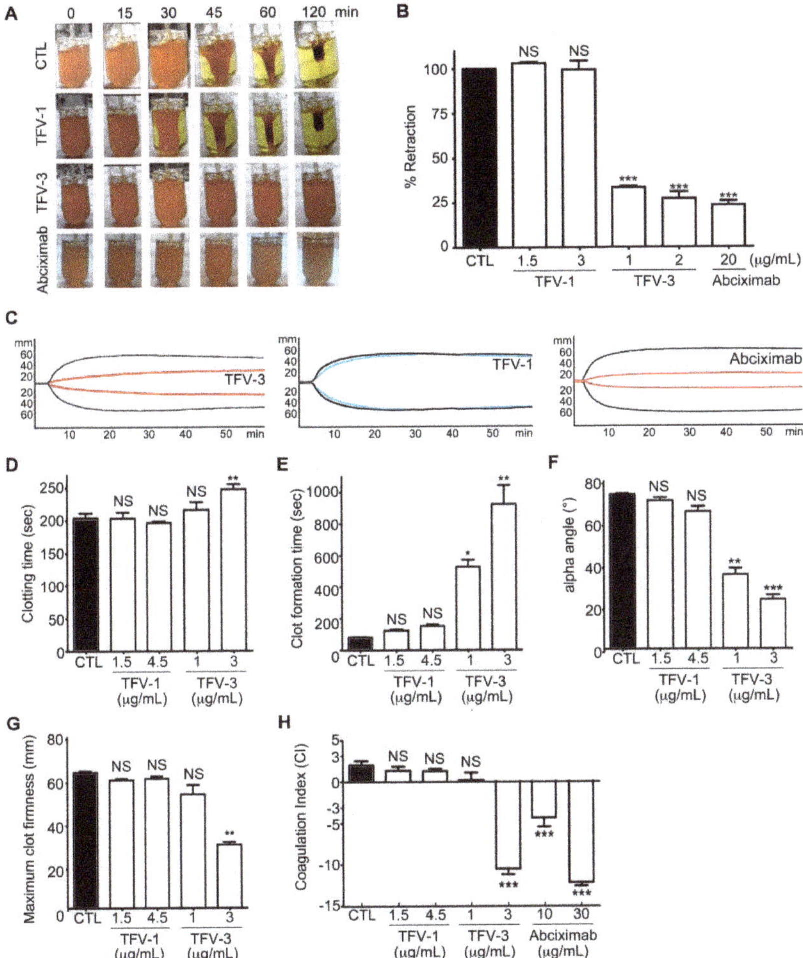

Figure 6. Effects of TFV-1 and TFV-3 on human hemostasis in vitro. (**A,B**) Effect of TFV-1 and TFV-3 on thrombin-induced clot retraction in human PRP. PRP was incubated with various concentrations of TFV-1 or TFV-3 at 37 °C for 3 min before addition of thrombin (4 U/mL). To observe the kinetics of clot retraction, photographs were taken at time 0 and every 15 min until 120 min. Percent retraction was measured by the volume of serum (test)/volume of serum (control). These data are presented as mean ± SEM ($n = 6$). *** $p < 0.001$ compared with the control group. (**C–H**) Physiologic platelet functions of TFV1 and TFV3 were evaluated by rotational thromboelastometry (ROTEM) assays. Human whole blood was incubated with TFV-1, TFV-3, or abciximab, and the ROTEM trace (**C**) of TFV-1 (1.5 μg/mL), TFV-3 (1 μg/mL) and abciximab (10 μg/mL) in human whole blood are shown. CTL (black line): in the absence of agents. Clotting time (**D**), clot formation time (**E**), α-angle (**F**), maximum clot firmness (**G**) and coagulation index (**H**) were evaluated with a ROTEM analyzer following recalcification of the blood, to determine clot and coagulation kinetics. Data were presented as means ± SEM ($n = 7$). * $p < 0.05$, ** $p < 0.01$, *** $p < 0.001$ compared with the control group.

To further mimic bleeding risk during perioperative scenarios, rotational thromboelastometry (ROTEM) [37,38] was used to evaluate platelet function in human whole blood by measuring the viscoelastic properties and kinetic changes of coagulation among these tested $\alpha_{IIb}\beta_3$ antagonists (Figure 6C–H). In the INTEM assay (intrinsically activated test using ellagic acid), four ROTEM variables were taken as a representation of hemostasis, including the clotting time (CT), the clot formation time (CFT), the α angle, and the maximum clot firmness (MCF). These were incorporated into a coagulation index (CI) to provide an overall assessment of coagulation and clot firmness for physiological hemostasis. CT is an indicator of the initial coagulation rate. CFT and α angle are correlated to the kinetics and rate of clot formation. MCF represents maximal platelet–fibrin interaction. In human whole blood, TFV-3 and abciximab dose-dependently repressed kinetics of clot formation (Figure 6C), progressively delayed the clotting time, and reduced clot firmness (Figure 6D–G), as well as reducing the CI (Figure 6H), leading to hypocoagulation. At similar concentrations, TFV-1 showed fewer repressive effects compared with controls and showing that TFV-1 has less influence on primary hemostasis.

2.8. TFV-1 Exhibits Anti-Thrombotic Activity, but Reduced Tendencies to Cause FcγRIIa-Mediated Immune Clearance and Hemorrhage In Vivo

We assessed the in vitro antiplatelet activity of TFV-1 and TFV-3 in mouse PRP. Pretreatment of TFV-1 and TFV-3 inhibited collagen-induced platelet aggregation of PRP in concentration-dependent fashion with IC50s of 1.24 and 0.08 µg/mL (169 and 11 nM), respectively (Figure 7A). Next we administered tested agents to mice intravenously for 10 min and collected blood samples by intracardiac puncture (Figure 7B,C). Compared with the ex vivo inhibitory efficacy of abciximab and eptifibatide (Figure 7D), TFV-1 and TFV-3 revealed favorable inhibitory potency on platelet aggregation of PRP induced by collagen (Figure 7B–D) or ADP (Figure S3). Additionally, TFV-1 and TFV-3 did not affect the initial platelet shape change caused by these inducers. Furthermore, we investigated the in vivo antithrombotic activities of TFV1 and TFV3 in a ferric chloride (FeCl₃)-induced arterial thrombosis model (Figure 7E,F). In response to 5% FeCl₃-induced carotid arterial thrombosis, complete occlusion occurred in the control group within 9 min. Prophylactic intravascular injections of TFV-1 or TFV-3 prevented FeCl₃-induced thrombus formation and delayed occlusion for over 80 min.

At an equally efficacious dose, we compared the in vivo effects of TFV-1 and TFV-3 with those of eptifibatide on platelet counts and bleeding times in a FcγRIIa-transgenic mouse model expressing FcγRIIa on platelets at equivalent levels to humans [39,40]. After intravenous injection of these tested agents into FcγRIIa transgenic mice, eptifibatide and TFV-3 not only caused descending platelet counts over 1–5 h (Figure 7G), but also prolonged tail bleeding time (Figure 7H) in a dose-dependent manner. In contrast, TFV-1 (0.25 or 0.5 mg/kg) did not affect platelet counts and bleeding times compared with the control group, suggesting that TFV-1 shows a lower tendency to cause FcγRIIa-mediated immune clearance of platelets and bleeding.

Since bleeding risks limit the use and doses of RGD-mimetics or $\alpha_{IIb}\beta_3$ antagonists, we used AP2 to assess safety margins of these agents (Table 2). We administered AP2 with a 10× higher minimum effective dose of TFV-1 (15 µg/mL) and found that unlike low doses of TFV-3, abciximab, eptifibatide, or rhodostomin, TFV-1/AP2 does not cause platelet aggregation (Table 2). At 3-times higher minimum effective antiplatelet doses of these agents (Table 2), TFV-1 neither prolonged the clot formation time in the ROTEM assay nor showed a defect in thrombin-induced clot retraction in human PRP, in contrast to TFV-3, abciximab, and eptifibatide. These findings are consistent with the results of bleeding side effects in vivo. Consistent with these observations, TFV-1 has a wider safety margin/index (Table 2) than TFV-3 and these clinical antithrombotic drugs.

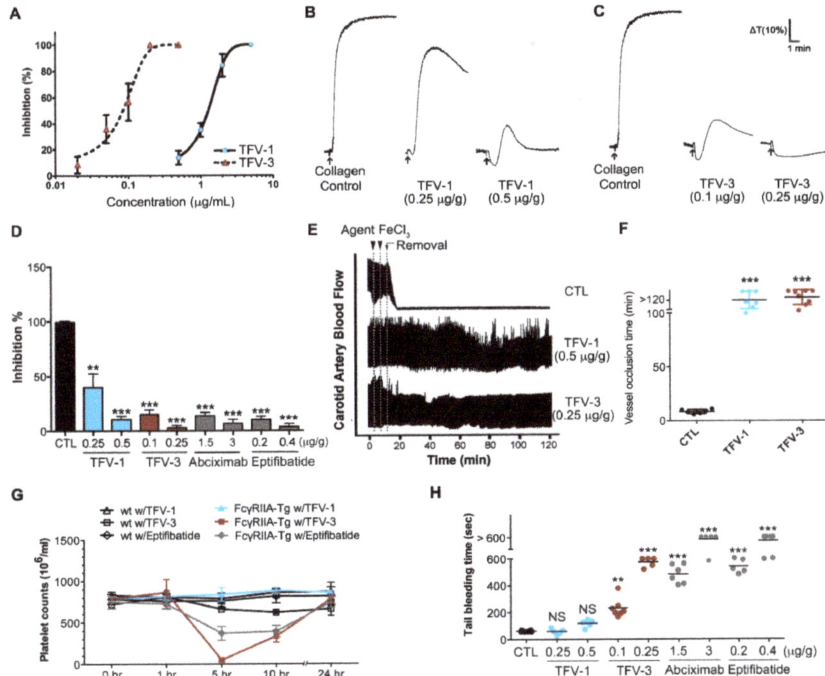

Figure 7. Ex vivo and in vivo effects of TFV-1: Selective inhibition of thrombosis, but not physiological hemostasis. (**A**) The in vitro collagen (10 µg/mL)-induced aggregation response of mouse PRP treated with TFV-1 or TFV-3. Concentration-dependent inhibition curves of TFV-1 and TFV-3 in mouse PRP are presented as mean ± SEM ($n = 3$). (**B–D**) Mice were intravenously treated with saline (control), TFV-1, TFV-3, abciximab, or eptifibatide for 10 min and then blood samples were collected by intracardiac puncture. PRP was obtained by centrifugation at 200× *g* for 4 min and then collagen (10 µg/mL) was added to trigger platelet aggregation. Platelet aggregation was measured by the turbidimetric method (ΔT) using a platelet aggregometer. Typical curves shown represent five independent experiments. Data are presented as a percent aggregation of control and mean ± SEM ($n = 8$). ** $p < 0.01$ and *** $p < 0.001$ compared with the control group (**E,F**) Antithrombotic activity of TFV-1 and TFV-3 in $FeCl_3$-induced thrombus formation in mouse carotid artery. Mice were intravenously administered TFV1 or TFV3. After 5 min, $FeCl_3$ injury was induced by a filter paper saturated with ferric chloride solution (10 %). After removal of the paper, carotid blood flow (mL/min) was monitored continuously until thromboembolism formation or for 60 min. Data are presented as the mean ± SEM ($n \geq 3$). *** $p < 0.001$ as compared with the vehicle control (saline). (**G**) Effect of TFV-1 and TFV-3 on immune clearance of platelets in FcγRIIa-transgenic mice. Wild-type WT and FcγRIIa-transgenic mice were intravenously treated with TFV-1, TFV-3, eptifibatide, or abciximab, and then whole blood (100 µL) was collected by puncture of the retro-orbital sinus with heparinized hematocrit tubes. Platelet counts were obtained at timed intervals after injection of antithrombotic agents. Mean platelet counts (±SEM) over time are shown. (**H**) Effect of TFV-1 and TFV-3 on tail bleeding time of FcγRIIa-transgenic mice. Bleeding times were measured 5 min after the intravenous injection of saline, TFV-1 or TFV-3 at the doses indicated. Bleeding times longer than 10 min were expressed as >10 min. The average bleeding time is indicated as (—). Each type of symbol represents the bleeding time of an individual mouse.

2.9. TFV-1 Does Not Affect the Interaction between Integrin $\alpha_{IIb}\beta_3$ and Its Mediator Talin Responsible for the Hemostatic Process

Upon endothelial injury, platelet cytoplasmic talin drives inside-out signaling of $\alpha_{IIb}\beta_3$, which is responsible for platelet adhesion to vessel walls and for clot formation in the initial hemostatic response [41]. Therefore, recent studies report that selectively targeting $G\alpha_{13}$-mediated outside-in signaling without affecting talin-mediated integrin processes may provide a strategy for preventing thrombosis with a higher safety margin [42,43]. Thus, we further investigated effects of TFV-1 on the mutually exclusive binding of the integrin mediator talin and $G\alpha_{13}$ to the β_3-domain in human PS. TFV-1 selectively inhibited $G\alpha_{13}$ binding to β_3, but did not affect talin binding to the β_3-domain, suggesting that TFV-1 acts as a selective inhibitor of $G\alpha_{13}$ binding to β_3 without perturbing talin-driven hemostatic processes (Figure 8).

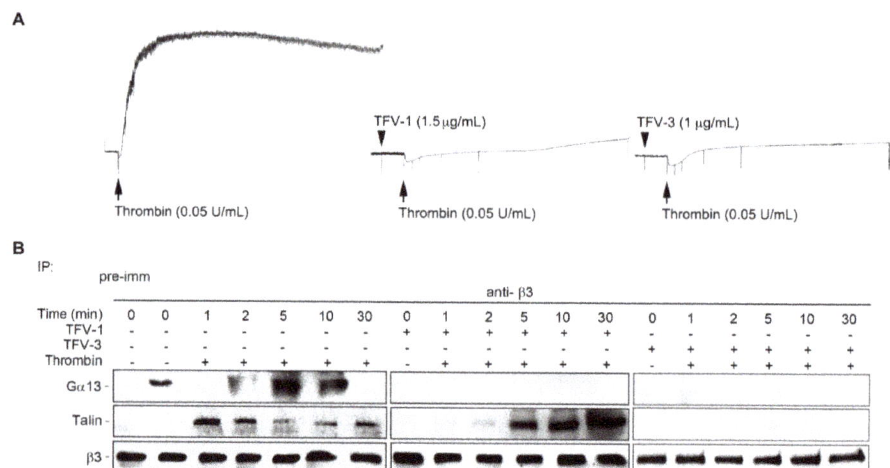

Figure 8. Comparison of the effect of TFV-1 and TFV-3 on the mutually exclusive binding of talin and $G\alpha_{13}$ to cytoplasmic β_3. Human platelets were stimulated with 0.1 U/mL α-thrombinin (Thr) in the presence of PBS, TFV-1 (2 µg/mL), or TFV-3 (1 µg/mL) with stirring (900 rpm) at 37 °C in an aggregometer and then solubilized at various time points. (**A**) Typical turbidity changes in human platelet suspension indicating integrin-dependent platelet aggregation. (**B**) Lysis of platelets by RIPA buffer that enables extraction and measurement of cytoplasmic proteins, including $G\alpha_{13}$ and talin. Lysed platelets were immunoprecipitated with anti-β_3 and immunoblotted for $G\alpha_{13}$, talin, and β_3. Tracings shown here were reproducible at least three times.

Table 2. IC_{50}, safety indices, platelet hemostatic functions, and tail-bleeding time of TFV-1, TFV-3, and clinical anti-thrombotic agents.

Antithrombotic Agents	TFV-1		TFV-3		Abciximab		Eptifibatide		Control
The Dose of Agents (Fold of IC_{50})	2×	6×	2×	6×	2×	6×	2×	6×	-
IC_{50} (µg/mL)	0.74		0.45		5		0.52		-
Safety index	20.27		2.22		2.00		1.92		-
Clot formation time (s)	135 (NS)	172 (NS)	540 (*)	967 (**)	569 (***)	1355 (***)	518 (**)	622 (**)	105
Inhibition of Clot retraction (%)	0	0	77.3	86.1	54.7	92.5	63.2	89.4	-
Tail-bleeding time (s)	66.8 (NS)	102.5 (NS)	233.3 (**)	574.0 (***)	373.4 (***)	580.8 (***)	538.6 (***)	1341.2 (***)	68.7

IC_{50} of collagen (10 µg/mL)-induced platelet aggregation in washed platelet suspension. Safety index is estimated as the lowest concentration of disintegrin to activate platelet (combining with 4 µg/mL AP2)/IC_{50} of disintegrin on collagen-induced platelet aggregation. Inhibition of thrombin-induced clot retraction of human PRP (%) was measured by the volume of serum (agent−control)/volume of serum (control). Tail bleeding times of mice were measured at increasing antithrombotic dosages as compared with the control group (68.7 s, $n = 28$). (These experiments were repeated at least three times and values were presented as means. * $p < 0.05$, ** $p < 0.01$, *** $p < 0.001$ compared with control group by Dunnett's test; NS, non-significance).

3. Discussion

Clinically available $\alpha_{IIb}\beta_3$ antagonists are highly efficacious antithrombotics, but their use is limited to percutaneous coronary intervention due to thrombocytopenia and bleeding [44]. Thrombocytopenia can occur on first exposure to RGD mimetic drugs and the platelet count declines abruptly within hours of commencement of drug administration [45], suggesting the presence of naturally occurring antiplatelet antibody [46]. Eptifibatide- or tirofiban-dependent antibodies (i.e., DDAbs) recognize ligand-induced binding sites (LIBS) in a drug-dependent manner [12,45,47], and subsequently recruit FcγRIIa and trigger a FcγRIIa-mediated platelet aggregation and consumption [19]. It has been reported that these intrinsic antibodies are capable of directly increasing in circulating procoagulant, suggesting that platelet activation might, in some instances, contribute to occasional thrombocytopenia [14]. However, the pathological mechanism by which such RGD mimetic drugs activate platelets remains unclear. Also, a potential antithrombotic agent with a higher safety profile is under active investigation.

We previously reported that the mAb, AP2, raised against conformation-dependent epitopes is used to mimic DDAb for predicting drug-induced platelet activation [29]. Here, we used AP2 as a platform to predict adverse reactions upon administration of RGD mimetic agents. Furthermore, we purified two small RGD-containing disintegrins, TFV-1 and TFV-3, both of which inhibited collagen-, ADP-, and thrombin-induced platelet aggregation in a concentration-dependent manner through $\alpha_{IIb}\beta_3$ receptor blockade. Especially, we found that TFV-1 inhibited ligand binding to $\alpha_{IIb}\beta_3$ by a mechanism different from that of clinically used $\alpha_{IIb}\beta_3$ antagonists and TFV-3. Our results revealed that TFV-1, with a binding motif near the α_{IIb} β-propeller domain, which is different from that of TFV-3 binding near the βA domain, does not prime resting platelets to bind fibrinogen and PAC-1 (Figure 3A,B). While eptifibatide, abciximab, and TFV-3 significantly prime platelets to bind ligands, suggesting that TFV-1 exhibits the minimal intrinsic property of causing the conformational change of $\alpha_{IIb}\beta_3$.

Upon conformational change, LIBS exposure occurs and DDAbs directly bind to conformation-altered $\alpha_{IIb}\beta_3$ [10,12]. Consistent with these findings, the treatment of resting platelets with eptifibatide or TFV-3 caused a significant increase in LIBS antibody, AP5 (Figure 3C,D) and mAb AP2 (Figure 3E,F) binding, while TFV-1 did not, suggesting that TFV-1 causes minor induction of DDAbs with minimal effect on conformational change. In support of data also shown in Figure 4, the combination of TFV-1 with mAb AP2 did not induce FcγRIIa-dependent platelet aggregation or the time-dependent downstream ITAM/Syk/PLCγ2 pathway, which is associated with FcγRIIa-mediated platelet consumption and thrombocytopenia [11].

Since bleeding complications impose a major limitation on the clinical use of current anti-integrin and anti-thrombotic therapies [7], we evaluated bleeding contraindications of TFV-1 and TFV-3 at their efficacious antithrombotic dosage in vivo (Figure 7). Since mouse platelets do not express Fcγ receptors and lack the genetic equivalent of human FcγRIIa [39], in this study, an FcγRIIa transgenic mouse model [40] was used to better understand tendencies of these agents to trigger FcγRIIa-mediated thrombocytopenia. At efficacious antithrombotic doses, TFV-1 neither decreased platelet counts (Figure 7G) nor prolonged tail-bleeding time (Figure 7H) of FcγRIIa transgenic mice, whereas eptifibatide, abciximab, and TFV-3 had this adverse effect in vivo. Consistent with these results, TFV-3 and abciximab not only inhibited thrombin-induced clot retraction of human PRP (Figure 6A,B), but also affected all values associated with platelet function and blood coagulation in human whole blood (Figure 6C–H; ROTEM assay), resulting in hypocoagulation states. By contrast, TFV-1 had minor effects on these coagulation indexes. Overall, our data suggest that TFV-1 is a unique disintegrin, with minimal conformational effects, which could potentially prevent thrombosis without affecting physiological hemostasis.

4. Conclusions

In summary, the lead $\alpha_{IIb}\beta_3$ antagonist, TFV1, with a specific binding motif, neither caused conformational changes in integrin $\alpha_{IIb}\beta_3$ nor triggered FcγRIIa-mediated activation of the downstream

ITAM/Syk/PLCγ2 pathway. Therefore, TFV-1 exhibited anti-thrombotic activity, but lower tendencies to cause FcγRIIa-mediated immune clearance of platelets and hemorrhage. Furthermore, subsequent structure–activity studies of TFV-1 and TFV-3, and their interactions with $\alpha_{IIb}\beta_3$ at a molecular level may provide valuable information for development of a second generation of $\alpha_{IIb}\beta_3$ antagonists.

5. Materials and Methods

5.1. Materials

mAb 7E3 and 10E5 were kindly supplied by Barry S. Coller (Rockefeller University, New York, NY, USA). The mAbs, AP2 and AP5, were kindly donated by Peter J. Newman (Blood Research Institute, Blood Center of Wisconsin, Milwaukee, WI, USA).

Lyophilized crude venom of *Protobothrops flavoviridis* (i.e., *Trimeresurus flavoviridis*) was purchased from Sigma Chemical Co. (St. Louis, MO, USA) and stored at 4 °C. Fractogel EMD TMAE was purchased from Merck (Darmstadt, Germany). Fast protein liquid chromatography (FPLC) Superdex 75HR 10/300 columns were from Amersham Pharmacia (Uppsala, Sweden). A high-performance liquid chromatography C_{18} column was from Waters (Milford, MA, USA). Acetonitrile and trifluoroacetic acid (TFA) were from Merck (Darmstadt, Germany). Molecular-mass standards for electrophoresis were from Bio-Rad (Hercules, CA, USA). Pierce™ Glu-C Protease, MS Grade (90054), and Sulfo-NHS-biotin reagents were from Pierce (Rockford, IL USA). Acrylamide, adenosine diphosphate (ADP), bovine serum albumin (BSA), collagen (bovine tendon type I), Coomassie blue R-250, human thrombin, prostaglandin E1 (PGE1), sodium dodecyl sulfate (SDS), and tris-(hydroxymethyl) aminoethane HCl (Tris-HCl) were obtained from Sigma Chemical Co. (St. Louis, MO, USA). Heparin was from Leo Pharmaceutical Product (Ballerup, Denmark). FITC-conjugated goat anti-mouse IgG was from Santa Cruz Biotechnology, Inc. (Santa Cruz, CA, USA). Alexa 488-conjugated fibrinogen was from Invitrogen (Thermo Fisher Scientific Inc., Waltham, MA, USA). FITC mouse anti-human PAC-1 was from BD Biosciences (Becton, Dickinson and Company, CA, USA).

5.2. Top-Down Analysis

Purified and dried TFV-1 and TFV-3 samples were dissolved in 8 M urea/50 mM ammonium bicarbonate buffer, and disulfide bonds were reduced in 10 mM dithiothreitol at 37 °C for 1 h. LC-MS/MS analyses were performed with a Q Exactive™ mass spectrometer coupled with an UltiMate™ 3000 RSLCnano system (Thermo Scientific) using an Acclaim PepMap RSLC C_{18} column (75 μm I.D. × 15 cm, 2 μm, 100 Å). The following gradient was used: 1 to 50% B for 19.5 min, 50 to 60% B for 3 min, 60 to 80% B for 2 min, and 80% B for 10 min (0.1% FA as mobile phase A and 95% acetonitrile/0.1% FA as mobile phase B). MS scan range was set to be m/z 400–2500 at a resolution of 140,000 (FWHM). The mass spectrum for each protein was deconvoluted with Protein Deconvolution 4.0 software (Thermo Scientific) to obtain molecular weights. Precursor ions, m/z 813.80, 915.52, 1046.03, 1220.20, 1464.03, and 1830.04 for TFV-1 and m/z 955.05, 1091.48, 1273.06, and 1527.47 for TFV-3, were selected for targeted-MS^2 analyses. Higher-energy collisional dissociation (HCD) at normalized collision energy of 32% was used to fragment these ions to generate multiplexed MS/MS spectrum for each protein. The mass spectra were inspected and relevant peaks were assigned manually.

5.3. Endoproteinase Asp-N or Glu-C Digestion and LC-MS/MS Analysis

Samples of TFV-1 and TFV-3 were denatured with 8 M urea and reduced in 10 mM DTT at 37 °C for 1 h. Alkylation was conducted using 50 mM iodoacetamide for 30 min in the dark at room temperature. Reduced TFV-1 and TFV-3 were diluted with 50 mM ammonium bicarbonate (Sigma-Aldrich Chemical Co.) and then digested with endoproteinase Asp-N (Roche) or Glu-C (Pierce) at 37 °C overnight, respectively. The same LC-MS/MS setup as described above was used for analyzing the peptide mixture. The following gradient was applied: 1 to 30% B for 39.5 min, 50 to 60% B for 3 min, 60 to 80% B for 2 min and 80% B for 10 min (0.1% FA as mobile phase A and 95% acetonitrile/0.1% FA as mobile phase B). Full MS scans were performed with a range of *m/z* 300–2000, and the 10 most intense ions from MS scans were subjected to fragmentation for MS/MS spectra. Raw data were processed using Proteome Discoverer 1.4 (Thermo Scientific) and a database search was performed using Mascot 2.4.1 (Matrix Science Inc., Boston, MA, USA) against the SwissProt database. Carbamidomethyl (C) was selected as a fixed modification and deamidation and oxidation were chosen as variable modifications. Mass tolerance windows were set as ±10 ppm and ±0.02 Da for peptide and fragment ions, respectively. Up to two missed cleavages were allowed for Asp-N digestion. All identified MS/MS spectra were manually confirmed to ensure quality.

5.4. Preparation of Human Platelets and Aggregation Assay

Blood was collected from healthy volunteers, who had not taken any medication within 2 weeks. All human participants provided informed consent, and this study was approved by the ethic committees and Joint Institutional Review Board (17-S-032-2), Medical Research Ethics Foundation, Taiwan. Preparation of human platelet-rich plasma (PRP) and platelet suspensions (PS) and platelet aggregation assay were performed as previously described [29].

5.5. Western Blotting and Immunoprecipitation

Washed platelets in an aggregometer cuvette (37 °C, 900 rpm) were treated with the tested agent in the presence of mAb AP2 or agonist thrombin. After incubation, platelets were lysed with lysis buffer (20 mM Tris-HCl buffer, pH7 .5, 150 mM NaCl, 1 mM EDTA, 1 mM EGTA, 1% Triton X-100, 2.5 mM sodium pyrophosphate, 1 mg/mL leupeptin, 1 mM β-glycerolphosphate, 1 mM Na_3VO_4, 1 mM PMSF). Insoluble materials were removed by centrifugation at 14,000 rpm for 15 min at 4 °C. Aliquots of cell lysates were resolved on 10% SDS-PAGE under reducing conditions and electrotransferred to Immobilon-PVDF membrane (Millipore). Western blotting and immunoprecipitation were conducted as described previously [19].

5.6. Binding Study

TFV-1 and TFV-3 were conjugated with FITC [48]. In brief, washed human platelets (3×10^8 platelets/mL) containing 2 μM PGE_1 were labeled with primary anti-$\alpha_{IIb}\beta_3$ integrin mAbs 7E3 or 10E5 at room temperature (RT) for 30 min. Labeled cells were washed and then incubated with secondary FITC-TFV-1 or FITC-TFV-3 at RT for 30 min with a continuous shaking. After incubation, cells were washed, resuspended in PBS, and analyzed immediately by FACS Calibur. In addition, washed human platelets were incubated with TFV-1, TFV-3, or abciximab at RT for 30 min. Following incubation, platelets were washed and then labeled with mAbs 7E3, 10E5 or AP2 at RT for 30 min. Labeled cells were incubated with secondary FITC-conjugated goat anti-mouse IgG at RT for 30 min and then analyzed by FACS Calibur (Becton Dickinson, Franklin Lakes, NJ, USA).

5.7. Priming Assay

The priming assay was performed as described previously, with minor modifications [25,49]. Washed platelets in HEPES-modified Tyrode's buffer were treated with eptifibatide, TFV-1, or TFV-3 for 30 min at room temperature. After incubation, washed human platelets were incubated with Alexa 488-conjugated fibrinogen or FITC-labelled PAC1 antibody for 30 min at 37 °C and analyzed by flow cytometry.

5.8. Definition of Safety Index

We defined a safety index = the lowest concentration of disintegrin to activate platelets in the presence of mAb AP2 (4 μg/mL)/ IC_{50} (μg/mL) of disintegrin for collagen-induced platelet aggregation in platelet suspension.

With eptifibatide, for example, in combination with 4 µg/mL AP2, the lowest concentration of eptifibatide to activate platelets was 1 µg/mL (Figure 4A). The IC_{50} of eptifibatide was about 0.52 µg/mL in collagen-induced platelet aggregation; therefore, we defined the safety index of eptifibatide as 1.92× to represent its safety margin in drug-induced platelet activation. The high safety index means that higher doses of disintegrin could cause platelet activation.

5.9. Clot Retraction

Clot retraction was conducted according to the method of Tucker [34] with minor modification. Briefly, agents were incubated with 200 µL PRP, 5 µL RBC (used to color the clot) at 37 °C for 1 min in glass tubes. After adding thrombin (4 U/mL), a sealed glass pipette was immediately placed in each tube and the kinetics of clot retraction were observed. Images were recorded at time 0 and every 15 min until 120 min. The ratio of clot retraction (%) was calculated by the volume of serum (test)/volume of serum (control).

5.10. Rotational Thromboelastometry (ROTEM)

Human whole blood was collected from healthy donors and the ROTEM assay was performed using the ROTEM® delta system. Briefly, whole blood was treated with agents at 37 °C. Blood mixtures were transferred to a ROTEM plastic cup, and samples were recalcified with star-TEM reagent (0.2 M $CaCl_2$) to initiate the INTEM assay (intrinsically activated test using ellagic acid). The speed at which a sample coagulates depends on the activity of plasma coagulation system, platelet function, fibrinolysis, and other factors. The clotting time (CT, sec), clot formation time (CFT, sec), alpha angle (α, o), and maximum clot firmness (MCF, mm) variables were taken as a representation of hemostasis, and could be incorporated into a coagulation index (CI) as defined by the equation: $CI = -0.6516CT - 0.3772CFT + 0.1224MCF + 0.0759\alpha - 7.7922$. The CI functions as an overall assessment of coagulation, with values less than -3.0 said to represent a hypo-coagulable state and values over $+3.0$ said to be hyper-coagulable.

5.11. Animal Preparation

FcγRIIa-transgenic mice obtained from The Jackson Laboratory [40] (weighing 24–30 g) and the male ICR mice (weighing 20–30 g) were used in all studies. Animals were given continuous access to food and water under controlled temperature (20 ± 1 °C) and humidity (55% ± 5%). Animal experimental protocols were approved by the Laboratory Animal Use Committee of Mackay Medical College (A1060020).

5.12. FeCl₃-Induced Arterial Thrombosis Model

Mice were anesthetized with sodium pentobarbital (50 mg/kg) by intraperitoneal injection, and then an incision was made with a scalpel directly over the right common carotid artery, and a 2-mm section of the carotid artery was exposed. A miniature Doppler flow probe was placed around the artery to monitor blood flow. Thrombus formation was induced by applying filter paper (diameter, 2 mm) saturated with 7.5% $FeCl_3$ solutions on the adventitia of the artery. After 3 min exposure, the filter paper was removed and carotid blood flow was continuously monitored for 80 min after $FeCl_3$ removal.

5.13. Tail-Bleeding Time

Mice were intravenously injected with agents via a lateral caudal vein. After injection for 5 min, a sharp cut 2 mm from the tip of the tail was made. The amputated tail was immediately placed in a tube filled with isotonic saline at 37 °C. Bleeding time was recorded for ≤10 min and the endpoint was the arrest of bleeding [50].

5.14. Statistical Analysis

Results were expressed as mean ± SEM. Statistical analysis was performed by one-way analysis of variance (ANOVA) and the Newman–Keuls multiple comparison test. A p-value less than 0.05 ($p < 0.05$) was considered as a significant difference.

Supplementary Materials: The following are available online at http://www.mdpi.com/2072-6651/12/1/11/s1, Figure S1: The top-down MS/MS spectra for TFV-1 and TFV-3; Figure S2: Effect of TFV-1 and TFV-3 on platelet aggregation of human platelet-rich plasma and washed platelet suspension; Figure S3: Effect of TFV-1 and TFV-3 on ADP-induced platelet aggregation of mouse platelet-rich plasma.

Author Contributions: Research concept and design: T.-F.H. and Y.-J.K.; Acquisition of data: Y.-J.K. and T.-Y.P.; Analysis and interpretation of data: Y.-J.K., T.-Y.P., C.-H.C., W.-J.C., and T.-F.H.; Writing of the manuscript: Y.-J.K.; Study supervision: T.-F.H. had full access to all the data in the review and take responsibility for the integrity of the manuscript. All authors have read and agreed to the published version of the manuscript.

Funding: This study was supported by the Ministry of Science and Technology of Taiwan (grant no. MOST 107-2320-B-715-004-MY3) and the Mackay Medical College (grant no. MMC-1071B25, MMC-1072B33, and MMC-1081B14).

Conflicts of Interest: The authors declare no conflict of interest.

Abbreviations

ADP	Adenosine 5′-diphosphate
BSA	Bovine serum albumin
FACS	Fluorescence activator cell sorter
FITC	Fluorescein isothiocyanate
FPLC	Fast protein liquid chromatography
GP	Glycoprotein
HPLC	High-performance liquid chromatography
IgG	Immunoglobulin G
IV	Intravenous
mAb	Monoclonal antibody
MALDI-TOF	Matrix-assisted laser desorption ionization-time of flight
LIBS	Ligand-induced binding site
PBS	Phosphate-buffered saline
PRP	Platelet-rich plasma
PS	Platelet suspension
RGD	Arg-Gly-Asp
SDS-PAGE	Sodium dodecyl sulfate-polyacrylamide gel electrophoresis
TFV	*Trimeresurus flavoviridis* venom

References

1. Bledzka, K.; Smyth, S.S.; Plow, E.F. Integrin alphaiibbeta3: From discovery to efficacious therapeutic target. *Circ. Res.* **2013**, *112*, 1189–1200. [CrossRef] [PubMed]

2. Pytela, R.; Pierschbacher, M.D.; Ginsberg, M.H.; Plow, E.F.; Ruoslahti, E. Platelet membrane glycoprotein iib/iiia: Member of a family of arg-gly-asp—Specific adhesion receptors. *Science* **1986**, *231*, 1559–1562. [CrossRef] [PubMed]

3. Huang, T.F.; Holt, J.C.; Lukasiewicz, H.; Niewiarowski, S. Trigramin. A low molecular weight peptide inhibiting fibrinogen interaction with platelet receptors expressed on glycoprotein iib-iiia complex. *J. Biol. Chem.* **1987**, *262*, 16157–16163. [PubMed]

4. Huang, T.F.; Holt, J.C.; Kirby, E.P.; Niewiarowski, S. Trigramin—Primary structure and its inhibition of vonwillebrand-factor binding to glycoprotein-iib/iiia complex on human-platelets. *Biochemistry* **1989**, *28*, 661–666. [CrossRef]

5. Huang, T.F.; Sheu, J.R.; Teng, C.M.; Chen, S.W.; Liu, C.S. Triflavin, an antiplatelet arg-gly-asp-containing peptide, is a specific antagonist of platelet membrane glycoprotein iib-iiia complex. *J. Biochem.* **1991**, *109*, 328–334.

6. Kuo, Y.J.; Chung, C.H.; Huang, T.F. From discovery of snake venom disintegrins to a safer therapeutic antithrombotic agent. *Toxins* **2019**, *11*, 372. [CrossRef]
7. Bassand, J.P. Current antithrombotic agents for acute coronary syndromes: Focus on bleeding risk. *Int. J. Cardiol.* **2013**, *163*, 5–18. [CrossRef]
8. Swieringa, F.; Kuijpers, M.J.; Heemskerk, J.W.; van der Meijden, P.E. Targeting platelet receptor function in thrombus formation: The risk of bleeding. *Blood Rev.* **2014**, *28*, 9–21. [CrossRef]
9. Berkowitz, S.D.; Harrington, R.A.; Rund, M.M.; Tcheng, J.E. Acute profound thrombocytopenia after c7e3 fab (abciximab) therapy. *Circulation* **1997**, *95*, 809–813. [CrossRef]
10. Gao, C.; Boylan, B.; Bougie, D.; Gill, J.C.; Birenbaum, J.; Newman, D.K.; Aster, R.H.; Newman, P.J. Eptifibatide-induced thrombocytopenia and thrombosis in humans require fcgammariia and the integrin beta3 cytoplasmic domain. *J. Clin. Investig.* **2009**, *119*, 504–511. [CrossRef]
11. Chong, B.H. Drug-induced thrombocytopenia: Mibs trumps libs. *Blood* **2012**, *119*, 6177–6178. [CrossRef] [PubMed]
12. Bougie, D.W.; Rasmussen, M.; Zhu, J.; Aster, R.H. Antibodies causing thrombocytopenia in patients treated with rgd-mimetic platelet inhibitors recognize ligand-specific conformers of alphaiib/beta3 integrin. *Blood* **2012**, *119*, 6317–6325. [CrossRef] [PubMed]
13. Padmanabhan, A.; Jones, C.G.; Bougie, D.W.; Curtis, B.R.; McFarland, J.G.; Wang, D.; Aster, R.H. Heparin-independent, pf4-dependent binding of hit antibodies to platelets: Implications for hit pathogenesis. *Blood* **2015**, *125*, 155–161. [CrossRef] [PubMed]
14. Morel, O.; Jesel, L.; Chauvin, M.; Freyssinet, J.M.; Toti, F. Eptifibatide-induced thrombocytopenia and circulating procoagulant platelet-derived microparticles in a patient with acute coronary syndrome. *J. Thromb. Haemost. JTH* **2003**, *1*, 2685–2687. [CrossRef] [PubMed]
15. Pedicord, D.L.; Dicker, I.; O'Neil, K.; Breth, L.; Wynn, R.; Hollis, G.F.; Billheimer, J.T.; Stern, A.M.; Seiffert, D. Cd32-dependent platelet activation by a drug-dependent antibody to glycoprotein iib/iiia antagonists. *Thromb. Haemost.* **2003**, *89*, 513–521. [CrossRef] [PubMed]
16. Zhi, H.; Rauova, L.; Hayes, V.; Gao, C.; Boylan, B.; Newman, D.K.; McKenzie, S.E.; Cooley, B.C.; Poncz, M.; Newman, P.J. Cooperative integrin/itam signaling in platelets enhances thrombus formation in vitro and in vivo. *Blood* **2013**, *121*, 1858–1867. [CrossRef] [PubMed]
17. Boylan, B.; Gao, C.; Rathore, V.; Gill, J.C.; Newman, D.K.; Newman, P.J. Identification of fcgammariia as the itam-bearing receptor mediating alphaiibbeta3 outside-in integrin signaling in human platelets. *Blood* **2008**, *112*, 2780–2786. [CrossRef]
18. Armstrong, P.C.; Peter, K. Gpiib/iiia inhibitors: From bench to bedside and back to bench again. *Thromb. Haemost.* **2012**, *107*, 808–814. [CrossRef]
19. Kuo, Y.J.; Chen, Y.R.; Hsu, C.C.; Peng, H.C.; Huang, T.F. An alphaiib beta3 antagonist prevents thrombosis without causing fc receptor gamma-chain iia-mediated thrombocytopenia. *J. Thromb. Haemost. JTH* **2017**, *15*, 2230–2244. [CrossRef]
20. Kawasaki, T.; Sakai, Y.; Taniuchi, Y.; Sato, K.; Maruyama, K.; Shimizu, M.; Kaku, S.; Yano, S.; Inagaki, O.; Tomioka, K.; et al. Biochemical characterization of a new disintegrin, flavostatin, isolated from trimeresurus flavoviridis venom. *Biochimie* **1996**, *78*, 245–252. [CrossRef]
21. Fujii, Y.; Okuda, D.; Fujimoto, Z.; Horii, K.; Morita, T.; Mizuno, H. Crystal structure of trimestatin, a disintegrin containing a cell adhesion recognition motif rgd. *J. Mol. Biol.* **2003**, *332*, 1115–1122. [CrossRef]
22. Byron, A.; Humphries, J.D.; Askari, J.A.; Craig, S.E.; Mould, A.P.; Humphries, M.J. Anti-integrin monoclonal antibodies. *J. Cell Sci.* **2009**, *122*, 4009–4011. [CrossRef] [PubMed]
23. Huang, T.F.; Liu, C.Z.; Ouyang, C.H.; Teng, C.M. Halysin, an antiplatelet arg-gly-asp-containing snake venom peptide, as fibrinogen receptor antagonist. *Biochem. Pharmacol.* **1991**, *42*, 1209–1219. [PubMed]
24. Artoni, A.; Li, J.; Mitchell, B.; Ruan, J.; Takagi, J.; Springer, T.A.; French, D.L.; Coller, B.S. Integrin beta3 regions controlling binding of murine mab 7e3: Implications for the mechanism of integrin alphaiibbeta3 activation. *Proc. Natl. Acad. Sci. USA* **2004**, *101*, 13114–13120. [CrossRef] [PubMed]
25. Negri, A.; Li, J.; Naini, S.; Coller, B.S.; Filizola, M. Structure-based virtual screening of small-molecule antagonists of platelet integrin alphaiibbeta3 that do not prime the receptor to bind ligand. *J. Comput. Aided Mol. Des.* **2012**, *26*, 1005–1015. [CrossRef]
26. Montalescot, G. Platelet biology and implications for antiplatelet therapy in atherothrombotic disease. *Clin. Appl. Thromb. Hemost.* **2011**, *17*, 371–380. [CrossRef]

27. Fullard, J.F. The role of the platelet glycoprotein iib/iiia in thrombosis and haemostasis. *Curr. Pharm. Des.* **2004**, *10*, 1567–1576. [CrossRef]

28. Pidard, D.; Montgomery, R.R.; Bennett, J.S.; Kunicki, T.J. Interaction of ap-2, a monoclonal antibody specific for the human platelet glycoprotein iib-iiia complex, with intact platelets. *J. Biol. Chem.* **1983**, *258*, 12582–12586.

29. Huang, T.F.; Chang, C.H.; Ho, P.L.; Chung, C.H. Fcgammarii mediates platelet aggregation caused by disintegrins and gpiib/iiia monoclonal antibody, ap2. *Exp. Hematol.* **2008**, *36*, 1704–1713. [CrossRef]

30. Ben Mkaddem, S.; Hayem, G.; Jonsson, F.; Rossato, E.; Boedec, E.; Boussetta, T.; El Benna, J.; Launay, P.; Goujon, J.M.; Benhamou, M.; et al. Shifting fcgammariia-itam from activation to inhibitory configuration ameliorates arthritis. *J. Clin. Investig.* **2014**, *124*, 3945–3959. [CrossRef]

31. Frenette, P.S.; Johnson, R.C.; Hynes, R.O.; Wagner, D.D. Platelets roll on stimulated endothelium in vivo: An interaction mediated by endothelial p-selectin. *Proc. Natl. Acad. Sci. USA* **1995**, *92*, 7450–7454. [CrossRef] [PubMed]

32. Merten, M.; Thiagarajan, P. P-selectin expression on platelets determines size and stability of platelet aggregates. *Circulation* **2000**, *102*, 1931–1936. [CrossRef] [PubMed]

33. Leytin, V.; Mody, M.; Semple, J.W.; Garvey, B.; Freedman, J. Quantification of platelet activation status by analyzing p-selectin expression. *Biochem. Biophys. Res. Commun.* **2000**, *273*, 565–570. [CrossRef] [PubMed]

34. Tucker, K.L.; Sage, T.; Gibbins, J.M. Clot retraction. *Methods Mol. Biol.* **2012**, *788*, 101–107.

35. Haling, J.R.; Monkley, S.J.; Critchley, D.R.; Petrich, B.G. Talin-dependent integrin activation is required for fibrin clot retraction by platelets. *Blood* **2011**, *117*, 1719–1722. [CrossRef]

36. Flevaris, P.; Li, Z.; Zhang, G.; Zheng, Y.; Liu, J.; Du, X. Two distinct roles of mitogen-activated protein kinases in platelets and a novel rac1-mapk-dependent integrin outside-in retractile signaling pathway. *Blood* **2009**, *113*, 893–901. [CrossRef]

37. Ganter, M.T.; Hofer, C.K. Coagulation monitoring: Current techniques and clinical use of viscoelastic point-of-care coagulation devices. *Anesth. Analg.* **2008**, *106*, 1366–1375. [CrossRef]

38. Lance, M.D. A general review of major global coagulation assays: Thrombelastography, thrombin generation test and clot waveform analysis. *Thromb. J.* **2015**, *13*, 1. [CrossRef]

39. Worth, R.G.; Chien, C.D.; Chien, P.; Reilly, M.P.; McKenzie, S.E.; Schreiber, A.D. Platelet fcgammariia binds and internalizes igg-containing complexes. *Exp. Hematol.* **2006**, *34*, 1490–1495. [CrossRef]

40. McKenzie, S.E.; Taylor, S.M.; Malladi, P.; Yuhan, H.; Cassel, D.L.; Chien, P.; Schwartz, E.; Schreiber, A.D.; Surrey, S.; Reilly, M.P. The role of the human fc receptor fc gamma riia in the immune clearance of platelets: A transgenic mouse model. *J. Immunol.* **1999**, *162*, 4311–4318.

41. Petrich, B.G.; Marchese, P.; Ruggeri, Z.M.; Spiess, S.; Weichert, R.A.; Ye, F.; Tiedt, R.; Skoda, R.C.; Monkley, S.J.; Critchley, D.R.; et al. Talin is required for integrin-mediated platelet function in hemostasis and thrombosis. *J. Exp. Med.* **2007**, *204*, 3103–3111. [CrossRef]

42. Shen, B.; Zhao, X.; O'Brien, K.A.; Stojanovic-Terpo, A.; Delaney, M.K.; Kim, K.; Cho, J.; Lam, S.C.; Du, X. A directional switch of integrin signalling and a new anti-thrombotic strategy. *Nature* **2013**, *503*, 131–135. [CrossRef]

43. Zhu, J.; Zhu, J.; Negri, A.; Provasi, D.; Filizola, M.; Coller, B.S.; Springer, T.A. Closed headpiece of integrin alphaiibbeta3 and its complex with an alphaiibbeta3-specific antagonist that does not induce opening. *Blood* **2010**, *116*, 5050–5059. [CrossRef] [PubMed]

44. Estevez, B.; Shen, B.; Du, X. Targeting integrin and integrin signaling in treating thrombosis. *Arterioscler. Thromb. Vasc. Biol.* **2015**, *35*, 24–29. [CrossRef]

45. Bougie, D.W.; Wilker, P.R.; Wuitschick, E.D.; Curtis, B.R.; Malik, M.; Levine, S.; Lind, R.N.; Pereira, J.; Aster, R.H. Acute thrombocytopenia after treatment with tirofiban or eptifibatide is associated with antibodies specific for ligand-occupied gpiib/iiia. *Blood* **2002**, *100*, 2071–2076. [CrossRef] [PubMed]

46. Billheimer, J.T.; Dicker, I.B.; Wynn, R.; Bradley, J.D.; Cromley, D.A.; Godonis, H.E.; Grimminger, L.C.; He, B.; Kieras, C.J.; Pedicord, D.L.; et al. Evidence that thrombocytopenia observed in humans treated with orally bioavailable glycoprotein iib/iiia antagonists is immune mediated. *Blood* **2002**, *99*, 3540–3546. [CrossRef] [PubMed]

47. Aster, R.H. Immune thrombocytopenia caused by glycoprotein iib/iiia inhibitors. *Chest* **2005**, *127*, 53S–59S. [CrossRef]

48. Liu, C.Z.; Wang, Y.W.; Shen, M.C.; Huang, T.F. Analysis of human platelet glycoprotein iib-iiia by fluorescein isothiocyanate-conjugated disintegrins with flow-cytometry. *Thromb. Haemost.* **1994**, *72*, 919–925.

49. Du, X.P.; Plow, E.F.; Frelinger, A.L., 3rd; O'Toole, T.E.; Loftus, J.C.; Ginsberg, M.H. Ligands "activate" integrin alpha iib beta 3 (platelet gpiib-iiia). *Cell* **1991**, *65*, 409–416. [CrossRef]
50. Chang, C.H.; Chung, C.H.; Kuo, H.L.; Hsu, C.C.; Huang, T.F. The highly specific platelet glycoprotein (gp) vi agonist trowaglerix impaired collagen-induced platelet aggregation ex vivo through matrix metalloproteinase-dependent gpvi shedding. *J. Thromb. Haemost. JTH* **2008**, *6*, 669–676. [CrossRef]

Article

Identification and Functional Characterization of a Novel Insecticidal Decapeptide from the Myrmicine Ant *Manica rubida*

John Heep [1], Marisa Skaljac [1], Jens Grotmann [1], Tobias Kessel [1], Maximilian Seip [1], Henrike Schmidtberg [2] and Andreas Vilcinskas [1,2,*]

[1] Branch for Bioresources, Fraunhofer Institute for Molecular Biology and Applied Ecology (IME), Winchesterstrasse 2, 35394 Giessen, Germany; John.Heep@ime.fraunhofer.de (J.H.); Marisa.Skaljac@ime.fraunhofer.de (M.S.); Jens.Grotmann@ime.fraunhofer.de (J.G.); Tobias.Kessel@ime.fraunhofer.de (T.K.); Maximilian.Seip@ime.fraunhofer.de (M.S.)

[2] Institute for Insect Biotechnology, Justus Liebig University of Giessen, Heinrich-Buff- Ring 26-32, 35392 Giessen, Germany; Henrike.Schmidtberg@agrar.uni-giessen.de

* Correspondence: Andreas.Vilcinskas@agrar.uni-giessen.de; Tel.: +49-641-99-37600

Received: 26 August 2019; Accepted: 23 September 2019; Published: 25 September 2019

Abstract: Ant venoms contain many small, linear peptides, an untapped source of bioactive peptide toxins. The control of agricultural insect pests currently depends primarily on chemical insecticides, but their intensive use damages the environment and human health, and encourages the emergence of resistant pest populations. This has promoted interest in animal venoms as a source of alternative, environmentally-friendly bio-insecticides. We tested the crude venom of the predatory ant, *Manica rubida*, and observed severe fitness costs in the parthenogenetic pea aphid (*Acyrthosiphon pisum*), a common agricultural pest. Therefore, we explored the *M. rubida* venom peptidome and identified a novel decapeptide U-MYRTX-MANr1 (NH_2-IDPKVLESLV-$CONH_2$) using a combination of Edman degradation and de novo peptide sequencing. Although this myrmicitoxin was inactive against bacteria and fungi, it reduced aphid survival and reproduction. Furthermore, both crude venom and U-MYRTX-MANr1 reversibly paralyzed injected aphids and induced a loss of body fluids. Components of *M. rubida* venom may act on various biological targets including ion channels and hemolymph coagulation proteins, as previously shown for other ant venom toxins. The remarkable insecticidal activity of *M. rubida* venom suggests it may be a promising source of additional bio-insecticide leads.

Keywords: mass spectrometry; LC-MS; Formicidae; Myrmicinae; *Myrmica rubra*; venom gland; bioinsecticide; antimicrobial peptide; aphids; *Acyrthosiphon pisum*

Key Contribution: Venom of the predatory ant, *Manica rubida*, contains a short peptidyl toxin (U-MYRTX-MANr1) that severely affects the fitness of a key agricultural pest (the pea aphid, *Acyrthosiphon pisum*), suggesting that ant venoms contain promising toxins for pest control.

---

1. Introduction

Ants (Hymenoptera: Formicidae) are a taxonomically diverse group of insects with more than 13,500 extant species [1]. They have evolved a venom apparatus derived from the ancestral reproductive system [2], but in contrast to other venomous phyla (e.g., snakes, spiders and scorpions) there have been few studies of ant venoms to functionally characterize their components [3]. This reflects the challenging taxonomy of ants, the limited amount of venom that can be extracted, and the common misconception that ant venoms are simple, consisting mainly of formic acid [4,5].

Recent transcriptomic and proteomic studies have shown that ant venoms are mixtures of many bioactive molecules with an impressive array of biological properties [2]. Although rich in short linear peptides (<5 kDa), ant venoms also contain complex peptides with disulfide bonds, as well as oligomeric proteins with a broad spectrum of activities [2,6–10]. Examples of linear peptides include dinoponeratoxins from the giant neotropical hunting ant, *Dinoponera australis* [11], ponericins from the predatory ant, *Neoponera goeldii* [12,13], and bicarinalin from *Tetramorium bicarinatum* [7]. These are classed as antimicrobial peptides (AMPs) because they show activity against microbial pathogens, but they may also possess paralytic, cytolytic, hemolytic and/or insecticidal properties [6,7,12,14,15], as ponericins do [12].

Insect pests reduce crop yields by feeding and transmitting plant pathogens, but they also act as vectors for many human and livestock pathogens [16,17]. Despite the increasing use of biological control methods, chemical insecticides remain the primary strategy for pest control in both agricultural and public health settings [18,19]. Over the past decade, the number of registered chemical insecticides has fallen due to the emergence of resistant pest populations and the de-registration of key insecticides by regulatory authorities, based on evidence of harm to the environment, to beneficial organisms, and to human health [20–23]. This has increased the demand for new and safe insecticidal lead compounds, novel insecticidal targets, and alternative methods for effective pest control.

Venom-derived peptidyl toxins have been fine-tuned by evolution to improve their selectivity and efficacy in the context of prey capture and defense against predators [2]. Insecticidal toxins derived from predatory arthropods therefore offer a promising source of novel bio-insecticides, and these frequently include peptidyl neurotoxins derived from venoms of arachnids, such as scorpions and spiders [24,25]. Ants also use venom for predation and/or defense against predators and competitors. For example, the linear peptide, poneratoxin, isolated from *Paraponera clavata*, targets the central nervous system of insects by blocking synaptic transmission [2,26,27]. In contrast, peptides with disulfide bonds often target ion channels, causing paralysis or incapacitation. For example, poneritoxin Ae1a from the predatory ant, *Anochetus emarginatus*, and MIITX$_1$-Mg1a from the giant red bull ant, *Myrmecia gulosa*, antagonize ion channels in sheep blowflies (*Lucilia cuprina*) and crickets (*Acheta domesticus*) [4,28].

Aphids (Hemiptera: Aphididae) are devastating insect pests that damage plants by feeding and by transmitting many important plant pathogens [19,29–31]. Aphid control relies predominantly on chemical insecticides such as carbamates, organophosphates, neonicotinoids and pyrethroids [18,29,32]. These insecticides mainly act on insect nerve and muscle targets, whereas other pesticides impair respiration, lipid synthesis, or cuticle formation during insect growth and development [18]. The long-term and frequent use of these insecticides have applied selective pressure to aphid populations, and multiple resistant forms have emerged, making some aphid species very difficult to control [18,33]. The polyphagous aphids, *Myzus persicae* and *Aphis gossypii*, are among the 12 most resistant insect species, with resistance to 75 and 48 chemical insecticidal compounds, respectively [18].

Recent studies have shown that both linear and disulfide-bonded toxins from venoms of spiders, scorpions, and ants are promising molecules for aphid control [34–39]. Some peptidyl toxins from scorpions (e.g., *Urodacus yaschenkoi* and *Urodacus manicatus*) and the ant *Myrmica rubra* act against aphids as stand-alone compounds [34,39]. Other scorpion and spider toxins are more active when fused to carrier proteins (e.g., *Galanthus nivalis* agglutinin) that mediate transport through the insect gut [37] or by engineering entomopathogenic fungi to express insecticidal proteins [36].

Manica rubida (Myrmicinae) is a stinging ant of moderate size (workers ~6–9 mm in length) that is prevalent in middle and southern Europe [40], favoring mountainous regions of 500–2000 m altitude [41]. Colony founding in *M. rubida* is semi-claustral [42], a strategy in which the founding queen actively preys upon other insects to enhance early colony development [42]. This species inflicts a painful sting and uses its potent venom to prevent nest invasions and to subdue prey [40]. Although *M. rubida* venom contains predominantly small, linear peptides (<5 kDa) [43], the peptide sequences are unknown and their activities and that of the crude venom have not been investigated in detail.

We explored the *M. rubida* venom peptidome to identify new peptides, using a combination of liquid chromatography/mass spectrometry (LC-MS) and Edman degradation. We also isolated one of the most abundant linear peptides in the crude venom (U-MYRTX-MANr1) and tested its antimicrobial activity and potency against the pea aphid (*Acyrthosiphon pisum*), a common agricultural pest.

2. Results

2.1. LC-MS Analysis of Crude Venom

Pooled *M. rubida* worker venom was analyzed by LC-MS. We identified 180 molecular features representing a set of 96 different peptides, most of which eluted between 25 and 55 min (23–50% acetonitrile) revealing low to moderate hydrophobicity (Figure 1). The mass range of the peptides was 368–3026 Da, representing sequences of 3–27 amino acids (Figure S1). The sequence length was calculated using an average amino acid molecular mass of 111.1254 Da [44].

Crude venom was reduced and alkylated to identify peptides with disulfide bonds. Among the most abundant peptides in the venom peptidome (relative peak intensity $\geq$ 3%), linear peptides were prevalent, but there were also some peptides with a single, intramolecular disulfide bond (Table 1).

Table 1. The most abundant peptides (relative peak intensity $\geq$ 3%) in the venom of *Manica rubida*. The venom peptidome is dominated by linear peptides, but contains a few peptides with one intramolecular disulfide bond. RT = retention time, Int. = relative peak intensity, MW = molecular weight, S-S = disulfide bond.

No.	RT [min]	Int. [%]	MW_{crude} [Da]	$MW_{red.}$ [Da]	$MW_{alk.}$ [Da]	S-S	Length [a]
1	26.3	8	1314.71	1314.71	1314.72	0	12
2	26.3	7	1352.66	1352.66	1352.66	0	12
3	26.3	4	1336.69	1336.70	1336.70	0	12
4	27.0	77	1434.80	1436.83	1550.87	1	13
5	27.1	8	1208.64	1210.66	1324.70	1	11
6	27.1	3	1472.75	1474.77	1588.81	1	13
7	32.1	15	1132.64	1132.65	1132.65	0	10
8	32.1	28	1093.64	1093.64	1093.64	0	10
9	32.1	17	1148.61	1148.61	1148.61	0	10
10	32.1	7	920.50	920.50	920.50	0	8
11	32.1	4	994.57	994.57	994.57	0	9
12	32.1	100	1110.67	1110.68	1110.66	0	10
13	38.9	4	1136.68	1136.68	1136.68	0	10
14	43.4	62	2739.63	2739.64	2739.64	0	25
15	43.4	12	2569.52	2569.53	2569.53	0	23
16	43.6	17	2978.60	2978.61	2978.61	0	27
17	44.7	5	2850.72	2850.72	2850.73	0	26
18	46.3	3	2987.79	2987.79	2987.79	0	27
19	51.5	3	2823.76	2823.77	2823.77	0	25
20	51.5	19	2840.79	2840.80	2840.80	0	26
21	54.5	92	2174.27	2174.27	2174.28	0	20
22	54.6	3	2218.22	2218.23	2218.23	0	20
23	54.6	45	2196.24	2196.25	2196.25	0	20
24	54.6	18	2212.21	2212.21	2212.21	0	20
25	54.6	4	2042.10	2042.11	2042.11	0	18
26	55.2	7	2840.79	2840.80	2840.80	0	26

[a] The amino acid sequence length was calculated using averagine (111.1254) [44].

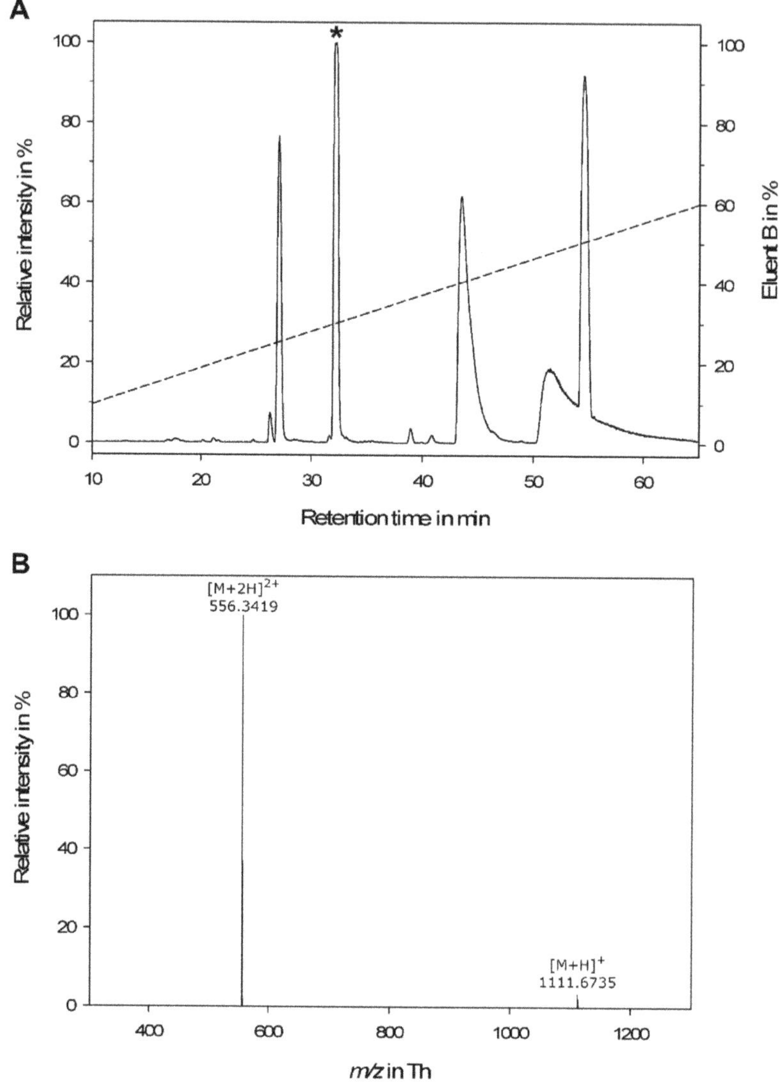

Figure 1. (**A**) Crude venom of the myrmicine ant, *Manica rubida*, was analyzed by LC-MS on a C$_{18}$ reversed-phase column with gradient elution (dashed line) using water and acetonitrile supplemented with 0.1% formic acid as mobile phases. The asterisk indicates the peptide that was isolated for further characterization. (**B**) The mass spectrum of the corresponding decapeptide U-MYRTX-MANr1 acquired on a high-resolution micrOTOF-QII instrument (Bruker Daltonics, Billerica, MA, USA).

2.2. Peptide Sequencing

A fraction from the crude venom containing a peptide with a molecular weight of 1110.67 Da, eluting at 32.1 min, was isolated and enriched. This fraction was selected for further analysis because it contained no co-eluting peptides.

The amino acid sequence of the peptide was determined using a combination of Edman degradation and de novo peptide sequencing. Stepwise Edman degradation revealed the decapeptide sequence

IDPKVLESLV (monoisotopic mass = 1111.65 Da). The difference of ~1 Da between the theoretical and experimental masses reflected C-terminal amidation. The structure of the novel peptide was confirmed by mass determination in the sub-3 ppm region and de novo peptide sequencing. The MS/MS fingerprint of the natural peptide and a synthetic analog achieved very high sequence coverage (Figure S2) and produced an exact match in terms of retention time.

The peptide was named U-MYRTX-MANr1, according to the proposed nomenclature system for ant venom peptides [2]. We added the tag "MAN" for *Manica* to avoid ambiguity, given the existence of several genera with similar names (e.g., *Myrmica*, *Malagidris* and *Mayriella*). The tag "r" is sufficient to denote the specific epithet, *rubida* within the genus *Manica*. The prefix "U" was added to indicate that the biological activity and pharmacological target of the peptide remain unidentified [45].

2.3. Antimicrobial Activity

Synthetic analogs of the decapeptide U-MYRTX-MANr1 (*M. rubida*) and the decapeptide U-MYRTX-MRArub1 from the myrmicine ant, *M. rubra* [39], were tested in antimicrobial assays against a range of bacteria and fungi. Neither decapeptide showed any activity against any strains we tested at concentrations up to 512 μg·mL^{-1}.

2.4. Effects of Crude Venom and Peptidyl Toxins on Aphid Survival and Reproduction

Insecticidal activities of crude *M. rubida* venom and the peptide U-MYRTX-MANr1 were determined by tracking *A. pisum* survival (Figure 2) and offspring production daily until 10 days post-injection (Table S1). In a previous study, the *M. rubra* peptide U-MYRTX-MYRrub1 was active against *A. pisum* [39]. Therefore, crude *M. rubra* venom and U-MYRTX-MYRrub1 were used as positive controls to determine the relative activity of *M. rubida* and *M. rubra* venom components on aphid fitness (Figure S3, Table S1).

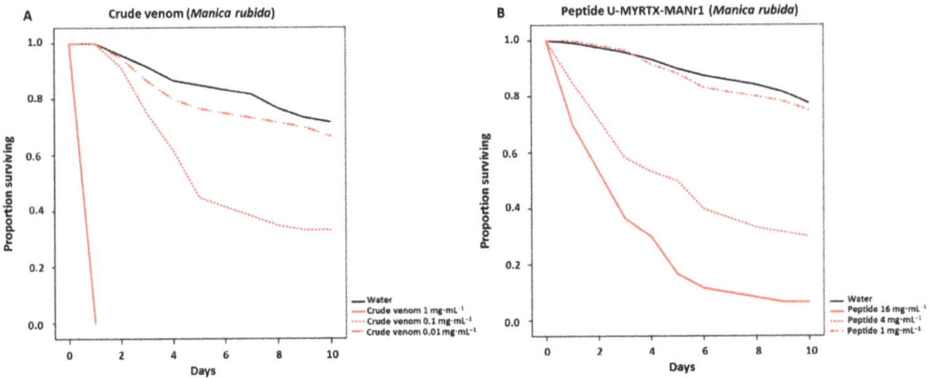

Figure 2. Low pea aphid survival reveals the strong insecticidal activity of crude *Manica rubida* venom (100% mortality) and peptide U-MYRTX-MANr1 at high (~93% mortality) and medium concentrations (~30% mortality) after 10 days. Survival (60 aphids in three biological replicates of 20 individuals per treatment) was monitored for 10 days following the injection of crude venom (**A**) or the peptide (**B**) into pea aphids, *Acyrthosiphon pisum*. Survival data were analyzed using Kaplan–Meier statistics and comparisons between the treatment and control were based on log-rank tests. Statistical data are shown in Table 2 and Table S2.

In most treatments, aphid survival was reduced in a concentration-dependent manner by the crude venom and peptidyl toxins from both species (Table 2). *M. rubida* venom was more potent than *M. rubra* venom (Table 2, Tables S2 and S3; Figure 2 and Figure S3). The highest concentration (1 mg·mL^{-1}) of *M. rubida* venom caused 100% mortality within 24 h, whereas ~20% of aphids were still alive 10 days after exposure to the same concentration of *M. rubra* venom (Figure 2A and Figure S3A). Furthermore, *M. rubida* and *M. rubra* venoms at a concentration of 0.1 mg·mL^{-1} reduced aphid survival to ~72% and ~30%, respectively (Table 2). All concentrations of the bovine serum albumin (BSA) control (1, 4 and 16 mg·mL^{-1}) and low concentrations of crude venom (0.01 mg·mL^{-1}) or peptide (1 mg·mL^{-1}) from either species had no significant effect on aphid survival (Table 2 and Table S4; Figure 2, Figures S3 and S4).

To determine the effect of venom/peptide injection on aphid reproduction, the number of offspring was scored for aphids that survived the treatments. Due to the strong effect of high concentrations of the crude venom/peptide on aphid survival, we only counted the number of offspring for aphids exposed to medium and low concentrations (Table S1). Medium concentrations of crude venom (0.1 mg·mL^{-1}) and peptidyl toxin (4 mg·mL^{-1}) from *M. rubida* both reduced the number of offspring by ~26%, whereas other treatments, including all treatments with the *M. rubra* venom/peptide, did not reduce aphid reproduction (Table S1).

Table 2. Crude ant venoms and associated peptidyl toxins strongly reduced the survival of injected aphids (*Acyrthosiphon pisum*) (60 aphids in three biological replicates of 20 individuals per treatment).

Treatment		Concentration (mg·mL^{-1})	% Survival	Significance [a]
Bovine serum albumin (BSA) [b]		16	78.3	ns
		4	84.7	ns
		1	91.7	$p < 0.05$
Crude venom	*Manica rubida*	1	0.0	$p < 0.0001$
		0.1	33.3	$p < 0.0001$
		0.01	66.7	ns
	Myrmica rubra	1	20.0	$p < 0.0001$
		0.1	71.7	ns
		0.01	66.7	ns
Peptides	U-MYRTX-MANr1 (*Manica rubida*)	16	6.7	$p < 0.0001$
		4	30.0	$p < 0.0001$
		1	75.0	ns
	U-MYRTX-MYRrub1 (*Myrmica rubra*)	16	41.7	$p < 0.0001$
		4	60.0	$p < 0.01$
		1	80.0	ns

[a] Compared to water (survival ~78%, 10 days post-injection); [b] BSA was used as peptide/protein control; ns = not significant.

2.5. Effects of Crude Venom and Peptide Toxins on Aphid Behavior and Injection Wound Healing

The strong toxicity of *M. rubida* crude venom and toxin U-MYRTX-MANr1 on aphid survival and reproduction persuaded us to investigate the effects of crude venom or toxin on the behavior of *A. pisum* immediately and 1 h after injection (Table 3). Interestingly, aphids placed on their backs immediately after injection with crude *M. rubida* venom displayed an inability or delayed ability (>3 min) to right themselves. Similar, but much less severe effects, were observed following the injection of U-MYRTX-MANr1 (Table 3). Aphids in the control group (water injection) showed normal movement and rapidly righted themselves when placed on their backs. Observations recorded for crude venom, peptide, and water immediately after injection remained unchanged after 1 h, but a massive loss of body fluids from the injection site occurred in aphids treated with *M. rubida* crude venom and U-MYRTX-MANr1 (Figure 3, B1–C3). This suggests that both the venom and the peptide impair wound healing and hemolymph coagulation.

Table 3. Aphid behavior and fitness after the injection of water, U-MYRTX-MANr1 (16 mg·mL^{-1}) or *Manica rubida* crude venom (1 mg·mL^{-1}) (30 individuals injected per treatment). Observations were recorded immediately after injection and 1 h later.

Treatment	Immediately Post-Injection	1 h Post-Injection
Control (water)	Immediate righting; responsiveness < 10 s; fast and active movement	No visible impact on vitality or fitness
U-MYRTX-MANr1 16 mg·mL^{-1}	Delayed righting Response; ~1 min; Disoriented and slow movement (mild paralysis)	Moderate reduction of vitality and fitness; continuous mild paralysis
Crude venom 1 mg·mL^{-1}	Delayed or no righting (>3 min); very limited movement (strong paralysis)	Extreme reduction of vitality and fitness (~40% with no physiological reactions); continuous strong paralysis; necrosis and extensive loss of hemolymph

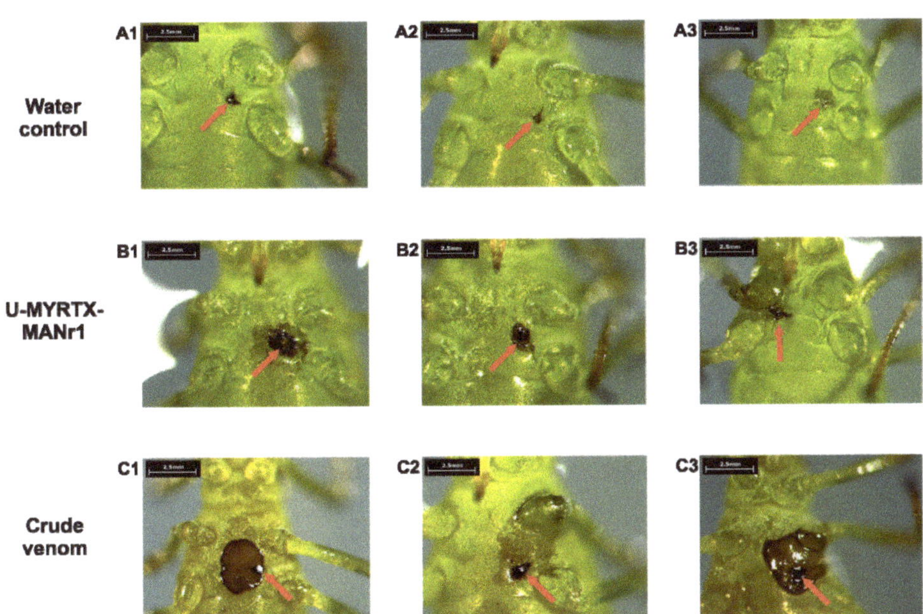

Figure 3. Injection wound healing in *Acyrthosiphon pisum* 1 h post-injection with water (A1–A3), U-MYRTX-MANr1 (16 mg·mL^{-1}) (B1–B3), or crude *Manica rubida* venom (1 mg·mL^{-1}) (C1–C3) (30 individuals injected per treatment). Hemolymph coagulation was impaired in aphids injected with U-MYRTX-MANr1, but the effects were more severe with crude venom, resulting in a massive loss of hemolymph. The injection site and melanization of exposed hemolymph are indicated with red arrows.

To understand the effectiveness of *M. rubida* venom components in more detail, injections using crude venom and toxin U-MYRTX-MYRrub1 from *M. rubra* were repeated. In contrast to injections of *M. rubida* venom, paralysis and fluid loss were significantly less severe for both crude venom and the *M. rubra* toxin (Table S5). Nevertheless, we observed differences between control aphids and those injected with crude *M. rubra* venom or peptidyl toxin (Figure S5, Table S5). Aphids injected with crude *M. rubra* venom experienced mild paralysis and required 10–30 s to right themselves. Paralytic effects were still evident 1 h post-injection, and were associated with mildly impaired wound healing/coagulation (Figure S5, C1–C3). In agreement with the *M. rubida* experiments, the effects on movement and wound healing/coagulation were weaker for toxin U-MYRTX-MYRrub1 than for crude venom. Aphids injected with the peptide were able to right themselves immediately and showed only mild paralysis (Table S5). Furthermore, the mild paralysis was maintained 1 h post-injection and wound healing/coagulation was only slightly impaired (Figure S5, B1–B3).

To further investigate the effectiveness of *M. rubida* venom components against aphids, histological investigations were also conducted. Histological sections of aphids injected with crude *M. rubida* venom showed the significant retardation of embryonic development, degradation of fat bodies, and dissociation of bacteriocytes (Figure S6). In addition, impaired wound healing following crude *M. rubida* venom injections was confirmed, as observed in other injection assays (Figure 3, C1–C3; Figure S6). Histological sections of aphids injected with water and peptide U-MYRTX-MANr1 did not show obvious effects (Figures S6–S8). Aphids injected with *M. rubra* venom components were not subjected to histological investigation.

2.6. Effects of Peptide U-MYRTX-MANr1 on Aphid Susceptibility to Chemical Insecticides

The oral activity of toxin U-MYRTX-MANr1 was determined by tracking aphid survival during 3 days of feeding (Figure S9). Aphid survival was not significantly affected following oral delivery of the peptide at a concentration of 500 $\mu g \cdot mL^{-1}$, as used in previous studies [34,39]. Aphids that survived 3 days on the control or peptide-supplemented diet were exposed to insecticide-treated bean plant leaf discs to determine the effect of U-MYRTX-MANr1 on aphid tolerance to three frequently used chemical insecticides: imidacloprid, spirotetramat, and methomyl. Each insecticide was tested at a sub-lethal concentration, determined in a previous study [46] (Table S6). Aphids previously exposed to peptide U-MYRTX-MANr1 were not significantly more sensitive to any of the chemical insecticides, compared to control aphids (Table S6). The peptide U-MYRTX-MYRrub1 from *M. rubra* was not included in this experiment because its effect on the susceptibility of *A. pisum* to chemical insecticides was previously reported [39].

3. Discussion

Manica rubida is not very aggressive, but workers respond promptly to disturbances and defend their colonies with powerful stings, which can be painful to humans. Although very little is known about the natural feeding preferences of *M. rubida*, N-isotope data have shown that this species is zoophagous [47]. This indicates that these ants are predominately scavengers and predators, using their potent venom to subdue prey. Our LC-MS analysis of *M. rubida* venom revealed significant heterogeneity of the venom peptidome, with numerous peptides varying in molecular weight, structure, and physocochemical properties. We also determined the first sequence of a peptidyl toxin isolated from *M. rubida* venom and characterized its biological activity.

Ant venoms are widely assumed to be simple mixtures, consisting mainly of formic acid, but this is only true for species of the subfamily *Formicinae* [4]. In contrast, species such as *M. gulosa* [4] and *Odontomanchus monticola* [48] produce venoms comprising complex mixtures of peptides with diverse biological activities, which is also true of other subfamilies (e.g., *Myrmeciinae*, *Ponerinae* and *Myrmicinae*) [7,49–51].

Touchard et al. [43] investigated venom characteristics of *M. rubida* and 81 other stinging ant species using MALDI-TOF-MS. Our LC-MS analysis of *M. rubida* crude venom is broadly consistent with these earlier results, revealing that its venom is a heterogeneous mixture of 96 distinct peptides. Interestingly, peptide lengths and molecular weights fall within a remarkably narrow window (Figure S1), ranging from 368 to 3026 Da, with the vast majority (82 peptides, 85.4%) between 1000 and 3000 Da and 9–27 amino acids (Figure S1). The average peptide in crude *M. rubida* venom has a molecular weight of 1855 (mean) or 1965 Da (median) and comprises 17 or 18 amino acids (Figure S1). Reduction and alkylation revealed that most peptides are linear and cysteine-free (Table 1). Surprisingly, the three most abundant peptides, all of which have a single intramolecular disulfide bond, co-eluted, suggesting that they have similar amino acid compositions. These findings support our earlier observations [39] and those of others [43] concerning the venom of the ruby ant, *M. rubra*, which consists predominantly of linear peptides with molecular weights of 1000–3500 Da. Indeed, small linear peptides in the 500–4000 Da range predominate in the venoms of many ants [43]. Furthermore, peptides from cone snails (*Conus* spp.) are also small (12–30 amino acids), but in contrast to those found in ant venoms, peptides from cone snails tend to be extremely rich in disulfide bonds [52]. Peptidyl toxins from snakes or scorpions are typically longer than those of ants, varying between 40 and 100 amino acids [53]. Even so, ant peptides could nevertheless possess interesting and useful biological activities.

Although the landscape of *M. rubida* venom peptides has been investigated, no studies have attempted to characterize individual peptide toxins. We used a combination of Edman degradation, accurate mass measurement, and de novo peptide sequencing to reveal the decapeptide sequence NH_2-IDPKVLESLV-$CONH_2$, and named this peptide, U-MYRTX-MANr1, according to current nomenclature rules for ant venom peptidyl toxins [2]. This peptide features mostly (60%) hydrophobic amino acids, but also contains both acidic (aspartic acid, glutamic acid) and basic (lysine) residues, resulting in amphipathic properties. Characteristics most often associated with anti-microbial peptides (AMPs) are amphipathicity, hydrophobicity and the presence of (multiple) positive charges [54], enabling them to pass through the bacterial outer membrane, initiated by attachment to anionic lipopolysaccharides [55]. U-MYRTX-MANr1 has a net neutral charge, which may explain its lack of antimicrobial activity in vitro. This result was supported by in silico predictions of negligible antimicrobial activity when the sequence was screened against the Database for Antimicrobial Activity and Structure of Peptides [56]. The absence of antimicrobial activity was also reported for the structurally similar peptide toxins U-MYRTX-MYRrub1 from the ruby ant *M. rubra* (sequence identity 80%, sequence similarity 100%) [39] and temporin-H, a so-called AMP isolated from the defensive skin secretions of the frog *Rana temporaria* (sequence identity 40%, sequence similarity 80%) [57].

A full understanding of the antimicrobial potential of U-MYRTX-MANr1 will require further studies to define its activity on bacterial membranes, as well as tests against a wider panel of microbes. A multiple alignment of U-MYRTX-MANr1 with other AMPs and peptide toxins, especially ant venom peptides, is shown in Figure S10. U-MYRTX-MANr1 and U-MYRTX-MYRrub1 are from different species of the Myrmicinae, but share a highly conserved domain, with ^{1}Ile ^{2}Asp ^{3}Pro ^{4}Lys ^{6}Leu ^{7}Glu ^{8}Ser ^{9}Leu as a common motif. They differ only at positions 5 (Val→Leu) and 10 (Val→Ala), and both substitutions are conservative (hydrophobic). Interestingly, this motif seems to be unique among peptide domains discovered thus far in ant venoms, with little structural conservation compared to other ant venom peptides such as pilosulins [51] and ponericins [12]. However, the recently discovered myrmicitoxin U_{12}-MYRTX-Tb1a from the myrmicine ant, *Tetramorium bicarinatum*, is an exception, in that it also shares the ^{3}Pro ^{6}Leu ^{8}Ser ^{9}Leu motif with U-MYRTX-MANr1 and U-MYRTX-MYRrub1. Such sequence alignments must be approached with some caution because only a limited number of peptide sequences from ant venom are known thus far. Further research is needed to characterize additional ant venom peptides to understand their structure–function relationships and modes of action.

Peptide toxins isolated from scorpion and spider venoms have exhibited diverse biological activities against aphids [34–38]. We tested *M. rubida* crude venom and the peptide U-MYRTX-MANr1 against *A. pisum*, expanding the portfolio of peptides that may be suitable for pest management applications. Both the crude venom and the peptide U-MYRTX-MANr1 significantly reduced aphid survival and reproduction (Table 2 and Table S1; Figure 2). Interestingly, the closely related peptide U-MYRTX-MYRrub1 from *M. rubra* [39] had lower potency against aphids than U-MYRTX-MANr1 (Table 2 and Table S1, Figure S3). These peptides differ at only two residues (Figure S10), but such small differences have a profound effect on bioactivity [58]. For example, the AMPs UyCT1 and D5 derived from the scorpion *Urodacus yaschenkoi* differ at only two residues, but likewise show vastly different activities against *A. pisum* [34].

Our previous study showed that U-MYRTX-MYRrub1 is orally active against *A. pisum* and that sublethal concentrations of this peptide increased the susceptibility of aphids to chemical insecticides [39]. Interestingly, although injected U-MYRTX-MANr1 was more potent than U-MYRTX-MYRrub1, it did not show any direct oral effect and it did not influence the insecticide susceptibility of *A. pisum* (Figure S6, Table S6). Orally delivered scorpion AMPs isolated from *U. yaschenkoi* and *U. manicatus* show remarkable activity against aphids, but this is probably associated with their net positive charge and non-selectivity for cell membranes [58,59]. The natural properties of U-MYRTX-MANr1 suggest that its insecticidal activity (Figure 2) does not involve membrane disruption, but has a different molecular target. Although considered rare in ant venoms, neurotoxic peptides are often found in animal venoms and their role is to achieve rapid prey immobilization [2]. These peptides typically block ion channels, with varying degrees of specificity and efficacy [2]. For example, the *P. clavata* peptide, poneratoxin, modulates voltage-gated sodium channels in both vertebrates and invertebrates, whereas the dimeric ectatomin Et-1 peptide from the neotropical ant, *Ectatomma tuberculatum*, blocks voltage-gated calcium channels and acts as a pore-forming peptide in eukaryotic cells [2,50]. Our behavioral assay revealed that crude *M. rubida* venom or the U-MYRTX-MANr1 peptide triggered severe paralysis after injection (Table 3 and Figure 3), with a much stronger effect than *M. rubra* venom components (Table S5 and Figure S5). This suggests that U-MYRTX-MANr1 may be a neurotoxin, although further studies are needed to confirm its activity against ion channels.

In addition to paralysis, we observed significant body fluid loss in aphids injected with either crude *M. rubida* venom or the isolated peptide (Figure 3 and Figure S6). This indicates that hemolymph coagulation may be impaired, which is an important component of wound healing [60,61]. Venom components of the endoparasitoid wasp, *Pimpla turionellae* (Hymenoptera: Ichneumonidae), suppress hemocyte-mediated immune responses at the cellular level [62]. Therefore, further studies should examine the correlation between ant venom components and the inactivation of hemocytes and/or clotting factors that may be essential for wound healing in aphids. Furthermore, it would be interesting to investigate the observed cessation of aphid embryonic development and tissue alterations after exposure to *M. rubida* venom (Figure S6).

Although potent peptide toxins can be integrated directly into pest management strategies by spraying the toxins onto plants (e.g., Spear-T developed by Vestaron), the more typical approach is the development of insect-resistant transgenic crops or engineered entomopathogens (e.g., fungi or baculoviruses) [24,25]. The development of bio-insecticides based on venom peptides frequently fails due to their lack of stability or low oral activity, as shown for U-MYRTX-MANr1 [63]. Therefore, natural peptides are usually replaced with synthetic versions that are more stable, as with the *U. yaschenkoi* AMPs discussed above [58,64,65]. Chemical modifications such as the replacement of individual disulfide bonds with diselenide bonds can improve the oral activity of venom peptides, making them more suitable for the development of bio-insecticides [63].

In summary, the proteomic analysis of *M. rubida* venom and functional characterization of the novel peptide U-MYRTX-MANr1 make a significant contribution to the underrepresented field of ant venom research. Our insect bioassays suggest that the *M. rubida* peptide toxin may act simultaneously against several molecular targets (e.g., ion channels or hemolymph coagulation proteins) as previously shown for the ant venom toxin, ectatomin Et-1 [2]. The remarkable insecticidal activity of crude *M. rubida* venom indicates that additional peptide toxins could also be suitable as leads for the development of novel bio-insecticides.

4. Materials and Methods

4.1. Ant Collection and Taxonomy

Ants identified as *M. rubida* workers based on their morphology, according to Seifert's identification key [66], were collected from a sunny open habitat with a rocky-loamy soil in a forest >500 meters above sea level, located near Tambach-Dietharz, Thuringia, Germany. In the laboratory, foraging workers were kept in a plastic box (180 × 135 × 60 mm) containing soil and plant litter from the environment as an artificial nest, fitted with a test tube water reservoir (160 × 16 mm). Ants were provided with 20% sucrose solution and mealworms (*Tenebrio molitor*) twice a week and maintained under constant conditions (~23 °C, 40% relative humidity, and a 16 h photoperiod).

4.2. Crude Venom Extraction

M. rubida workers were anesthetized with CO_2 and the sting apparatus was removed from the abdominal segment. Venom glands and reservoirs of five foraging workers were gently transferred to a 1.5 mL microcentrifuge tube and pooled in 100 µL of chilled methanol. The crude extract was immediately centrifuged for 30 min at 18,000× *g*. The supernatant was transferred to a new tube for crude venom analysis and fractionation. We used different solvents for crude venom extraction, for reduction and alkylation, and for injection.

4.3. Reduction and Alkylation

Crude venom was extracted in 100 µL of 50 mM ammonium bicarbonate containing 5% acetonitrile and was supplemented with 5 mM dithiothreitol before incubation for 30 min at 56 °C. After cooling the mixture to room temperature, cysteine residues were alkylated with 15 mM iodoacetamide for 30 min in the dark.

4.4. LC-MS Analysis of Venom and Peptides

Conditions and settings for reversed-phase LC and MS data acquisition were modified slightly from our previous study [39]. Peptides were separated with 0.1% formic acid in water (eluent A) and 0.1% formic acid in acetonitrile (eluent B). The starting concentration (5% B) was maintained for 5 min, increased to 60% B in 60 min, then steeply increased to 95% B in 5 min, held for 9 min, then reduced to the starting concentration (5% B) in 1 min, followed by a re-equilibration step (10 min). The following MS instrument settings were used: capillary voltage 3200 V, Hexapole RF 200 Vpp, ion energy 5.0 eV, and collision RF 460 Vpp.

4.5. Edman Degradation

Automated N-terminal sequencing was performed by stepwise Edman degradation (Proteome Factory AG, Berlin, Germany) using a Procise Model 492 cLc protein sequencer (Applied Biosystems, Foster City, CA, USA) according to the manufacturer's protocol.

4.6. Peptide Synthesis

The U-MYRTX-MANr1 peptide was synthesized by Romer Labs (Butzbach, Germany) according to the manufacturer's protocol (purity > 98%).

4.7. Determination of Total Protein Concentration for Injection Assays

Peptides from six *M. rubida* venom glands or 12 *M. rubra* venom glands were extracted in 50 µL water as described above. Total protein concentration was determined using a NanoDrop 2000 spectrophotometer (Thermo Fisher Scientific, Waltham, MA, USA). The absorbance of a 1.5 µL droplet of crude venom was determined at 205 nm and the total protein concentration was calculated using an extinction coefficient (ε_{205}) of 31 mL·mg^{-1}·cm^{-1} [67] and Scopes' method [68]. We used 1 mg·mL^{-1} solutions of BSA and the synthetic *M. rubra* decapeptide NH$_2$-IDPKLLESLA-CONH$_2$ [39] as controls. Crude extracts were further diluted to 1.0, 0.1 and 0.01 mg·mL^{-1} for aphid injection assays.

4.8. Antimicrobial Assay

The minimal inhibition concentration (MIC) was determined by broth dilution assay in a 384-well plate with a final working volume of 20 µL as previously described [69]. Antibacterial and antifungal activity were screened against the following test strains: *Bacillus megaterium* ATCC 14945, *Bacillus subtilis* DSM 10, *Escherichia coli* D31, *Listeria fleischmannii* DSM 24998, *Listeria monocytogenes* ATCC 15313, *Micrococcus luteus* DSM 20030, Staphylococcus aureus ATCC 25923, *Moraxella catarrhalis* DSM 9143, *Pseudomonas aeruginosa* ATCC 27853, *Staphylococcus epidermidis* ATCC 35984, and *Candida albicans* ATCC 90028. We also used the synthetic analog of the *M. rubra* decapeptide U-MYRTX-MRArub1 [39].

Microbial strains were cultivated in Mueller–Hinton II or brain heart infusion broth, as appropriate. Peptides were serially diluted (two-fold), resulting in a concentration range from 512 to 0.016 $\mu g \cdot mL^{-1}$. Endpoint MIC values were read after 16 h of incubation at 37 °C (*C. albicans* 28 °C) with continuous shaking on a LUMIstar Omega plate reader (BMG Labtech, Ortenberg, Germany). We measured absorbance at 600 nm (*B. megaterium, B. subtilis, E. coli, P. aeruginosa* and *S. epidermidis*) or luminescence (*L. fleischmanii, L. monocytogenes, M. luteus, M. catarrhalis, S. aureus* and *C. albicans*) using BacTiter-Glo (Promega, Mannheim, Germany). All experiments were conducted in technical and biological triplicates, including sterility and growth controls (gentamycin and rifampicin).

4.9. Maintenance of Aphids, Injection and Incapacitation Assays

The parthenogenetic *A. pisum* clone LL01 was reared under constant conditions on broad beans, *Vicia faba* var. *minor*, as previously described [34,70]. Age-synchronized aphids (5 days old) were used in all experiments [71]. Aphids were injected laterally, between the middle and hind legs, with 25 nL of crude venom extract or peptides, using glass capillaries held on a M3301 micromanipulator (World Precision Instruments, Hitchin, UK). Crude *M. rubida* and *M. rubra* venoms were tested at concentrations of 0.01, 0.1 and 1 $mg \cdot mL^{-1}$, whereas peptide toxins U-MYRTX-MANr1 and U-MYRTX-MYRrub1 were tested at concentrations of 1, 4 and 16 $mg \cdot mL^{-1}$. The *M. rubra* peptide U-MYRTX-MYRrub1 was used as a control because its insecticidal activity had already been confirmed [39]. Water and BSA (1, 4 and 16 $mg \cdot mL^{-1}$) were used as negative controls. We injected 60 aphids in three biological replicates of 20 individuals per treatment. Injected aphids were reared individually for 10 days in Petri dishes with *V. faba* leaves on 1% agarose gel [71,72]. Aphid survival and offspring production were monitored daily [46]. Newly emerged nymphs were counted daily and removed. Fresh Petri dishes with *V. faba* leaves were provided every 5 days to ensure the aphids were maintained in an optimal environment.

Aphid behavior was observed after injections with crude venom (1 $mg \cdot mL^{-1}$) or the pure peptides (16 $mg \cdot mL^{-1}$) compared to a water control, with 30 individuals injected per treatment group. Injected aphids were placed on their backs and the time they took to correct their position was recorded [4]. Aphid fitness and behavior were assessed immediately after and 1 h after injection and images were acquired using a Leica M125 C stereomicroscope.

4.10. Feeding Assays and Insecticide Bioassays

The oral activity of toxin U-MYRTX-MANr1 and its effect on aphid susceptibility to chemical insecticides were tested as previously described [39] with minor modifications. Briefly, we fed *A. pisum* nymphs (5 days old) for 3 days in modified chambers [73] on an artificial diet [74] mixed with U-MYRTX-MANr1 (500 $\mu g \cdot mL^{-1}$) or a negative control diet in which the peptide solution was replaced with water. Survival was scored daily over the 3 days of feeding. We tested ~1300 aphids per treatment in three biological replicates.

Aphids that survived the 3 day feeding treatment with U-MYRTX-MANr1 or the control were transferred to insecticide bioassays. We tested three chemical insecticides: imidacloprid (neonicotinoid), methomyl (carbamate) and spirotetramat (tetramic acid derivative) [18,32]. All were acquired from Chem Service Inc. (West Chester, PA, USA). For each insecticide, a stock solution of 1000 $\mu g \cdot mL^{-1}$ was prepared in acetone and working solutions (imidacloprid = 0.0975 $\mu g \cdot mL^{-1}$; methomyl = 6.25 $\mu g \cdot mL^{-1}$; spirotetramat = 1.56 $\mu g \cdot mL^{-1}$) were prepared in water. Sub-lethal concentrations of each compound were used, based on previous studies [46].

Insecticide bioassays were carried out as previously described [39,46]. Briefly, *V. faba* stems with roots (2–3 weeks old) were dipped into Falcon tubes (50 mL) containing each insecticide working solution for 24 h. Petri dishes were then prepared with bean leaf discs prepared from the treated stems, as recommended by the Insecticide Resistance Action Committee (IRAC) [75]. Ten aphids, previously treated with the peptide or control diet, were transferred to each leaf disc in 6–8 replicates for each insecticide. Each experiment was conducted with three biological replicates, along with corresponding solvent and water controls. Aphid mortality was scored after exposure for 3 days.

4.11. Histological Preparations

For light microscopy, 30 aphids (5 days old) 12 h after each treatment were pre-fixed in 2.5% glutaraldehyde in 0.1 M phosphate buffer (pH 7.4) for 1 h. Treatment groups were assigned as follows: (I) water injection, (II) peptide toxin U-MYRTX-MANr1 (16 mg·mL^{-1}), and (III) crude *M. rubida* venom (1 mg·mL^{-1}). After washing in phosphate buffer, aphids were post-fixed in 1% OsO_4 in the same buffer for 1 h. After dehydration in a graded ethanol series, aphids were embedded in Araldite epoxy resin (Plano, Germany). Semi-thin sections were prepared from five randomly chosen specimens representing each treatment using a Reichert Om/U3 ultramicrotome. Sections were stained with 1% toluidine blue in 1% sodium borate and examined using a Leica DM 4 B microscope.

4.12. Data Analysis

Molecular features were extracted from acquired mass spectra using Compass DataAnalysis v4.2 (Bruker Daltonics, Billerica, MA, USA) as previously described [39]. MS data were analyzed and visualized using SigmaPlot v12.5 (Systat Software, San Jose, CA, USA). We analyzed aphid fitness data using IBM SPSS Statistics v17 (Armonk, New York, NY, USA). Statistical significance was defined as $p < 0.05$. Survival data were analyzed by Kaplan–Meier survival analysis and comparisons between groups were based on log-rank tests. The total number of offspring was analyzed using the Mann–Whitney U test for non-parametric data and Student's t-test for normally distributed data. For insecticide bioassays, total mortality for each insecticide treatment was corrected according to Abbott's formula, based on mortality scored in control groups [76]. The mortality in solvent control groups was 4–12%. We used the Mann–Whitney U test to compare mortality between the two feeding treatments (peptide toxin and diet control). For sequence alignment, we used Clustal Omega v1.2.4 from the European Bioinformatics Institute (Hinxton, UK) [77] and the online tool Sequence Identity And Similarity (SIAS) [78].

Supplementary Materials: The following are available online at http://www.mdpi.com/2072-6651/11/10/562/s1, Figure S1: Distribution of peptides in the venom of *Manica rubida* (n = 96 peptides) according to amino acid sequence length and molecular weight. Figure S2: Fragmentation pattern of the peptide U-MYRTX-MANr1. Table S1: Effect of crude venom and peptide toxins on the total number of offspring in *Acyrthosiphon pisum*. Table S2: Statistical data for the Kaplan–Meier survival analysis shown in Figure 2. Figure S3: Insecticidal activity of *Myrmica rubra* crude venom and peptide U-MYRTX-MRArub1 in *Acyrthosiphon pisum*. Table S3: Statistical data for Kaplan–Meier survival analysis shown in Figure S3. Figure S4: Insecticidal activity of bovine serum albumin (BSA) control in *Acyrthosiphon pisum*. Table S4: Statistical data for Kaplan–Meier survival analysis shown in Figure S4. Table S5: Aphid behavior and fitness after the injection of water, U-MYRTX-MYRrub1, or *Myrmica rubra* crude venom. Figure S5: Injection wound healing in *Acyrthosiphon pisum* 1 h post-injection of water, U-MYRTX-MYRrub1, or crude *Myrmica rubra* venom. Figure S6: Semi-thin cross-section through the abdomen of *Acyrthosiphon pisum* injected with crude *Manica rubida* venom. Figure S7: Semi-thin cross-section through abdomen of *Acyrthosiphon pisum* injected with *Manica rubida* peptide U-MYRTX-MANr1. Figure S8: Semi-thin cross-section through the abdomen of *Acyrthosiphon pisum* injected with water. Figure S9: Insecticidal activity of orally delivered *Manica rubida* peptide U-MYRTX-MANr1. Table S6: Summary of statistical data for the insecticidal activity of orally delivered *Manica rubida* peptide U-MYRTX-MANr1 and its effect on the susceptibility of aphids to chemical insecticides. Figure S10: Alignment of U-MYRTX-MANr1 with peptide sequences found in the venom of other arthropods or defensive skin secretions of frogs.

Author Contributions: Conceptualization, J.H., M.S.(Marisa Skaljac) and A.V.; Formal analysis, J.H. and M.S.(Marisa Skaljac); Funding acquisition, A.V.; Investigation, J.H., J.G., T.K., M.S.(Maximilian Seip) and H.S.; Methodology, J.H., M.S.(Marisa Skaljac) and H.S.; Project administration, M.S.(Marisa Skaljac) and A.V.; Resources, A.V.; Software, J.H. and M.S.(Marisa Skaljac); Supervision, M.S.(Marisa Skaljac) and A.V.; Validation, J.H., M.S.(Marisa Skaljac), H.S. and A.V.; Visualization, J.H., M.S.(Marisa Skaljac) and H.S.; Writing—original draft, J.H. and M.S.(Marisa Skaljac); Writing—review and editing, A.V.

Funding: The authors acknowledge generous funding by the Hessen State Ministry of Higher Education, Research and the Arts (HMWK) via the "LOEWE Center for Insect Biotechnology and Bioresources".

Acknowledgments: We would like to thank Olga Lang, Svenja Thöneböhn, Benedikt Leis and Sanja Mihajlovic from Fraunhofer IME (Giessen, Germany) for their valuable help and support in this study. We thank the "Imaging

Unit" of the Biomedical Research Center Senckenberg (Justus-Liebig University of Giessen) for providing the ultramicrotome equipment. We are grateful to Richard M. Twyman for editing the manuscript.

Conflicts of Interest: The authors declare no conflict of interest.

References

1. AntWeb. Available online: https://www.antweb.org (accessed on 1 July 2019).
2. Touchard, A.; Aili, R.S.; Fox, G.E.; Escoubas, P.; Orivel, J.; Nicholson, M.G.; Dejean, A. The Biochemical Toxin Arsenal from Ant Venoms. *Toxins* **2016**, *8*, 30. [CrossRef] [PubMed]
3. Aili, S.R.; Touchard, A.; Petitclerc, F.D.R.; Dejean, A.; Orivel, J.R.M.; Padula, M.P.; Escoubas, P.; Nicholson, G.M. Combined peptidomic and proteomic analysis of electrically stimulated and manually dissected venom from the South American bullet ant *Paraponera clavata*. *J. Proteome Res.* **2017**, *16*, 1339–1351. [CrossRef] [PubMed]
4. Robinson, S.D.; Mueller, A.; Clayton, D.; Starobova, H.; Hamilton, B.R.; Payne, R.J.; Vetter, I.; King, G.F.; Undheim, E.A.B. A comprehensive portrait of the venom of the giant red bull ant, *Myrmecia gulosa*, reveals a hyperdiverse hymenopteran toxin gene family. *Sci. Adv.* **2018**, *4*, eaau4640. [CrossRef] [PubMed]
5. Osman, M.F.H.; Brander, J. Weitere Beiträge zur Kenntnis der chemischen Zusammensetzung des Giftes von Ameisen aus der Gattung Formica. *Z. Naturforsch. B* **1961**, *16*, 749–753. [CrossRef]
6. Pluzhnikov, K.A.; Kozlov, S.A.; Vassilevski, A.A.; Vorontsova, O.V.; Feofanov, A.V.; Grishin, E.V. Linear antimicrobial peptides from *Ectatomma quadridens* ant venom. *Biochimie* **2014**, *107*, 211–215. [CrossRef] [PubMed]
7. Rifflet, A.; Gavalda, S.; Tene, N.; Orivel, J.; Leprince, J.; Guilhaudis, L.; Genin, E.; Vetillard, A.; Treilhou, M. Identification and characterization of a novel antimicrobial peptide from the venom of the ant *Tetramorium bicarinatum*. *Peptides* **2012**, *38*, 363–370. [CrossRef] [PubMed]
8. Aili, S.R.; Touchard, A.; Escoubas, P.; Padula, M.P.; Orivel, J.; Dejean, A.; Nicholson, G.M. Diversity of peptide toxins from stinging ant venoms. *Toxicon* **2014**, *92*, 166–178. [CrossRef] [PubMed]
9. Kazuma, K.; Masuko, K.; Konno, K.; Inagaki, H. Combined Venom Gland Transcriptomic and Venom Peptidomic Analysis of the Predatory Ant *Odontomachus monticola*. *Toxins* **2017**, *9*, 323. [CrossRef] [PubMed]
10. Bouzid, W.; Klopp, C.; Verdenaud, M.; Ducancel, F.; Vétillard, A. Profiling the venom gland transcriptome of *Tetramorium bicarinatum* (Hymenoptera: Formicidae): The first transcriptome analysis of an ant species. *Toxicon* **2013**, *70*, 70–81. [CrossRef]
11. Johnson, S.R.; Copello, J.A.; Evans, M.S.; Suarez, A.V. A biochemical characterization of the major peptides from the venom of the giant neotropical hunting ant *Dinoponera australis*. *Toxicon* **2010**, *55*, 702–710. [CrossRef]
12. Orivel, J.; Redeker, V.; Le Caer, J.P.; Krier, F.; Revol-Junelles, A.M.; Longeon, A.; Chaffotte, A.; Dejean, A.; Rossier, J. Ponericins, new antibacterial and insecticidal peptides from the venom of the ant *Pachycondyla goeldii*. *J. Biol. Chem.* **2001**, *276*, 17823–17829. [CrossRef] [PubMed]
13. Schmidt, C.A.; Shattuck, S.O. The higher classification of the ant subfamily Ponerinae (Hymenoptera: Formicidae), with a review of ponerine ecology and behavior. *Zootaxa* **2014**, *3817*, 1–242. [CrossRef] [PubMed]
14. Lima, D.B.; Torres, A.F.C.; Mello, C.P.; de Menezes, R.; Sampaio, T.L.; Canuto, J.A.; da Silva, J.J.A.; Freire, V.N.; Quinet, Y.P.; Havt, A.; et al. Antimicrobial effect of Dinoponera quadriceps (Hymenoptera: Formicidae) venom against Staphylococcus aureus strains. *J. Appl. Microbiol.* **2014**, *117*, 390–396. [CrossRef] [PubMed]
15. Schluns, H.; Crozier, R.H. Molecular and chemical immune defenses in ants (Hymenoptera: Formicidae). *Myrmecol. News* **2009**, *12*, 237–249.
16. Pimentel, D. Pesticides and Pest Control. In *Integrated Pest Management: Innovation-Development Process: Volume 1*; Peshin, R., Dhawan, A.K., Eds.; Springer: Dordrecht, The Netherlands, 2009; pp. 83–87.
17. Oerke, E.C. Crop losses to pests. *J. Agric. Sci.* **2005**, *144*, 31–43. [CrossRef]
18. Sparks, T.C.; Nauen, R. IRAC: Mode of action classification and insecticide resistance management. *Pestic. Biochem. Physiol.* **2015**, *121*, 122–128. [CrossRef]
19. King, G.F.; Hardy, M.C. Spider-venom peptides: Structure, pharmacology, and potential for control of insect pests. *Annu. Rev. Entomol.* **2013**, *58*, 475–496. [CrossRef] [PubMed]
20. Herzig, V.; Ikonomopoulou, M.; Smith, J.J.; Dziemborowicz, S.; Gilchrist, J.; Kuhn-Nentwig, L.; Rezende, F.O.; Moreira, L.A.; Nicholson, G.M.; Bosmans, F.; et al. Molecular basis of the remarkable species selectivity of an insecticidal sodium channel toxin from the African spider *Augacephalus ezendami*. *Sci. Rep.* **2016**, *6*, 29538. [CrossRef]

21. Jørgensen, P.S.; Aktipis, A.; Brown, Z.; Carrière, Y.; Downes, S.; Dunn, R.R.; Epstein, G.; Frisvold, G.B.; Hawthorne, D.; Gröhn, Y.T.; et al. Antibiotic and pesticide susceptibility and the Anthropocene operating space. *Nat. Sustain.* **2018**, *1*, 632–641. [CrossRef]

22. Barzman, M.; Bàrberi, P.; Birch, A.N.E.; Boonekamp, P.; Dachbrodt-Saaydeh, S.; Graf, B.; Hommel, B.; Jensen, J.E.; Kiss, J.; Kudsk, P.; et al. Eight principles of integrated pest management. *Agron. Sustain. Dev.* **2015**, *35*, 1199–1215. [CrossRef]

23. European Parliament. Think Thank. Farming without Plant Protection Products. Available online: https: //www.europarl.europa.eu/RegData/etudes/IDAN/2019/634416/EPRS_IDA(2019)634416_EN.pdf (accessed on 1 July 2019).

24. Windley, M.J.; Herzig, V.; Dziemborowicz, S.A.; Hardy, M.C.; King, G.F.; Nicholson, G.M. Spider-Venom Peptides as Bioinsecticides. *Toxins* **2012**, *4*, 191–227. [CrossRef] [PubMed]

25. King, G.F. Tying pest insects in knots: The deployment of spider-venom-derived knottins as bioinsecticides. *Pest Manag. Sci.* **2019**. [CrossRef] [PubMed]

26. Piek, T.; Duval, A.; Hue, B.; Karst, H.; Lapied, B.; Mantel, P.; Nakajima, T.; Pelhate, M.; Schmidt, J.O. Poneratoxin, a novel peptide neurotoxin from the venom of the ant, *Paraponera clavata*. *Comp. Biochem. Physiol. C Comp. Pharmacol.* **1991**, *99*, 487–495. [CrossRef]

27. Piek, T.; Hue, B.; Mantel, P.; Terumi, N.; Schmidt, J.O. Pharmacological characterization and chemical fractionation of the venom of the ponerine ant, *Paraponera clavata* (F.). *Comp. Biochem. Physiol. C Comp. Pharmacol.* **1991**, *99*, 481–486. [CrossRef]

28. Touchard, A.; Brust, A.; Cardoso, F.C.; Chin, Y.K.Y.; Herzig, V.; Jin, A.H.; Dejean, A.; Alewood, P.F.; King, G.F.; Orivel, J.; et al. Isolation and characterization of a structurally unique β-hairpin venom peptide from the predatory ant *Anochetus emarginatus*. *Biochim. Biophys. Acta* **2016**, *1860*, 2553–2562. [CrossRef] [PubMed]

29. Van Emden, H.F.; Harrington, R. *Aphids as Crop Pests*; CABI: Wallingford, UK, 2017.

30. Will, T.; Vilcinskas, A. Aphid-proof plants: Biotechnology-based approaches for aphid control. In *Yellow Biotechnology II*; Springer: Berlin/Heidelberg, Germany, 2013; pp. 179–203.

31. Vilcinskas, A. *Biology and Ecology of Aphids*; CRC Press: Boca Raton, FL, USA, 2016.

32. IRAC—Insecticide Resistance Action Committee. Available online: http://www.irac-online.org/documents/ sucking-pests-moa-poster/ (accessed on 1 July 2019).

33. Bass, C.; Puinean, A.M.; Zimmer, C.T.; Denholm, I.; Field, L.M.; Foster, S.P.; Gutbrod, O.; Nauen, R.; Slater, R.; Williamson, M.S. The evolution of insecticide resistance in the peach potato aphid, *Myzus persicae*. *Insect Biochem. Mol. Biol.* **2014**, *51*, 41–51. [CrossRef] [PubMed]

34. Luna-Ramirez, K.; Skaljac, M.; Grotmann, J.; Kirfel, P.; Vilcinskas, A. Orally delivered scorpion antimicrobial peptides exhibit activity against Pea Aphid (*Acyrthosiphon pisum*) and its bacterial symbionts. *Toxins* **2017**, *9*, 261. [CrossRef] [PubMed]

35. Pal, N.; Yamamoto, T.; King, G.F.; Waine, C.; Bonning, B. Aphicidal efficacy of scorpion-and spider-derived neurotoxins. *Toxicon* **2013**, *70*, 114–122. [CrossRef]

36. Xie, M.; Zhang, Y.J.; Zhai, X.M.; Zhao, J.J.; Peng, D.L.; Wu, G. Expression of a scorpion toxin gene BmKit enhances the virulence of *Lecanicillium lecanii* against aphids. *J. Pest Sci.* **2015**, *88*, 637–644. [CrossRef]

37. Yang, S.; Fitches, E.; Pyati, P.; Gatehouse, J.A. Effect of insecticidal fusion proteins containing spider toxins targeting sodium and calcium ion channels on pyrethroid-resistant strains of peach-potato aphid (*Myzus persicae*). *Pest Manag. Sci.* **2015**, *71*, 951–956. [CrossRef]

38. Bonning, B.C.; Pal, N.; Liu, S.; Wang, Z.; Sivakumar, S.; Dixon, P.M.; King, G.F.; Miller, W.A. Toxin delivery by the coat protein of an aphid-vectored plant virus provides plant resistance to aphids. *Nat. Biotechnol.* **2014**, *32*, 102. [CrossRef] [PubMed]

39. Heep, J.; Klaus, A.; Kessel, T.; Seip, M.; Vilcinskas, A.; Skaljac, M. Proteomic Analysis of the Venom from the Ruby Ant *Myrmica rubra* and the Isolation of a Novel Insecticidal Decapeptide. *Insects* **2019**, *10*, 42. [CrossRef] [PubMed]

40. Wheeler, G.C.; Wheeler, J. The Natural History of Manica (Hymenoptera: Formicidae). *J. Kans. Entomol. Soc.* **1970**, *43*, 129–162.

41. Lenoir, A.; Devers, S.; Marchand, P.; Bressac, C.; Savolainen, R. Microgynous queens in the Paleartic ant, *Manica rubida*: Dispersal morphs or social parasites? *J. Insect Sci.* **2010**, *10*. [CrossRef] [PubMed]

42. Brown, M.J.F.; Bonhoeffer, S. On the evolution of claustral colony founding in ants. *Evol. Ecol. Res.* **2003**, *5*, 305–313.

43. Touchard, A.; Koh, J.M.; Aili, S.R.; Dejean, A.; Nicholson, G.M.; Orivel, J.; Escoubas, P. The complexity and structural diversity of ant venom peptidomes is revealed by mass spectrometry profiling. *Rapid Commun. Mass Spectrom.* **2015**, *29*, 385–396. [CrossRef]

44. Senko, M.W.; Beu, S.C.; McLaffertycor, F.W. Determination of monoisotopic masses and ion populations for large biomolecules from resolved isotopic distributions. *J. Am. Soc. Mass Spectrom.* **1995**, *6*, 229–233. [CrossRef]

45. King, G.F.; Gentz, M.C.; Escoubas, P.; Nicholson, G.M. A rational nomenclature for naming peptide toxins from spiders and other venomous animals. *Toxicon* **2008**, *52*, 264–276. [CrossRef]

46. Skaljac, M.; Kirfel, P.; Grotmann, J.; Vilcinskas, A. Fitness costs of infection with *Serratia symbiotica* are associated with greater susceptibility to insecticides in the pea aphid *Acyrthosiphon pisum*. *Pest Manag. Sci.* **2018**, *74*, 1829–1836. [CrossRef]

47. Fiedler, K.; Kuhlmann, F.; Schlick-Steiner, B.C.; Steiner, F.M.; Gebauer, G. Stable N-isotope signatures of central European ants—Assessing positions in a trophic gradient. *Insectes Soc.* **2007**, *54*, 393–402. [CrossRef]

48. Tani, N.; Kazuma, K.; Ohtsuka, Y.; Shigeri, Y.; Masuko, K.; Konno, K.; Inagaki, H. Mass Spectrometry Analysis and Biological Characterization of the Predatory Ant *Odontomachus monticola* Venom and Venom Sac Components. *Toxins* **2019**, *11*, 50. [CrossRef] [PubMed]

49. Pan, J.; Hink, W.F. Isolation and characterization of myrmexins, six isoforms of venom proteins with anti-inflammatory activity from the tropical ant, *Pseudomyrmex triplarinus*. *Toxicon* **2000**, *38*, 1403–1413. [CrossRef]

50. Szolajska, E.; Poznanski, J.; Ferber, M.L.; Michalik, J.; Gout, E.; Fender, P.; Bailly, I.; Dublet, B.; Chroboczek, J. Poneratoxin, a neurotoxin from ant venom: Structure and expression in insect cells and construction of a bio-insecticide. *Eur. J. Biochem.* **2004**, *271*, 2127–2136. [CrossRef] [PubMed]

51. Wanandy, T.; Gueven, N.; Davies, N.W.; Brown, S.G.A.; Wiese, M.D. Pilosulins: A review of the structure and mode of action of venom peptides from an Australian ant *Myrmecia pilosula*. *Toxicon* **2015**, *98*, 54–61. [CrossRef] [PubMed]

52. Terlau, H.; Olivera, B.M. Conus venoms: A rich source of novel ion channel-targeted peptides. *Physiol. Rev.* **2004**, *84*, 41–68. [CrossRef] [PubMed]

53. Olivera, B.M.; Rivier, J.; Clark, C.; Ramilo, C.A.; Corpuz, G.P.; Abogadie, F.C.; Mena, E.E.; Hillyard, D.R.; Cruz, L.J. Diversity of Conus neuropeptides. *Science* **1990**, *249*, 257–263. [CrossRef]

54. Chen, Y.; Guarnieri, M.T.; Vasil, A.I.; Vasil, M.L.; Mant, C.T.; Hodges, R.S. Role of Peptide Hydrophobicity in the Mechanism of Action of α-Helical Antimicrobial Peptides. *Antimicrob. Agents Chemother.* **2007**, *51*, 1398. [CrossRef]

55. Hale, J.D.F.; Hancock, R.E.W. Alternative mechanisms of action of cationic antimicrobial peptides on bacteria. *Expert Rev. Anti-Infect. Ther.* **2007**, *5*, 951–959. [CrossRef]

56. Pirtskhalava, M.; Gabrielian, A.; Cruz, P.; Griggs, H.L.; Squires, R.B.; Hurt, D.E.; Grigolava, M.; Chubinidze, M.; Gogoladze, G.; Vishnepolsky, B.; et al. DBAASP v.2: An enhanced database of structure and antimicrobial/cytotoxic activity of natural and synthetic peptides. *Nucleic Acids Res.* **2015**, *44*, D1104–D1112. [CrossRef]

57. Simmaco, M.; Mignogna, G.; Canofeni, S.; Miele, R.; Mangoni, M.L.; Barra, D. Temporins, Antimicrobial Peptides from the European Red Frog *Rana temporaria*. *Eur. J. Biochem.* **1996**, *242*, 788–792. [CrossRef]

58. Luna-Ramirez, K.; Tonk, M.; Rahnamaeian, M.; Vilcinskas, A. Bioactivity of Natural and Engineered Antimicrobial Peptides from Venom of the Scorpions *Urodacus yaschenkoi* and *U. manicatus*. *Toxins* **2017**, *9*, 22. [CrossRef] [PubMed]

59. Sani, M.A.; Separovic, F. How membrane-active peptides get into lipid membranes. *Acc. Chem. Res.* **2016**, *49*, 1130–1138. [CrossRef] [PubMed]

60. Dushay, M.S. Insect hemolymph clotting. *Cell. Mol. Life Sci.* **2009**, *66*, 2643–2650. [CrossRef] [PubMed]

61. Altincicek, B.; Gross, J.; Vilcinskas, A. Wounding-mediated gene expression and accelerated viviparous reproduction of the pea aphid *Acyrthosiphon pisum*. *Insect Mol. Biol.* **2008**, *17*, 711–716. [CrossRef]

62. Er, A.; Sak, O.; Ergin, E.; Uçkan, F.; Rivers, D.B. Venom-induced immunosuppression: An overview of hemocyte-mediated responses. *Psyche A J. Entomol.* **2011**, *2011*, 276376. [CrossRef]

63. Herzig, V.; De Araujo, D.A.; Greenwood, P.K.; Chin, K.Y.; Windley, J.M.; Chong, Y.; Muttenthaler, M.; Mobli, M.; Audsley, N.; Nicholson, M.G.; et al. Evaluation of Chemical Strategies for Improving the Stability and Oral Toxicity of Insecticidal Peptides. *Biomedicines* **2018**, *6*, 90. [CrossRef] [PubMed]

64. Keymanesh, K.; Soltani, S.; Sardari, S. Application of antimicrobial peptides in agriculture and food industry. *World J. Microbiol. Biotechnol.* **2009**, *25*, 933–944. [CrossRef]

65. Marcos, J.F.; Muñoz, A.; Pérez-Payá, E.; Misra, S.; López-García, B. Identification and Rational Design of Novel Antimicrobial Peptides for Plant Protection. *Annu. Rev. Phytopathol.* **2008**, *46*, 273–301. [CrossRef]

66. Seifert, B. *Die Ameisen Mittel-und Nordeuropas*; Lutra Verlag-u. Vertriebsges: Görlitz/Tauer, Germany, 2007.

67. Anthis, N.J.; Clore, G.M. Sequence-specific determination of protein and peptide concentrations by absorbance at 205 nm. *Protein Sci.* **2013**, *22*, 851–858. [CrossRef]

68. Scopes, R.K. Measurement of protein by spectrophotometry at 205 nm. *Anal. Biochem.* **1974**, *59*, 277–282. [CrossRef]

69. Hirsch, R.; Wiesner, J.; Marker, A.; Pfeifer, Y.; Bauer, A.; Hammann, P.E.; Vilcinskas, A. Profiling antimicrobial peptides from the medical maggot Lucilia sericata as potential antibiotics for MDR Gram-negative bacteria. *J. Antimicrob. Chemother.* **2018**, *74*, 96–107. [CrossRef] [PubMed]

70. Will, T.; Schmidtberg, H.; Skaljac, M.; Vilcinskas, A. Heat shock protein 83 plays pleiotropic roles in embryogenesis, longevity, and fecundity of the pea aphid *Acyrthosiphon pisum*. *Dev. Genes Evol.* **2017**, *227*, 1–9. [CrossRef] [PubMed]

71. Sapountzis, P.; Duport, G.; Balmand, S.; Gaget, K.; Jaubert-Possamai, S.; Febvay, G.; Charles, H.; Rahbé, Y.; Colella, S.; Calevro, F. New insight into the RNA interference response against cathepsin-L gene in the pea aphid, *Acyrthosiphon pisum*: Molting or gut phenotypes specifically induced by injection or feeding treatments. *Insect Biochem. Mol. Biol.* **2014**, *51*, 20–32. [CrossRef] [PubMed]

72. Will, T.; Vilcinskas, A. The structural sheath protein of aphids is required for phloem feeding. *Insect Biochem. Mol. Biol.* **2015**, *57*, 34–40. [CrossRef] [PubMed]

73. Sadeghi, A.; Van Damme, E.J.M.; Smagghe, G.; Cullen, E. Evaluation of the susceptibility of the pea aphid, *Acyrthosiphon pisum*, to a selection of novel biorational insecticides using an artificial diet. *J. Insect Sci.* **2009**, *9*. [CrossRef] [PubMed]

74. Akey, D.H.; Beck, S.D. Continuous Rearing of the Pea Aphid, *Acyrthosiphon pisum*, on a Holidic Diet. *Ann. Entomol. Soc. Am.* **1971**, *64*, 353–356. [CrossRef]

75. IRAC—Insecticide Resistance Action Committee. Susceptibility Test Method 019. Available online: https://www.irac-online.org/methods/aphids-adultnymphs/ (accessed on 1 July 2019).

76. Abbott, W.S. A method of computing the effectiveness of an insecticide. *J. Econ. Entomol.* **1925**, *18*, 265–267. [CrossRef]

77. Madeira, F.; Park, Y.M.; Lee, J.; Buso, N.; Gur, T.; Madhusoodanan, N.; Basutkar, P.; Tivey, A.R.N.; Potter, S.C.; Finn, R.D.; et al. The EMBL-EBI search and sequence analysis tools APIs in 2019. *Nucleic Acids Res.* **2019**, *47*, W636–W641. [CrossRef]

78. Immunomedicine Group. Sequence Identity and Similarity (SIAS). Available online: http://bio.med.ucm.es/Tools/sias.html (accessed on 1 July 2019).

Article

ACP-TX-I and ACP-TX-II, Two Novel Phospholipases A$_2$ Isolated from Trans-Pecos Copperhead *Agkistrodon contortrix pictigaster* Venom: Biochemical and Functional Characterization

Salomón Huancahuire-Vega [1,*], Luciana M. Hollanda [2], Mauricio Gomes-Heleno [3], Edda E. Newball-Noriega [1] and Sergio Marangoni [3]

[1] Departamento de Ciencias Básicas, Facultad de Ciencias de la Salud, Escuela de Medicina Humana, Universidad Peruana Unión (UPeU), Lima 15, Peru; eddanewball@upeu.edu.pe
[2] Instituto de Tecnologia e Pesquisa, Universidade Tiradentes (UNIT), Aracaju 49032-490, SE, Brazil; luciana.hollanda@gmail.com
[3] Departamento de Bioquímica, Instituto de Biologia, Universidade Estadual de Campinas (UNICAMP), Campinas 13083-970, SP, Brazil; maoheleno@yahoo.com.br (M.G.-H.); marango@unicamp.br (S.M.)
* Correspondence: salomonhv77@yahoo.com.br; Tel.: +51-9-9757-4011

Received: 26 September 2019; Accepted: 6 November 2019; Published: 14 November 2019

Abstract: This work reports the purification and biochemical and functional characterization of ACP-TX-I and ACP-TX-II, two phospholipases A$_2$ (PLA$_2$) from *Agkistrodon contortrix pictigaster* venom. Both PLA$_2$s were highly purified by a single chromatographic step on a C$_{18}$ reverse phase HPLC column. Various peptide sequences from these two toxins showed similarity to those of other PLA$_2$ toxins from viperid snake venoms. ACP-TX-I belongs to the catalytically inactive K49 PLA$_2$ class, while ACP-TX-II is a D49 PLA$_2$, and is enzymatically active. ACP-TX-I PLA$_2$ is monomeric, which results in markedly diminished myotoxic and inflammatory activities when compared with dimeric K49 PLA$_2$s, confirming the hypothesis that dimeric structure contributes heavily to the profound myotoxicity of the most active viperid K49 PLA$_2$s. ACP-TX-II exhibits the main pharmacological actions reported for this protein family, including in vivo local myotoxicity, edema-forming activity, and in vitro cytotoxicity. ACP-TX-I PLA$_2$ is cytotoxic to A549 lung carcinoma cells, indicating that cytotoxicity to these tumor cells does not require enzymatic activity.

Keywords: snake venom; *Agkistrodon contortrix pictigaster*; D49 PLA$_2$; homologous K49 PLA$_2$; myotoxin; edema-forming activity and cytotoxicity

Key Contribution: This study reports the first isolation and functional characterization of two basic PLA$_2$ from *A. c. pictigaster* venom.

---

1. Introduction

Phospholipases A$_2$ (PLA$_2$), which hydrolyze 2-acyl ester bonds of 3-sn-phospholipids, releasing lysophospholipids and fatty acids, are widespread in snake venoms, facilitating the immobilization and digestion of prey [1,2]. Venom PLA$_2$s belong to the secretory PLA$_2$ (sPLA$_2$) family (groups IA, elapids and IIA, viperids) [3,4]. PLA$_2$s are small proteins (13–15 kDa) with 115–122 residues, and seven conserved disulfide bonds [5,6]. They are subdivided into two main subgroups: (1) D49 PLA$_2$s, which have an aspartate residue at position 49, and typically have high catalytic activity [7,8], and (2) K49 PLA$_2$s, with a lysine residue at position 49. These have little or no catalytic activity, but still induce various biological effects [9,10]. Both types of proteins show significant similarity in their three-dimensional structures, although they exhibit different pharmacological actions, such as myotoxicity, neurotoxicity, anticoagulant activity, platelet aggregation inhibition/activation, hemolysis, edematogenicity, hypotension, bactericidal action, proinflammatory cytotoxicity, and antitumor activity [2,11].

A great number of sequences and crystal structures of viperid PLA_2s have been determined [11–13]. Despite their overarching similarities, sequence differences result in diverse biological functions [14]. Accordingly, characterization of individual toxins is important to better understand the pathophysiology of envenomation and to potentially improve therapeutic procedures [15].

Agkistrodon is a genus of pit vipers ranging from the southern United States to northern Costa Rica [16]. Currently, this genus comprises four species: *A. contortrix* (copperheads), *A. piscivorus* (cottonmouth), *A. bilineatus* (cantils), and *A. taylori* (Taylor's cantils) [17,18]. Copperheads comprise several subspecies: *A. c. contortrix* (southern copperhead), *A. c. laticinctus* (broad-banded copperhead), *A. c. mokasen* (northern copperhead), *A. c. phaeogaster* (Osage copperhead), and *A. c. pictigaster* (Trans-Pecos copperhead). Subspecific taxonomy is based largely on gross morphology, color pattern, and scale counts [18]. Cottonmouths and copperheads are among the most common venomous snakes in the southeastern United States. Cottonmouths frequent streams, rivers, ponds, marshes, and swamps, whereas copperheads are found in deciduous hardwood forests with moist leaf litter, large logs, scattered rocks, and high levels of vegetative cover. These snakes account for ~30% of the non-lethal human envenomations in this region [19,20].

The Trans-pecos copperhead (*A. c. pictagaster*) is found in western Texas, northern Chihuahua, and Coahuila (Mexico) [16]. Juveniles usually prey on invertebrates (spiders, millipedes, and insects), frogs, and small lizards, whereas adults primarily prey on vertebrates, including amphibians (salamanders and anurans), reptiles (lizards and snakes), birds, and small mammals (rodents) [16].

Studies on the biochemical composition and toxic activities of copperhead venoms, including D49 PLA_2 and homologous K49 PLA_2, have been almost exclusively restricted to *A. c. contortrix* and *A. c. latincictus*, [5,21–23]. In contrast, little information is available on *A. c. pictigaster*. Partial characterizations of two disintegrins from this venom have been described by Lucena et al. [24]. A comparison of venom proteome variation in the genus *Agkistrodon* found that *A. c. pictigaster* venom contains ten protein families, dominated by PLA_2s (38.2%) and metalloproteinases (30.2%). The venom showed proteolytic, hemorrhagic, and myotoxic activities [25].

This work is the first report of two basic PLA_2s isolated from *A. c. pictigaster* venom, with their identification and structural characterization found by biochemical and enzymatic experiments. Furthermore, we describe their biological activities and cytotoxic properties upon an A549 tumor cell line. The results of this study illuminate structure-function relationships of ACP-TX-I and ACP-TX-II PLA_2, and improve our understanding of the chemistry of this venom.

2. Results

2.1. Purification and Biochemical Characterization of ACP-TX-I and ACP-TX-II

Chromatographic separation of *A. c. pictigaster* venom by reversed phase high-performance liquid chromatography (RP-HPLC) on a C_{18} column resulted in 29 fractions, with two prominent peaks (16 and 18) eluting in more than 30% acetonitrile (Figure 1). These peaks, representing about 30% of total venom protein, were collected and screened for PLA_2 activity. Fractions 16 and 18 were named ACP-TX-I and ACP-TX-II, respectively. Both ACP-TX-I and ACP-TX-II exhibited high purity when re-chromatographed using the same chromatography system, each showing only one peak (Figure S1). These peaks were also analyzed by SDS-PAGE, which manifested a single electrophoretic band with an *Mr* of approximately 14 kDa under reducing and non-reducing conditions (Figure 1 insert).

ESI-MS analysis demonstrated that the proteins were homogeneous, with molecular masses of 12,209.7 and 14,041.1 Da for ACP-TX-I and ACP-TX-II, respectively (Figure 2A,B).

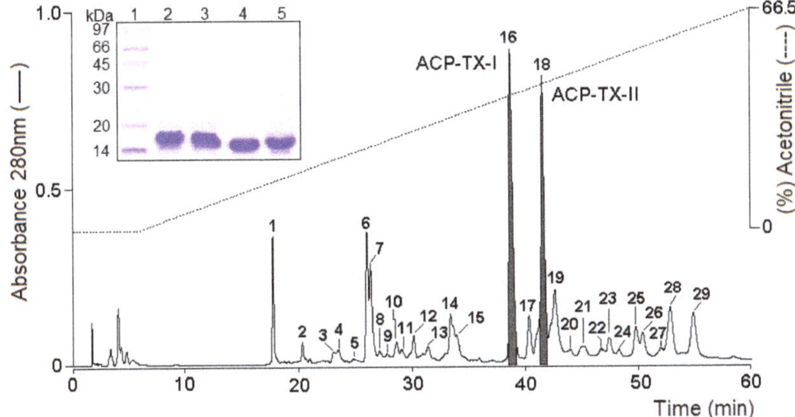

Figure 1. When *A. c. pictigaster* venom was fractionated on a C_{18} μ-Bondapak column, two phospholipases A_2 dominated the elution profile. Fraction 18 (ACP-TX-II) possessed PLA_2 activity, while fraction 16 (ACP-TX-I) showed cytotoxic activity, despite a lack of enzymatic activity. Insert: Electrophoretic profile in Tricine SDS-PAGE. (1) Molecular mass markers; (2) ACP-TX-I not reduced; (3) ACP-TX-I reduced with DTT (1M); (4) ACP-TX-II not reduced; and (5) ACP-TX-II reduced with DTT (1M).

2.2. Determination of the Amino Acid Sequences of ACP-TX-I and ACP-TX-II

In order to identify the purified proteins, they were digested with trypsin, and tryptic peptides were detected and characterized by mass spectrometry. Amino acid sequences of several tryptic peptides were obtained (Table 1). ACP-TX-I and ACP-TX-II shared 7 and 6 peptides with other viperid PLA_2s, respectively.

Assembly of partial protein sequences by similarity, using BLAST and multiple alignment, demonstrated that ACP-TX-I is a K49 PLA_2 (Figure 3A), and is quite similar to well-characterized members of this family such as ACL PLA_2 from *A. c. laticinctus*, MjTX-II from *Bothrops moojeni*, BnSP-7 from *B. neuwiedi pauloensis*, etc. Sequenced peptides accounted for 64 amino acids, assuming that the number of residues is identical to that of homologous viperid toxins. This represented an estimated 53% of the protein (Figure 3A). ACP-TX-II shares conserved domain sequences common to catalytically active D49 PLA_2s. Peptide 1, having the sequence DATDRCCFVHDCCYGQ/K, contains an Asp (aspartic acid) residue that corresponds to position 49 in the complete amino acid sequence (Figure 3B).

Table 1. Tryptic peptides of ACP-TX-I and ACP-TX-II PLA_2. Peptides were separated and sequenced by mass spectrometry. Molecular masses are monoisotopic.

Peptides	Mass (Da) Expected	Amino Acid Sequence	Mass (Da) Calculated
ACP-TX-I			
1	9.864.857	GQ/KPK/QDATDR	9.864.781
2	14.075.661	DATDRCCFVHQ/K	14.076.024
3	7.753.587	VTGCDPK	7.753.535
4	14.737.322	AI/LI/LCEEK/QNPCL/IQ/K	14.737.319
5	17.537.551	MCECDK/QAVAI/LCL/IRE	17.537.619
6	11.235.574	ENL/IDTYNQ/KQ/K	11.235.509
7	9.844.505	TYWK/QYPQ/K	9.845.069
ACP-TX-II			
1	20.627.936	DATDRCCFVHDCCYGQ/K	20.627.754
2	15.045.434	CCFVHDCCYGQ/K	15.045.356
3	22.618.772	CCFVHDCCYGK/QI/LTACSPQ/K	22.619.149
4	17.687.631	Q/KI/LCECDRAAAI/LCFR	17.687.807
5	10.494.556	DNI/L/I/LTYDSQ/K	10.494.666
6	9.845.087	TYWKYPQ/K	9.845.069

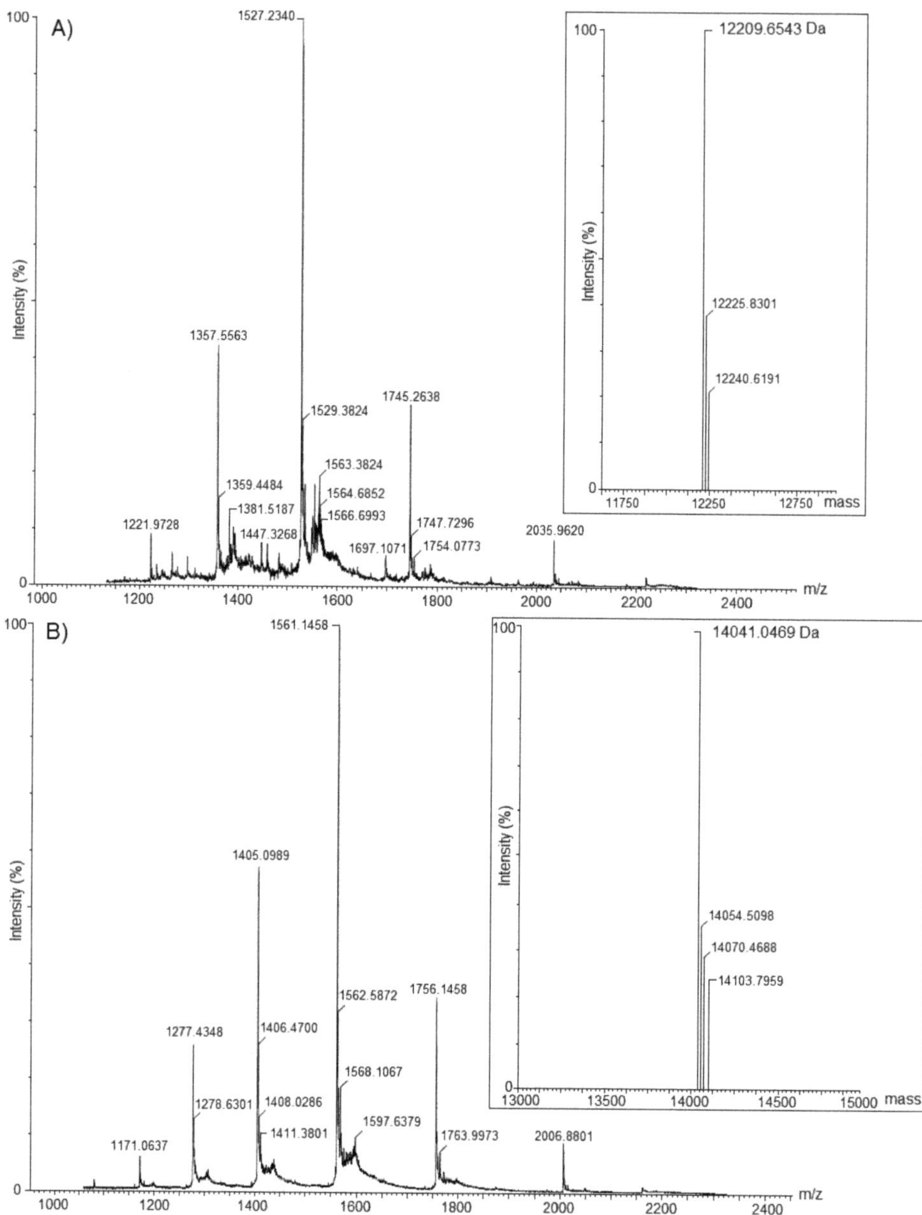

Figure 2. Molecular mass determination of ACP-TX-I (**A**) and ACP-TX-II (**B**) by nanoelectrospray tandem mass spectrometry using a Quadrupole Time-of-flight (Q-Tof) Ultima API mass spectrometer (MicroMass/Waters) with an output mass range of 6000–20,000 Da at a "resolution" of 0.1 Da/channel. Raw and deconvoluted electrospray mass spectra are shown (inserts).

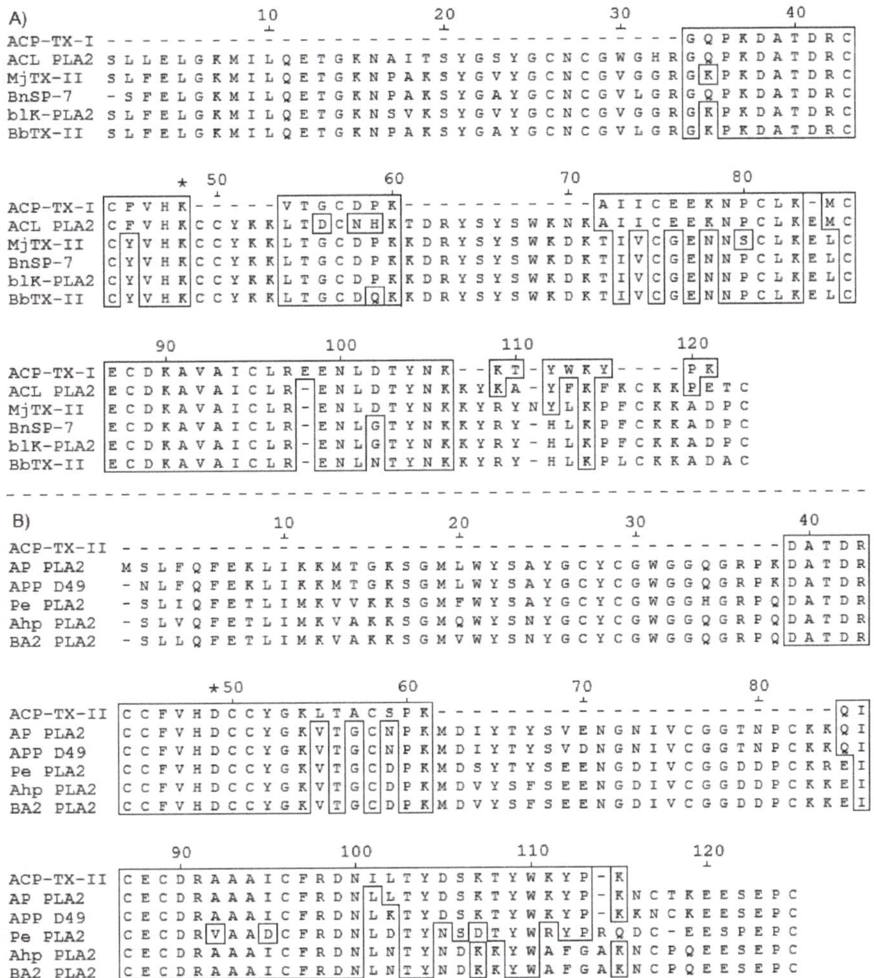

Figure 3. ACP-TX-I (**A**) and ACP-TX-II (**B**) show significant similarity to K49 and D49 PLA$_2$s, respectively (Edit Seq version 5.01© Program, DNASTAR Inc., Madison, WI, USA, 2001). ACL from *A. c. laticinctus* [26]; MjTX-II from *B. moojeni* [27]; BnSP-7 from *B. neuwiedi pauloensis* [28]; blK PLA$_2$ from *B. leucurus* [29]; BbTX-II from *B. brazili* [6]; AP PLA$_2$ from *A. piscivorus* [30]; APP PLA$_2$ from *A. p. piscivorus* [31]; Pe PLA$_2$ from *Protobothrops elegans* [32]; Ahp and BA2 PLA$_2$ from *A. halys pallas* [33,34]. Hyphens indicate gaps generated by the alignment software.

2.3. Activity Measurements of ACP-TX-II

ACP-TX-I did not show PLA$_2$ activity, but possessed a mass of ~14 kDa. ACP-TX-II displayed specific PLA$_2$ activity of 29.31 ± 1.62 nmol/min (Figure 4A). The pH optimum was 8.0 (Figure 4B) and this protein was stable at temperatures from 35 to 40 °C (Figure 4D). At low concentrations, ACP-TX-II showed a sigmoidal relationship with temperature (Figure 4C) and a strict dependence on calcium ions (10 mM) for full activity. Substitution of Ca^{2+} with Mg^{2+}, Mn^{2+}, Cd^{2+}, or Zn^{2+} (10 mM) significantly reduced enzyme activity (Figure 4E). Enzymatic activity of ACP-TX-II was abolished by EDTA and treatment with *p*-BPB. Incubation with crotapotins, F5 and F6, from *C. d. collilineatus* venom also diminished activity, while heparin did not significantly inhibit catalysis (Figure 4F).

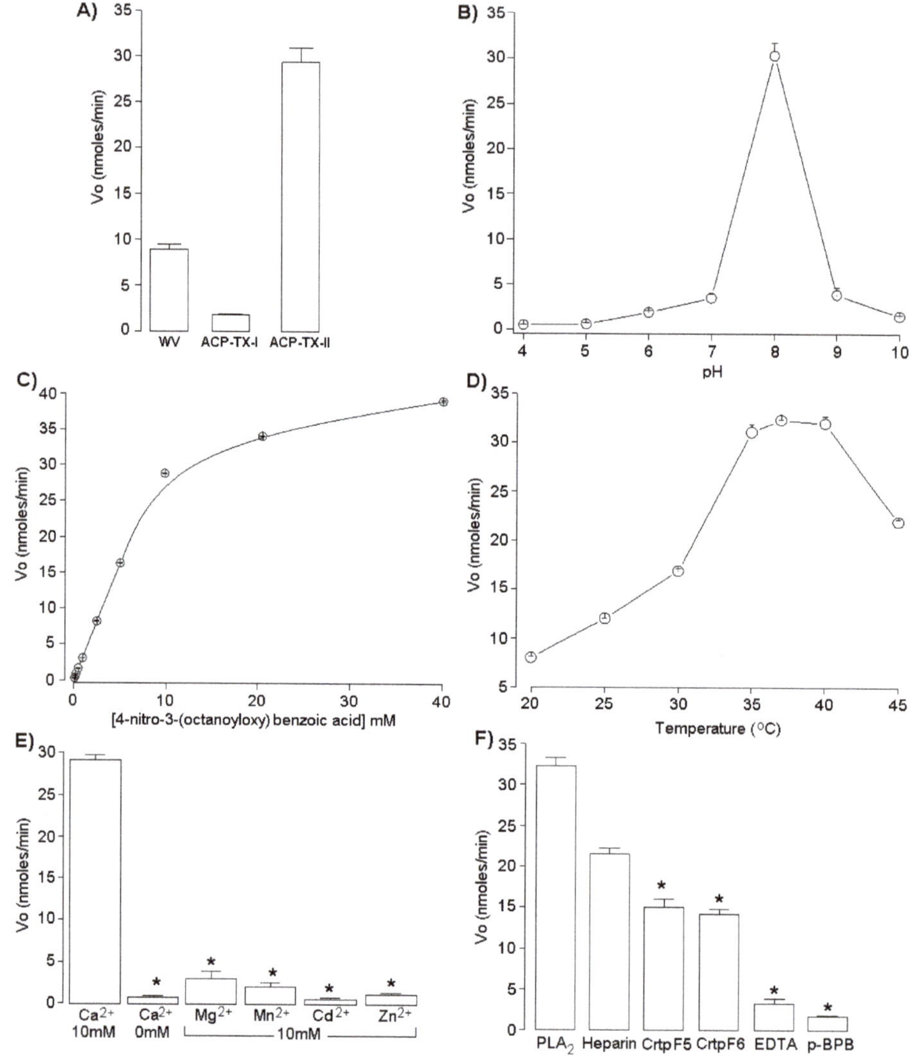

Figure 4. (**A**) PLA$_2$ activity of *A. c. pictigaster* venom, ACP-TX-I and ACP-TX-II PLA$_2$; (**B**) Effect of pH on the PLA$_2$ activity of ACP-TX-II; (**C**) Effect of substrate concentration on the PLA$_2$ activity of ACP-TX-II; (**D**) Effect of temperature on the PLA$_2$ activity of ACP-TX-II; (**E**) Influence of ions (10 mM each) on PLA$_2$ activity of ACP-TX-II; (**F**) Effect of heparin, EDTA crotapotins (F5 and F6) and chemical modification with BPB on PLA$_2$ activity of ACP-TX-II. The results are the mean ± SEM of three determinations (* $p < 0.05$).

2.4. Pharmacological Activities of ACP-TX-I and ACP-TX-II

In vivo, ACP-TX-II PLA$_2$ (20 and 50 μg), injected intramuscularly (IM), induced local myonecrosis, and time-course analysis showed a maximum increase in plasma CK 3 h after injection, returning to normal by 24 h (Figure 5B). In contrast, ACP-TX-I PLA$_2$ showed no local myotoxic effect, even at high concentrations (Figure 5A). ACP-TX-I and ACP-TX-II PLA$_2$ did not show systemic myotoxicity after intravenous (IV) injection (Figure S2).

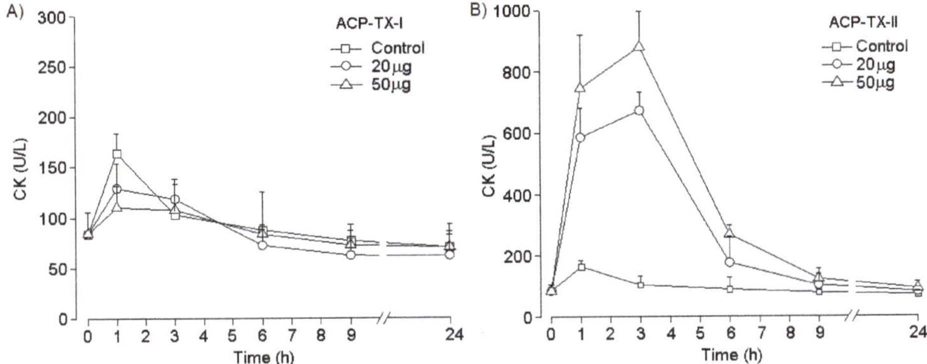

Figure 5. Myotoxic activity of ACP-TX-I PLA$_2$ (**A**) and ACP-TX-II PLA$_2$ (**B**) in mice. Time-course of the increments in plasma CK activity after an intramuscular injection of 20 and 50 μg of toxins compared to the injection of vehicle alone (PBS). Points represent means ± SD of four mice per group.

ACP-TX-I PLA$_2$ had little edematogenic effect, since a 50-μg injection was necessary in order to observe 23.4% edema after 3 h (Figure 6A). In contrast, ACP-TX-II PLA$_2$ presented marked paw edema, with maximal activity (40.54%) 3 h after a 50-μg injection. Edema returned to normal levels after 24 h, showing dose-dependent activity.

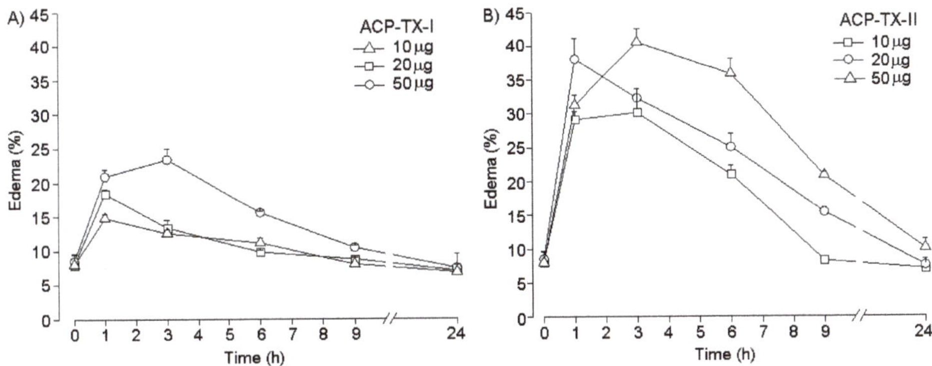

Figure 6. Edema-forming activity of ACP-TX-I PLA$_2$ (**A**) and ACP-TX-II PLA$_2$ (**B**) in mice. Induction of edema by toxins (10, 20 and 50 μg), injected SC in the footpads of mice. At various time intervals, the increase in footpad volume, compared to controls, was expressed as percent edema. Each point represents the mean ± SD of four animals per group.

ACP-TX-I was cytotoxic to cultured NIH/3T3 (non-tumor fibroblasts) and A549 (human lung carcinoma) cells treated with different concentrations of ACP-TX-I and ACP-TX-II PLA$_2$ (5–500 μg/mL) during a period of 24 h (Figure 7). ACP-TX-I PLA$_2$ cytotoxicity was dose-dependent on both cell types, causing a 50% decrease in cell viability at doses ≥20 μg/mL (Figure 7A). ACP-TX-II PLA$_2$ did not show toxicity in non-tumor cells used in this study, and 250–500 μg were required to decrease A549 tumor cell viability by 20 to 25% compared to controls, but the difference was not statistically significant (Figure 7B).

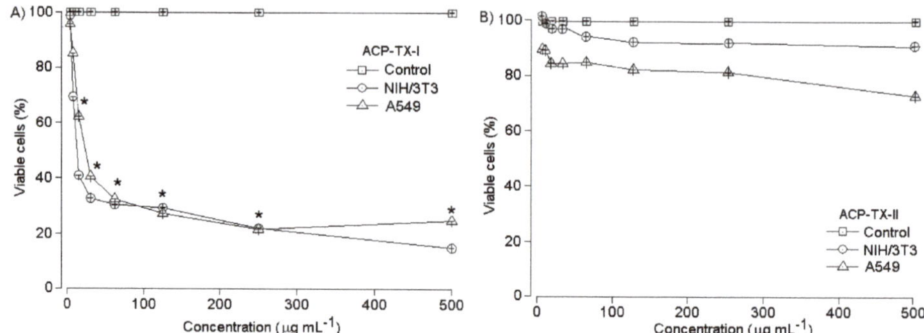

Figure 7. In vitro cytotoxic activity of ACP-TX-I PLA$_2$ (**A**) and ACP-TX-II PLA$_2$ (**B**) in NIH/3T3 (non-tumor fibroblasts) and A549 (human lung carcinoma) cells. Cell viability (%) was estimated by neutral red uptake assay. Experiments were performed in triplicate (* $p < 0.05$).

3. Discussion

Crude venom of the Trans-Pecos copperhead, *A. c. pictigaster*, was fractionated by HPLC. Fractions of interest were analyzed by MS and screened for diverse bioactivities. This work reports isolation and characterization of two basic PLA$_2$s, ACP-TX-I K49 and ACP-TX-II D49, the main components of this venom (Figure 1). Using C$_{18}$ reversed phase HPLC, 29 peaks were obtained. The two most prominent peaks, 16 and 18, named ACP-TX-I and ACP-TX-II, respectively, were selected because of PLA$_2$ activity and/or molecular mass in SDS-PAGE (Figure 1 insert). Multiple PLA$_2$ isoforms in the same venom were derived from accelerated microevolution, in which high mutation rates in gene coding regions, mainly associated with amino acids exposed to the solvent, allowed development of new functions [6,14].

Both ACP-TX-I and ACP-TX-II were obtained in high purity. Re-chromatography using the same chromatography system showed only one peak for each fraction (Figure S1). On SDS-PAGE, a single band was seen with an Mr of approximately 14 kDa under reducing and non-reducing conditions (Figure 1 insert). Using this purification method, several other PLA$_2$ (Bp13 PLA$_2$, PhTX-I, -II,-III, Bleu-PLA$_2$, Bbil-TX, BbTX-II, -III, etc.) from different venoms have been purified, showing that it is rapid and efficient for the purification of these proteins in one step [6–8,35–38].

Molecular masses determined for ACP-TX-I (12,209.7 Da) and ACP-TX-II (14,041.1 Da), obtained with ESI mass spectrometry (Figure 2), are very close to those of PLA$_2$s isolated from other snake venoms [26,34]. Partial amino acid sequencing of ACP-TX-I suggested that this protein probably has a Lys residue at position 49 (Figure 3A), based upon its high similarity to well characterized K49 PLA$_2$s such as ACL, MjTX-II, BnSP-7, bl-K, and BbTX-II from *A. c. latincictus*, *B. moojeni*, *B. neuwiedi pauloensis*, *B. leucurus*, and *B. brazili*, [6,26–29]. In addition, ACP-TX-I showed negligible catalytic activity compared to crude venom and ACP-TX-II (Figure 4A).

In contrast to various homologous K49 PLA$_2$s, ACP-TX-I migrated as a monomer in SDS-PAGE when analyzed under reducing and non-reducing conditions (Figure 1 insert). Recently, it was demonstrated that SDS induces oligomerization of the Lys49 PLA$_2$, BPII, from *Protobothrops flavoviridis* [39], although under the same conditions, ACP-TX-I behaves as a monomer. K49 PLA$_2$s isolated from three cottonmouths (*A. p. piscivorus*, *A. p. leucostoma*, and *A. p. conanti*) also behaved as monomers [14].

Partial amino acid sequencing of ACP-TX-II PLA$_2$ showed that it belongs to the D49 PLA$_2$ family, with an Asp residue at position 49 (Figure 3B). The catalytic site formed by H48, D49, Y52, and D89 is conserved, as reflected by the high enzymatic activity of the toxin (Figure 4A). Comparison of the amino acid sequence of ACP-TX-II shows a high degree of homology with other myotoxic D49 PLA$_2$ from viperid venoms (Figure 3B). It was not possible to determine the amino acid sequence corresponding to the Ca^{2+}-binding loop, comprising of residues Y24 to G35, and containing glycine residues at positions 26, 30, 32, and 33, and cysteine residues at positions 27 and 29 [40]. This domain is important to maintain Ca^{2+} in the correct position for a nucleophilic attack on the substrate.

Optimum enzymatic activity of ACP-TX-II PLA_2 occurred at pH 8 and 37 °C (Figure 4B,D), and Ca^{2+} was an obligatory co-factor [38,41]. Ca^{2+} replacement with other divalent ions (Mg^{2+}, Mn^{2+}, Cd^{2+}, and Zn^{2+}) resulted in a loss of catalytic activity (Figure 4E) [42]. Since ACP-TX-II requires Ca^{2+} for activity, chelators such as EDTA inhibited the enzymatic activity (Figure 4F). Histidine residue alkylation also abolished enzymatic activity of ACP-TX-II PLA_2 (Figure 4F), although alkylation does not completely disrupt the 3D structure of PLA_2 enzymes and their capacity to bind phospholipids, but it could modify their ability to interact with specific ligands or proteins [43].

Crotapotin is the acidic moiety of crotoxin that specifically chaperones the basic PLA_2 subunit, inhibiting its catalytic activity [44]. F5 and F6 crotapotins from *C. d. collilineatus* inhibited the enzymatic activity of ACP-TX-II PLA_2 by approximately 55% (Figure 4F). These results are consistent with PhTX-II PLA_2 from *Porthidium hyoprora*, which was inhibited ~60% by F2 and F3 crotapotins from *C. d. collilineatus* [8]. These results suggest that crotapotins may bind to *Agkistrodon* PLA_2, much like how they bind to crotaline PLA_2s.

Although snake venom PLA_2s exhibit diverse pharmacological activities [2], myotoxicity is one of the most common effects [45]. Necrosis induced in vivo in skeletal musculature by intramuscular injection or ex vivo by incubation with differentiated skeletal muscle cells [46] is evidenced by increased plasma CK levels. ACP-TX-II PLA_2 increased serum CK levels when injected intramuscularly (Figure 5B). Local myotoxicity is a characteristic of viperid envenomations, in which PLA_2s affect predominantly muscles near the injection site [47]. This is consistent with clinical examinations of *Agkistrodon* bites in the United States, where the main clinical manifestations are local effects that are in some cases associated with permanent dysfunction, while systemic effects are generally absent [19].

ACP-TX-II PLA_2 showed no systemic myotoxicity when injected intravenously. CK levels were similar to those of controls (Figure S2). Perhaps it lacks specificity and attaches to tissues at the site of injection, consistent with the hypothesis that myotoxins may act locally or systemically, proposed by Gutierrez and Ownby [46] in order to explain the pharmacological specificity of venom PLA_2s. Myotoxic PLA_2s bind predominantly to different cell types, as well as to muscle fibers, and are quickly sequestered after injection. On the other hand, systemic myotoxic PLA_2s, such as PLA_2s F6 and F6a from *Crotalus durissus collilineatus* [47], have high selectivity for skeletal muscle fibers and do not attach to other cells. This specificity allows systemic myotoxins to spread beyond the injection site, reaching the bloodstream and distant muscle cells and causing rhabdomyolysis.

Homologous K49 PLA_2s, despite lacking catalytic activity, also cause myonecrosis when injected intramuscularly in mice [48]. Basic/hydrophobic amino acid residues located in the C-terminal region are considered one of the structural determinants for K49 myotoxicity [49,50]. To exert myotoxicity, these toxins function as obligate dimers [11]. According to this model, catalytically inactive K49 dimers undergo structural rearrangement when a membrane fatty acid enters the hydrophobic channel of one of the monomers. This reorientation aligns the C-terminal regions of the two monomers in the same plane and facilitates membrane destabilization when specific hydrophobic amino acids (Leu121, Phe125) penetrate the membrane. Once the membrane becomes disorganized, cells lose ionic control, resulting in cell death [11].

Unlike most *Bothrops* K49 PLA_2s, ACP-TX-I does not display detectable local myotoxicity in mice (Figure 5A). Its lack of myotoxicity may reflect its monomeric structure (Figure 1 insert), preventing it from adopting different oligomeric configurations, appropriate to the physicochemical environment. These data corroborate findings with MjTX-I, which is a monomeric K49 PLA_2 in solution isolated from *Bothrops moojeni* venom, and which shows markedly decreased myotoxic activity [51].

ACP-TX-I and ACP-TX-II induce edema, but even at high concentrations, ACP-TX-I does not reach 30% edema (Figure 6A). The marked edema induced by ACP-TX-II is likely due to phospholipid hydrolysis. This possibility is supported by studies showing that chemical modification of D49 PLA_2s with *p*-BPB, which nullifies catalytic activity, dramatically reduces edematogenic activity of MT-III PLA$_2$ from *B. asper* as well as other D49 PLA$_2$s [52,53].

In addition to in vivo myotoxic and edematogenic activities, ACP-TX-I and ACP-TX-II were assayed for cytotoxicity in vitro on NIH/3T3 human fibroblasts and A549 lung cancer cells. ACP-TX-I showed high cytotoxicity in both cell lines (Figure 7A). On the other hand, even at concentrations as high as 500 µg/mL, ACP-TX-II had no inhibitory effect on NIH/3T3 fibroblasts. However, the A549 lung cancer cell line was sensitive to this myotoxin, losing ~30% viability after 24 h of incubation (Figure 7B).

The exact molecular mechanism by which snake venom PLA$_2$s decrease cell viability is unknown. Some authors have proposed that cytotoxic activity on tumor cells is associated with apoptosis induction [54], and propose that PLA$_2$ activity accelerates the rate of phospholipid renewal, which induces membrane changes that occur during apoptosis [55]. However, other mechanisms have been proposed: CC-PLA$_2$-1 and CC-PLA$_2$-2 from *Cerastes cerastes* inhibit cancerous cell adhesion and migration, as well as angiogenesis [56]. Crotoxin B interferes with signaling at epidermal growth factor receptors [57]. *Bothrops* myotoxins promote fatty acid-dependent lysis by interacting with a receptor able to activate intracellular lipase [58], etc. A K49 PLA$_2$ from *Protobothrops flavoviridis* induces cell death by caspase-independent apoptosis, accompanied by rapid plasma membrane disruption in human leukemia cells. Some homologous PLA$_2$s, such as MTX-II, exert cytotoxicity regardless of catalytic activity [59]. It is noteworthy that the cytotoxic mechanism depends on activities exerted by molecular regions other than the catalytic site.

4. Materials and Methods

4.1. Venom and Reagents

Venom was obtained from the National Natural Toxins Research Center (Texas A&M University-Kingsville). Solvents and reagents used were HPLC grade, sequence grade, or high purity, obtained from Sigma, Aldrich Chemicals, Merck and BioRad.

4.2. Purification of ACP-TX-I and ACP-TX-II

PLA$_2$ enzymes, ACP-TX-I and ACP-TX-II, from *A. c. pictigaster* venom were isolated by RP- HPLC, following Huancahuire-Vega et al. [7]. A measure of 20 mg of whole venom was dissolved in 250 µL of 0.1% TFA (buffer A) and centrifuged at 4500 g. Supernatant was then applied to an analytical RP-HPLC µ-Bondapak C$_{18}$ column (0.78 × 30 cm; Waters 991-PDA system Milford, MA., USA), and equilibrated with buffer A for 15 min. Protein elution employed a linear gradient (0–100%, *v/v*) of 66.5% acetonitrile in 0.1% TFA (buffer B) at a flow rate of 1.0 mL/min. Elution was monitored at 280 nm and PLA$_2$ activity was assayed in each fraction. Active PLA$_2$ fractions (ACP-TX-I and II) were collected, lyophilized, and used for subsequent biochemical/functional characterization.

4.3. Electrophoresis

Molecular masses of ACP-TX-I and II were determined under reducing and non-reducing conditions using Tricine SDS-PAGE in a discontinuous gel and buffer system [60]. Lysozyme, soybean trypsin inhibitor, carbonic anhydrase, ovalbumin, albumin, and phosphorylase B were used as molecular mass markers.

4.4. Determination of Molecular Masses of the Purified Proteins by Mass Spectrometry

A measure of 4.5 µL aliquots of ACP-TX-I and ACP-TX-II PLA$_2$ was injected using a C$_{18}$ (100 µm × 100 mm) ultra-high-performance reversed phase column (nanoAcquity UPLC, Waters) coupled with nanoelectrospray tandem mass spectrometry on a Quadrupole Time-of-flight (Q-Tof) Ultima API mass spectrometer (MicroMass/Waters Milford, MA., USA) at a flow rate of 600 nL/min. The spectrometer was operated in MS continuum mode, and data acquisition was from *m/z* 100–3000 at a scan rate of 1 s and an interscan delay of 0.1 s. The gradient used was 0–50% acetonitrile in 0.1%

formic acid over 45 min. A number of m/z mass spectra were accumulated over about 300 scans, and data were converted to molecular masses using maximum-entropy-based software from Masslynx 4.1 (Waters, Milford, MA., USA, 2005). Output masses ranged from 6000–20,000 Da at a 0.1 Da/channel "resolution." The simulated isotope pattern model was used with the spectrum blur width parameter set to 0.2 Da, and minimum intensity ratios between successive peaks were 20% (left and right). After the smoothing of deconvoluted spectra, mass centroid values were obtained using 80% of the peak top and a minimum peak width at half the height of 4 channels [8].

4.5. Analysis of Tryptic Digests

Prior to trypsin addition (Promega sequencing grade modified), ACP-TX-I and II PLA$_2$ were reduced (DTT 5 mM for 25 min to 56 °C) and alkylated (iodoacetamide, 14 mM for 30 min). After trypsin addition (20 ng/µL in Ambic 0.05 M), samples were incubated 16 h at 37 °C. The reaction was stopped with 0.4% formic acid and samples were centrifuged at 2500 rpm for 10 min. Pellets were discarded and supernatants were dried in a Speed Vac. Peptides were separated by C$_{18}$ reverse phase chromatography (100 µm × 100 mm) (nanoAcquity UPLC, Waters, Milford, MA., USA) coupled with nanoelectrospray tandem mass spectrometry on a Q-Tof Ultima API mass spectrometer (MicroMass/Waters) at a flow rate of 600 nL/min. In order to select ions of interest, an ESI/MS mass spectrum (TOF MS mode) was acquired for each HPLC fraction before performing tandem mass spectrometry, over the mass range of 100–2000 m/z. Then, these ions were fragmented in the collision cell (TOF MS/MS mode). Raw data files from LC-MS/MS runs were processed using Masslynx 4.1 (Waters) and analyzed using the MASCOT search engine, version 2.3 (Matrix Science Ltd. London, UK) against the snake database, using the following parameters: trypsin as the enzyme, fragment mass tolerance of ±0.1 Da, peptide mass tolerance of ±0.1 Da, and oxidation as a variable modification for methionine. Sequence alignments of ACP-TX-I and ACP-TX-II with K49 and D49 PLA$_2$s, respectively, were made using ClustalW in Edit Seq 5.01 © DNASTAR. (Madison, WI., USA, 2001).

4.6. PLA$_2$ Activity

PLA$_2$ activity was assayed as described by Holzer and Mackessy [61], and adapted for 96-well plates. The final volume of the standard assay mixture was 260 µL and contained 20 µL of substrate 4-nitro-3-(octanoyloxy) benzoic acid (3 mM), 200 µL of buffer (10 mM Tris–HCl, 10 mM CaCl$_2$, and 100 mM NaCl, pH 8.0), 20 µL of water, and 20 µL of ACP-TX-I or ACP-TX-II (1 mg/mL). The mixture was incubated at 37 °C for 40 min, measuring absorbances at intervals of 10 min. The initial velocity (Vo) was calculated based on the value of absorbance after 20 min of reaction. ACP-TX-II was chosen by studying kinetic parameters. Different substrate concentrations (40, 20, 10, 5, 2.5, 1.0, 0.5, 0.3, 0.2, and 0.1 mM) incubated in in Tris–HCl buffer, pH 8.0 at 37 °C, were used. The optimal temperature was determined by incubating the enzyme at different temperatures. Similarly, buffers of different pHs (4–10) were used in order to determine the optimal pH. All assays were done in triplicate and absorbances at 425 nm were measured with a VersaMax 190 multiwell plate reader (Molecular Devices, Sunnyvale, CA, USA).

4.7. Inhibition and Chemical Modifications

The effects of EDTA and low molecular weight heparin (Mr 6.000 Da) on enzymatic activity of ACP-TX-II PLA$_2$ were performed by incubating the enzyme with a 1 mM solution of EDTA or a molar ratio of 2:1 (heparin:toxin) at 37 °C for 30 min. Similarly, the effect of crotapotins F5 and F6 (1 mg/mL) from *Crotalus durissus collilineatus* upon enzymatic activity of ACP-TX-II was evaluated under the same conditions. Additionally, modification of His residues of ACP-TX-II with *p*-bromophenacyl bromide (BPB) (1.5 mg/mL in ethanol) was performed [43].

4.8. Myotoxic Activity

Different amounts (20 and 50 µg) of ACP-TX-I and II PLA$_2$ dissolved in 100 µL of PBS were injected IM or IV, in groups of four Swiss mice (18–20 g). The control group received 100 µL of PBS. Blood was collected from the tail into the heparinized capillary tubes at different intervals (1, 3, 6, 9, and 24 h), and plasma creatine kinase (CK; EC 2.7.3.2) activity was determined by a kinetic assay (Sigma 47-UV). Activity was expressed in U/L, where one unit is defined as the phosphorylation of 1 mmol of creatine/min at 25 °C.

4.9. Edema-Forming Activity

Fifty µL of phosphate-buffered saline (PBS; 0.12 M NaCl, 0.04 M sodium phosphate, pH 7.2) with ACP-TX-I and II PLA$_2$ (10, 20, and 50 µg/paw) were injected into the subplantar region of the right footpad of five Swiss mice (18–20 g). The left footpad received 50 µL of PBS, as a control. Immediately before inoculation (basal) and at different time intervals (1, 3, 6, 9, and 24 h) paw volume was evaluated by plethysmography (Model 7140 Plethysmometer Ugo Basile, VA., Italy). Edema-forming activity was expressed as the percentage increase in volume of the right foot pad in comparison to the left foot pad (control). The percentage of edema in toxin-inoculated paws was calculated with the equation: % edema = $[((Tx \times 100)/T_0) - 100]$. T_0 is the paw volume before toxin injection. Tx is the edema (volume) measured at each time interval. The percentage of edema calculated was subtracted from the matched values at each time point in the saline injected hind paw (control) [62].

4.10. Cytotoxic Activity

Cytotoxic activity was assayed on NIH 3T3 fibroblasts (ATCC®CCL-1658™) and A549 lung cancer (ATCC®CCL-185™) cells, grown in plastic flasks (25 cm^2) with RPMI 1640 medium (Cultilab, Campinas, SP, Brazil), was supplemented with 2% L-glutamine, 120 µg/mL garamycin, and 13% inactivated fetal bovine serum (complete medium). Cultures were incubated at 37 °C in an atmosphere containing 5% CO$_2$. The medium was changed every 48 h, and when the culture reached confluence, subculturing was performed by treatment with trypsin and versene (Adolfo Lutz, São Paulo, SP, Brazil). Variable amounts of both ACP-TX-I and II were diluted in the assay medium and added to cells in 96-well plates. Experiments were carried out in triplicate. Cellular viability was assayed by the neutral red uptake assay of Ates et al. [63]. After treatment with toxins, the medium was removed and the culture was washed with PBS. For each well, 0.2 mL RPMI medium containing 50 µg/mL Neutral Red dye was added. The plate was incubated for 3 h at 37 °C to capture dye by viable cell lysosomes. After incubation, the medium containing the dye was removed and the wells rapidly washed with formalin-calcium to remove unincorporated dye from the cells. Then 0.1 mL of a solution of 1% (*v/v*) acetic acid: 50% (*v/v*) ethanol was added to each well to extract the dye. After shaking for 10 min on a microtitre plate shaker, absorbance was read at 540 nm. Cell viability was expressed as percentages compared to control and untreated cells.

4.11. Statistical Analyses

Results were reported as mean ± SEM. Dunnett's test was used to determine the significance of differences among means by analysis of variance when several experimental groups were compared with the control group. Differences were considered statistically significant if $p < 0.05$.

Supplementary Materials: The following are available online at http://www.mdpi.com/2072-6651/11/11/661/s1, Figure S1: Re-chromatography on an analytical RP-HPLC C18 analytical column of ACP-TX-I and ACP-TX-II. Figure S2: ACP-TX-II produces local myotoxicity when injected intramuscularly, but little systemic myotoxicity when injected intravenously, whereas ACP-TX-I and crude venom injected intravenously produce no systemic myotoxicity.

Author Contributions: All authors approved the submitted version. Authors contributions were: conceptualization, S.H.-V. and S.M.; Methodology, L.M.H., M.G.-H., E.E.N.-N. and S.H.-V.; Formal Analysis, S.H.-V. and S.M.; Investigation, L.M.H., M.G.-H., E.E.N.-N. and S.H.-V.; Writing—Original Draft Preparation, M.G.-H., E.E.N.-N.;

Writing—Review & Editing, L.M.H., S.H.-V. and S.M.; Supervision, S.H.-V., and S.M.; Project Administration, S.M.; Funding Acquisition, S.H.-V.

Funding: This work was supported by the National Council for Scientific and Technological Development (CNPq, process 163536/2013-9) and is part of Postdoctoral research of Salomón Huancahuire-Vega.

Acknowledgments: We acknowledge the Mass spectrometry Laboratory at Brazilian Biosciences National laboratory, CNPEM-ABTLUS, Campinas, Brazil, for its support with mass spectrometric analyses.

Conflicts of Interest: The authors have no conflict of interest.

Ethical statement: The animals and research protocols used in this study followed the guidelines of the Ethical Committee for use of animals of ECAE-IB-UNICAMP SP, Brazil and international law and policies.

References

1. Kini, R.M. Excitement ahead: Structure, function and mechanism of snake venom phospholipase A2 enzymes. *Toxicon* **2003**, *42*, 827–840. [CrossRef] [PubMed]
2. Gutiérrez, J.M.; Lomonte, B. Phospholipases A2: Unveiling the secrets of a functionally versatile group of snake venom toxins. *Toxicon* **2013**, *62*, 27–39. [CrossRef] [PubMed]
3. Schaloske, R.H.; Dennis, E.A. The phospholipase A2 superfamily and its group numbering system. *Biochim. Biophys. Acta-Mol. Cell Biol. Lipids* **2006**, *1761*, 1246–1259. [CrossRef] [PubMed]
4. Six, D.A.; Dennis, E.A. The expanding superfamily of phospholipase A2 enzymes: Classification and characterization. *Biochim. Biophys. Acta-Mol. Cell Biol. Lipids* **2000**, *1488*, 1–19. [CrossRef]
5. Oliveira, A.; Bleicher, L.; Schrago, C.G.; Silva Junior, F.P. Conservation analysis and decomposition of residue correlation networks in the phospholipase A2 superfamily (PLA2s): Insights into the structure-function relationships of snake venom toxins. *Toxicon* **2018**, *146*, 50–60. [CrossRef]
6. Huancahuire-Vega, S.; Ponce-Soto, L.A.; Martins-de-Souza, D.; Marangoni, S. Structural and functional characterization of brazilitoxins II and III (BbTX-II and -III), two myotoxins from the venom of Bothrops brazili snake. *Toxicon* **2009**, *54*, 818–827. [CrossRef]
7. Huancahuire-Vega, S.; Ponce-Soto, L.A.; Martins-De-Souza, D.; Marangoni, S. Biochemical and pharmacological characterization of PhTX-I a new myotoxic phospholipase A 2 isolated from Porthidium hyoprora snake venom. *Comp. Biochem. Physiol. Toxicol. Pharmacol. CBP* **2011**, *154*, 108–119. [CrossRef]
8. Huancahuire-Vega, S.; Ponce-Soto, L.A.; Marangoni, S. PhTX-II a basic myotoxic phospholipase A_2 from Porthidium hyoprora snake venom, pharmacological characterization and amino acid sequence by mass spectrometry. *Toxins* **2014**, *6*, 3077–3097. [CrossRef]
9. de Roodt, A.; Fernández, J.; Solano, D.; Lomonte, B. A myotoxic Lys49 phospholipase A2-homologue is the major component of the venom of Bothrops cotiara from Misiones, Argentina. *Toxicon* **2018**, *148*, 143–148. [CrossRef]
10. Grabner, A.N.; Alfonso, J.; Kayano, A.M.; Moreira-Dill, L.S.; Dos Santos, A.P.A.; Caldeira, C.A.S.; Sobrinho, J.C.; Gome, A.; Grabner, F.P.; Cardoso, F.F.; et al. BmajPLA2-II, a basic Lys49-phospholipase A2 homologue from Bothrops marajoensis snake venom with parasiticidal potential. *Int. J. Biol. Macromol.* **2017**, *102*, 571–581. [CrossRef]
11. Fernandes, C.A.H.; Comparetti, E.J.; Borges, R.J.; Huancahuire-Vega, S.; Ponce, L.A.; Marangoni, S.; Soares, A.M.; Fontes, M.R.M. Structural bases for a complete myotoxic mechanism: Crystal structures of two non-catalytic phospholipases A_2-like from Bothrops brazili venom. *Biochim. Biophys. Acta-Proteins Proteom.* **2013**, *1834*, 2772–2781. [CrossRef] [PubMed]
12. de Lima, L.F.G.; Borges, R.J.; Viviescas, M.A.; Fernandes, C.A.H.; Fontes, M.R.M. Structural studies with BnSP-7 reveal an atypical oligomeric conformation compared to phospholipases A_2-like toxins. *Biochimie* **2017**, *142*, 11–21. [CrossRef] [PubMed]
13. Salvador, G.H.M.; dos Santos, J.I.; Lomonte, B.; Fontes, M.R.M. Crystal structure of a phospholipase A_2 from Bothrops asper venom: Insights into a new putative myotoxic cluster. *Biochimie* **2017**, *133*, 95–102. [CrossRef] [PubMed]
14. Jia, Y.; Villarreal, J. Phospholipases A_2 purified from cottonmouth snake venoms display no antibacterial effect against four representative bacterial species. *Toxicon* **2018**, *151*, 1–4. [CrossRef] [PubMed]
15. Angulo, Y.; Lomonte, B. Biochemistry and toxicology of toxins purified from the venom of the snake Bothrops asper. *Toxicon* **2009**, *54*, 949–957. [CrossRef]

16. Campbell, J. *The Venomous Reptiles of the Western Hemisphere*; Cornell University: New York, NY, USA, 2004.

17. Guiher, T.J.; Burbrink, F.T. Demographic and phylogeographic histories of two venomous North American snakes of the genus Agkistrodon. *Mol. Phylogenet. Evol.* **2008**, *48*, 543–553. [CrossRef]

18. Knight, R.A. *Molecular Systematics of the Agkistrodon Complex*; Texas Tech University: Lubbock, TX, USA, 1991.

19. Domanski, K.; Kleinschmidt, K.C.; Greene, S.; Ruha, A.M.; Berbata, V.; Onisko, N.; Campleman, S.; Brent, J.; Wax, P. Cottonmouth snake bites reported to the ToxIC North American snakebite registry 2013–2017. *Clin. Toxicol.* **2019**, 1–5. [CrossRef]

20. Walter, F.G.; Stolz, U.; Shirazi, F.; Walter, C.M.; McNally, J. Epidemiology of the reported severity of copperhead (Agkistrodon contortrix) snakebite. *South. Med. J.* **2012**, *105*, 313–320. [CrossRef]

21. Johnson, E.K.; Ownby, C.L. Isolation of a myotoxin from the venom of Agkistrodon contortrix laticinctus (broad-banded copperhead) and pathogenesis of myonecrosis induced by it in mice. *Toxicon* **1993**, *31*, 243–255. [CrossRef]

22. Bocian, A.; Urbanik, M.; Hus, K.; Łyskowski, A.; Petriaal, V.; Andrejčáková, Z.; Petrillová, M.; Legáth, J. Proteomic analyses of Agkistrodon contortrix contortrix venom using 2D electrophoresis and MS techniques. *Toxins* **2016**, *8*, 372. [CrossRef]

23. Fořtová, H.; Suttnar, J.; Dyr, J.E.; Pristach, J. Simultaneous isolation of protein C activator, fibrin clot promoting enzyme (fiprozyme) and phospholipase A$_2$ from the venom of the southern copperhead snake. *J. Chromatogr. B Biomed. Appl.* **1997**, *694*, 49–53. [CrossRef]

24. Lucena, S.; Rodríguez-Acosta, A.; Grilli, E.; Alfonso, A.; Goins, A.; Ogbata, I.; Walls, R.; Suntravat, M.; Uzcátegui, N.L.; Guerrero, B.; et al. The characterization of trans-pecos copperhead (Agkistrodon contortrix pictigaster) venom and isolation of two new dimeric disintegrins. *Biologicals* **2016**, *44*, 191–197. [CrossRef] [PubMed]

25. Lomontea, B.; Tsai, W.-C.; Ureña-Diaz, J.M.; Sanz, L.; Mora-Obando, D.; Sánchez, E.E.; Fry, B.G.; Gutiérreza, J.M.; Gibbs, H.L.; Sovic, G.M.; et al. Venomics of New World pit vipers: Genus-wide comparisons of venom proteomes across Agkistrodon. *J. Proteom.* **2014**, *96*, 103–116. [CrossRef] [PubMed]

26. De Araujo, H.S.S.; White, S.P.; Ownby, C.L. cDNA cloning and sequence analysis of a lysine-49 phospholipase A 2 myotoxin from Agkistrodon contortrix laticinctus snake venom. *Arch. Biochem. Biophys.* **1996**, *326*, 21–30. [CrossRef]

27. Soares, A.M.; Rodrigues, V.M.; Homsi-Brandeburgo, M.I.; Toyama, M.H.; Lombardi, F.R.; Armo, R.K.; Giglio, J.R. A rapid procedure for the isolation of the LYS-49 myotoxin II from bothrops moojeni (caissaca) venom: Biochemical characterization, crystallization, myotoxic and edematogenic activity. *Toxicon* **1998**, *36*, 503–514. [CrossRef]

28. Soares, A.M.; Guerra-Sá, R.; Borja-Oliveira, C.R.; Rodrigues, V.M.; Rodrigues-Simioni, L.; Rodrigues, V.; Fontes, M.R.M.; Lomonte, B.; Gutiérrez, J.M.; Giglio, J.R. Structural and functional characterization of BnSP-7, a Lys49 myotoxic phospholipase A$_2$ homologue from Bothrops neuwiedi pauloensis venom. *Arch. Biochem. Biophys.* **2000**, *378*, 201–209. [CrossRef]

29. Higuchi, D.A.; Barbosa, C.M.V.; Bincoletto, C.; Chagas, J.R.; Magalhaes, A.; Richardson, M.; Sanchez, E.F.; Pesquero, J.B.; Araujo, R.C.; Pesquero, J.L. Purification and partial characterization of two phospholipases A2 from Bothrops leucurus (white-tailed-jararaca) snake venom. *Biochimie* **2007**, *89*, 319–328. [CrossRef]

30. Lathrop, B.K.; Burack, W.R.; Biltonen, R.L.; Rule, G.S. Expression of a group II phospholipase A2 from the venom of Agkistrodon piscivorus piscivorus in Escherichia coli: Recovery and renaturation from bacterial inclusion bodies. *Protein Expr. Purif.* **1992**, *3*, 512–517. [CrossRef]

31. Welches, W.; Reardon, I.; Heinrikson, R.L. An examination of structural interactions presumed to be of importance in the stabilization of phospholipase A$_2$ dimers based upon comparative protein sequence analysis of a monomeric and dimeric enzyme from the venom of Agkistrodon p. piscivorus. *J. Protein Chem.* **1993**, *12*, 187–193. [CrossRef]

32. Chijiwa, T.; Tokunaga, E.; Ikeda, R.; Terada, K.; Ogawa, T.; Oda-Ueda, N.; Hattori, S.; Nozaki, M.; Ohno, M. Discovery of novel [Arg49] phospholipase A$_2$ isozymes from *Protobothrops elegans* venom and regional evolution of Crotalinae snake venom phospholipase A$_2$ isozymes in the southwestern islands of Japan and Taiwan. *Toxicon* **2006**, *48*, 672–682. [CrossRef]

33. Wang, Y.; Cui, G.; Zhao, M.; Yang, J.; Wang, C.; Giese, R.; Peng, S. Bioassay-directed purification of an acidic phospholipase A 2 from Agkistrodon halys pallas venom. *Toxicon* **2008**, *51*, 1131–1139. [CrossRef] [PubMed]

34. Pan, H.; Liu, X.; Ou-Yang, L.; Yang, G.; Zhou, Y.; Li, Z.; Wu, X. Diversity of cDNAs encoding phospholipase A 2 from Agkistrodon halys Pallas venom, and its expression in E. coli. *Toxicon* **1998**, *36*, 1155–1163. [CrossRef]

35. Marques, P.P.; Esteves, A.; Lancellotti, M.; Ponce-Soto, L.A.; Marangoni, S. Novel acidic phospholipase A_2 from Porthidium hyoprora causes inflammation with mast cell rich infiltrate. *Biochem. Biophys.* **2015**, *1*, 78–84. [CrossRef] [PubMed]

36. Marangoni, F.A.; Ponce-Soto, L.A.; Marangoni, S.; Landucci, E.C.T. Unmasking snake venom of bothrops leucurus: Purification and pharmacological and structural characterization of new PL A2 Bleu TX-III. *Biomed. Res. Int.* **2013**, *2013*. [CrossRef] [PubMed]

37. Victor, C.C.; Floriano, R.S.; Rodrigues-Simioni, L.; Winck, F.V.; Baldasso, P.A.; Ponce-Soto, L.A.; Marangoni, S. Biochemical, pharmacological, and structural characterization of new basic PLA2 Bbil-TX from bothriopsis bilineata snake venom. *Biomed. Res. Int.* **2013**, *2013*. [CrossRef]

38. Sucasaca-Monzón, G.; Randazzo-Moura, P.; Rocha, T.; Torres-Huaco, F.D.; Vilca-Quispe, A.; Alberto Ponce-Soto, L.; Marangoni, S.; da Cruz-Höfling, M.A.; Rodrigues-Simioni, L. Bp-13 PLA2: Purification and neuromuscular activity of a new Asp49 toxin isolated from bothrops pauloensis snake venom. *Biochem. Res. Int.* **2015**, *2015*. [CrossRef] [PubMed]

39. Matsui, T.; Kamata, S.; Ishii, K.; Maruno, T.; Ghanem, N.; Uchiyama, S.; Kato, K.; Suzuki, A.; Oda-Ueda, N.; Ogawa, T.; et al. SDS-induced oligomerization of Lys49-phospholipase A_2 from snake venom. *Sci. Rep.* **2019**, *9*, 1–8. [CrossRef]

40. Singh, G.; Gourinath, S.; Sharma, S.; Paramasivam, M.; Srinivasan, A.; Singh, T.P. Sequence and crystal structure determination of a basic phospholipase A2 from common krait (Bungarus caeruleus) at 2.4 A resolution: Identification and characterization of its pharmacological sites. *J. Mol. Biol.* **2001**, *307*, 1049–1059. [CrossRef]

41. Diz Filho, E.B.S.; Marangoni, S.; Toyama, D.O.; Fagundes, F.H.R.; Oliveira, S.C.B.; Fonseca, F.V.; Calgarotto, A.K.; Joazeiro, P.P.; Toyama, M.H. Enzymatic and structural characterization of new PLA2 isoform isolated from white venom of Crotalus durissus ruruima. *Toxicon* **2009**, *53*, 104–114. [CrossRef]

42. Yu, B.Z.; Jain, M.K.; Berg, O.G. The Divalent Cation Is Obligatory for the Binding of Ligands to the Catalytic Site of Secreted Phospholipase A_2. *Biochemistry* **1993**, *32*, 6485–6492. [CrossRef]

43. Huancahuire-Vega, S.; Correa, D.H.A.; Hollanda, L.M.; Lancellotti, M.; Ramos, C.H.I.; Ponce-Soto, L.A. Chemical modifications of PhTX-I myotoxin from porthidium hyoprora snake venom: Effects on structural, enzymatic, and pharmacological properties. *Biomed. Res. Int* **2013**, *2013*. [CrossRef] [PubMed]

44. Aird, S.D.; Kaiser, I.I.; Lewis, R.V.; Kruggel, W.G. Rattlesnake presynaptic neurotoxins: Primary structure and evolutionary origin of the acidic subunit. *Biochemistry* **1985**, *24*, 7054–7058. [CrossRef] [PubMed]

45. Montecucco, C.; Gutiérrez, J.M.; Lomonte, B. Cellular pathology induced by snake venom phospholipase A2 myotoxins and neurotoxins: Common aspects of their mechanisms of action. *Cell. Mol. Life Sci.* **2008**, *65*, 2897–2912. [CrossRef] [PubMed]

46. Gutiérrez, J.M.; Ownby, C.L. Skeletal muscle degeneration induced by venom phospholipases A_2: Insights into the mechanisms of local and systemic myotoxicity. *Toxicon* **2003**, *42*, 915–931. [CrossRef]

47. Gutiérrez, J.M.; Alberto Ponce-Soto, L.; Marangoni, S.; Lomonte, B. Systemic and local myotoxicity induced by snake venom group II phospholipases A_2: Comparison between crotoxin, crotoxin B and a Lys49 PLA2 homologue. *Toxicon* **2008**, *51*, 80–92. [CrossRef]

48. Gutiérrez, J.M.; Lomonte, B.; Chaves, F.; Moreno, E.; Cerdas, L. Activities of a Toxic a Isolated From the Venom of the Snake Bothrops. *Comp. Biochem. Physiol. Part C: Comp. Pharmacol.* **1986**, *84*, 159–164.

49. Lomonte, B.; Angulo, Y.; Calderón, L. An overview of lysine-49 phospholipase A2 myotoxins from crotalid snake venoms and their structural determinants of myotoxic action. *Toxicon* **2003**, *42*, 885–901. [CrossRef]

50. Lomonte, B.; Rangel, J. Snake venom Lys49 myotoxins: From phospholipases A_2 to non-enzymatic membrane disruptors. *Toxicon* **2012**, *60*, 520–530. [CrossRef]

51. Salvador, G.H.M.; Fernandes, G.A.H.; Magro, A.J.; Marchi-Salvador, D.P.; Cavalcante, W.L.G.; Fernandez, R.M.; Gallacci, M.; Soares, A.M.; Oliveira, C.L.P.; Fontes, M.R.M. Structural and Phylogenetic Studies with MjTX-I Reveal a Multi-Oligomeric Toxin—a Novel Feature in Lys49-PLA2s Protein Class. *PLoS ONE* **2013**, *8*, 4. [CrossRef]

52. Chaves, F.; León, G.; Alvarado, V.H.; Gutiérrez, J.M. Pharmacological modulation of edema induced by Lys-49 and Asp-49 myotoxic phospholipases A_2 isolated from the venom of the snake Bothrops asper (terciopelo). *Toxicon* **1998**, *36*, 1861–1869. [CrossRef]

53. Zuliani, J.P.; Fernandes, C.M.; Zamuner, S.R.; Gutiérrez, J.M.; Teixeira, C.F.P. Inflammatory events induced by Lys-49 and Asp-49 phospholipases A$_2$ isolated from Bothrops asper snake venom: Role of catalytic activity. *Toxicon* **2005**, *45*, 335–346. [CrossRef] [PubMed]

54. Chen, K.C.; Kao, P.H.; Lin, S.R.; Sen Chang, L. p38 MAPK activation and mitochondrial depolarization mediate the cytotoxicity of Taiwan cobra phospholipase A$_2$ on human neuroblastoma SK-N.-SH cells. *Toxicol. Lett.* **2008**, *180*, 53–58. [CrossRef] [PubMed]

55. Panini, S.R.; Yang, L.; Rusinol, A.E.; Sinensky, M.S.; Bonventre, J.V.; Leslie, C.C. Arachidonate metabolism and the signaling pathway of induction of apoptosis by oxidized LDL/oxysterol. *J. Lipid Res.* **2001**, *42*, 1678–1686. [PubMed]

56. Zouari-Kessentini, R.; Luis, J.; Karray, A.; Kallech-Ziri, O.; Srairi, N.; Bazaa, A.; Loret, E.; Sofiane, B.; Mohamed, E.; Marrakchi, N. Two purified and characterized phospholipases A2 from Cerastes cerastes venom, that inhibit cancerous cell adhesion and migration. *Toxicon* **2009**, *53*, 444–453. [CrossRef] [PubMed]

57. Donato, N.J.; Martin, C.A.; Perez, M.; Newman, R.A.; Vidal, J.C.; Etcheverry, M. Regulation of epidermal growth factor receptor activity by crotoxin, a snake venom phospholipase A$_2$ toxin: A novel growth inhibitory mechanism. *Biochem. Pharmacol.* **1996**, *51*, 1535–1543. [CrossRef]

58. Gebrim, L.C.; Marcussi, S.; Menaldo, D.; de Menezes, C.S.R.; Nomizo, A.; Hamaguchi, A.; Silveira-Lacerda, E.P.; Homsi-Brandeburgo, A.I.; Sampaio, S.V.; Soares, A.M.; et al. Antitumor effects of snake venom chemically modified Lys49 phospholipase A2-like BthTX-I and a synthetic peptide derived from its C-terminal region. *Biologicals* **2009**, *37*, 222–229. [CrossRef]

59. Costa, T.R.; Menaldo, D.L.; Oliveira, C.Z.; Santos-Filho, N.A.; Teixeira, S.S.; Nomizo, A.; Fuly, A.L.; Monteiro, M.C.; Souza, B.M.; Palma, M.S.; et al. Myotoxic phospholipases A2 isolated from Bothrops brazili snake venom and synthetic peptides derived from their C-terminal region: Cytotoxic effect on microorganism and tumor cells. *Peptides* **2008**, *29*, 1645–1656. [CrossRef]

60. Schägger, H.; von Jagow, G. Tricine-sodium dodecyl sulfate-polyacrylamide gel electrophoresis for the separation of proteins in the range from 1 to 100 kDa. *Anal. Biochem.* **1987**, *166*, 368–379. [CrossRef]

61. Holzer, M.; Mackessy, S.P. An aqueous endpoint assay of snake venom phospholipase A$_2$. *Toxicon* **1996**, *34*, 1149–1155. [CrossRef]

62. Ferreira, T.; Camargo, E.A.; Ribela, M.T.C.P.; Damico, D.C.; Marangoni, S.; Antunes, E.; Nucci, G.D.; Landucci, E.C.T. Inflammatory oedema induced by Lachesis muta muta (Surucucu) venom and LmTX-I in the rat paw and dorsal skin. *Toxicon* **2009**, *53*, 69–77. [CrossRef]

63. Ates, G.; Rogiers, T.V.; Rodrigues, R.M. Assaying Cellular Viability Using the Neutral Red Uptake Assay. *Cell Viability Assays* **2017**, *1601*, 19–26.

Article

Lebetin 2, a Snake Venom-Derived B-Type Natriuretic Peptide, Provides Immediate and Prolonged Protection against Myocardial Ischemia-Reperfusion Injury via Modulation of Post-Ischemic Inflammatory Response

Bochra Tourki [1,2], Anais Dumesnil [3], Elise Belaidi [4], Slim Ghrir [1], Diane Godin-Ribuot [4], Naziha Marrakchi [1], Vincent Richard [3], Paul Mulder [3] and Erij Messadi [1,*]

[1] Laboratoire des Venins et Biomolécules Thérapeutiques (LR11IPT08) et Plateforme de Physiologie et de Physiopathologie Cardiovasculaires (P2C), Institut Pasteur de Tunis, Université Tunis El Manar, 1068 Tunis, Tunisia

[2] Université Carthage Tunis, 1054 Bizerte, Tunisia

[3] Normandie Univ, UNIROUEN, Inserm U1096, FHU REMOD-VHF, 76000 Rouen, France

[4] Université Grenoble Alpes, Inserm U1042, Laboratoire HP2, 38000 Grenoble, France

* Correspondence: erij.messadi@pasteur.tn; Tel.: +216-53-617-425; Fax: +216-71-791-833

Received: 19 July 2019; Accepted: 16 August 2019; Published: 10 September 2019

Abstract: Myocardial infarction (MI) followed by left ventricular (LV) remodeling is the most frequent cause of heart failure. Lebetin 2 (L2), a snake venom-derived natriuretic peptide, exerts cardioprotection during acute myocardial ischemia-reperfusion (IR) ex vivo. However, its effects on delayed consequences of IR injury, including post-MI inflammation and fibrosis have not been defined. Here, we determined whether a single L2 injection exerts cardioprotection in IR murine models in vivo, and whether inflammatory response to ischemic injury plays a role in L2-induced effects. We quantified infarct size (IS), fibrosis, inflammation, and both endothelial cell and cardiomyocyte densities in injured myocardium and compared these values with those induced by B-type natriuretic peptide (BNP). Both L2 and BNP reduced IS, fibrosis, and inflammatory response after IR, as evidenced by decreased leukocyte and proinflammatory M1 macrophage infiltrations in the infarcted area compared to untreated animals. However, only L2 increased anti-inflammatory M2-like macrophages. L2 also induced a higher density of endothelial cells and cardiomyocytes. Our data show that L2 has strong, acute, prolonged cardioprotective effects in post-MI that are mediated, at least in part, by the modulation of the post-ischemic inflammatory response and especially, by the enhancement of M2-like macrophages, thus reducing IR-induced necrotic and fibrotic effects.

Keywords: natriuretic peptide; myocardial infarction; ischemia-reperfusion injury; inflammation; fibrosis

Key Contribution: The natriuretic-like peptide L2 exerted acute, prolonged post-ischemic effects, in vivo, after a single injection prior to the onset of myocardial reperfusion. L2 alleviated heart damage by attenuating necrosis and fibrosis, while promoting anti-inflammatory M2-macrophages in the infarcted myocardium, thus providing novel insights into the mechanism of action of natriuretic peptides.

---

1. Introduction

The gold standard of therapy for acute myocardial infarction (MI) is percutaneous intervention, with the aim of restoring blood flow to ischemic myocardium as quickly as possible. Nevertheless, reperfusion itself exacerbates myocardial injury and delays the functional recovery of the ischemic heart. This phenomenon is known as ischemia-reperfusion (IR) injury. Deleterious consequences of myocardial IR are not limited to myocytes but also include coronary endothelial cells [1]. Despite advances in mechanical and pharmacological reperfusion therapy, MI still leads to a higher risk for developing heart failure (HF), a process known as ventricular remodeling, which involves structural lesions (i.e., cardiomyocyte growth and death, inflammation, collagen matrix alterations and microvascular rarefaction) in infarcted and non-infarcted myocardium. Left ventricular (LV) remodeling is governed by complex interrelated mechanisms. Among them, post-MI inflammation plays a critical role, since it triggers wound healing and scar formation, including interstitial fibrosis, a major determinant of ventricular function impairment after MI [2]. The inflammatory response is mainly characterized by neutrophil infiltration, followed by monocyte/macrophage and lymphocyte influx in ischemic myocardium. Infiltrating monocytes express a proinflammatory (M1) phenotype, followed by a switch to a healing phenotype (M2). Both phenotypes are involved in resolution of the inflammatory process [3]. Thus, new therapies to optimize temporal and spatial regulation of the inflammatory response or a direct resolution of the imbalance between pro- and anti-inflammatory components offer interesting strategies to prevent or reverse post-MI LV dysfunction.

Several pharmacological postconditioning strategies have been used to prevent detrimental IR effects [4]. Recently, the B-type natriuretic peptide (BNP) has emerged as an important therapeutic tool in patients with HF [5]. In clinical practice, pharmacological agents that enhance the biological actions of this peptide, such as nesiritide, neprilysin inhibitors, or more recently, the angiotensin receptor antagonist-neprilysin inhibitor LCZ696, have shown potential for translational research to improve HF patient care [6,7]. However, some issues related to their efficacy, benefits, cardiovascular risks, or mechanisms of action are controversial [8,9]. In particular, although BNP's effect on cardiac remodeling and fibrosis is established [10,11], its anti-inflammatory activity is still under debate [9,12] and its direct effect on modulation of inflammatory cell subsets during IR has not yet been demonstrated [13].

Clinical results have shown that snake venom-derived compounds, such as *Dendroaspis* natriuretic peptide (DNP), may offer superior therapeutic benefits in chronic HF [14]. This is likely due to greater potency and increased stability as compared to human family members [15,16], while displaying similar benefits in cardiac ischemia through natriuretic receptor-mediated signaling [17,18]. Recent studies focused on Lebetin 2 (L2), a 38-amino acid peptide (4 kDa) isolated from *Macrovipera lebetina* venom [19,20], that shares structural homology with natriuretic peptide (NP) family members, BNP, atrial natriuretic peptide (ANP), and DNP [20] (ranked by decreasing order of homology). Interestingly, L2 exerts cardioprotection in an IR ex vivo murine model, with additional effects compared to those of BNP under the same conditions [18]. These cardiac effects are mediated through a BNP-like mechanism of action, involving the NP receptor (NPR)/cyclic guanosine monophosphate (cGMP)-mediated pathway, downstream activation of mitochondrial K_{ATP} channels, and inhibition of mitochondrial permeability transition pore (mPTP) at the time of reperfusion [18].

In the current study, we extended the reperfusion period to investigate the effect of L2 on delayed consequences of IR, in vivo, including cardiomyocyte death, collagen matrix alterations, endothelial cell rarefaction, and post-MI inflammatory response, since these parameters are determinants for tissue healing. We focused particularly on L2/BNP-induced inflammatory-cell modulation by examining M1/M2 macrophage recruitment in the infarcted heart. L2 proved effective against MI with acute and prolonged effects, after a single injection administered prior to the onset of reperfusion. To the best of our knowledge, this report describes novel insights into mechanisms of NPs in myocardial repair, since L2, but not BNP, induced an increase in M2-macrophage subtype after MI, contributing to the resolution of the inflammatory process, and subsequently reducing IR-induced necrotic and fibrotic effects.

2. Results

2.1. L2 Effect on Blood Pressure and Heart Rate

To define effective doses of L2 and BNP, we investigated their influence on blood pressure and heart rate (HR, see Materials and Methods). Mean baseline values for blood pressure and HR did not differ statistically among experimental groups in either rats or mice (Table 1). BNP or L2 induced a dose-dependent decrease in the mean arterial pressure (MAP, Figure 1a,c, Table 1). The maximal hypotensive response to BNP or L2 was further documented by comparing areas under curves (AUCs, Figure 1b,d). The HR was not statistically different among experimental groups before or after treatment (Table 1). In rats, the effect of 100 ng/g L2 was equivalent to the effect of BNP at 50 ng/g (Figure 1a,b, AUCs NS). In mice, 25 ng/g L2 was equivalent to 20 ng/g BNP at inducing hypotensive response (Figure 1c,d, AUCs NS). The doses selected significantly decreased blood pressure; however, the maximal hypotensive responses to these doses, occurring within 30 min after bolus injection, were les than 30% in all cases (Figure 1a,c). Therefore, these doses were used in subsequent IR experiments, based on their ability to elicit a mild decrease in blood pressure, which minimized the deleterious effect of hypotension.

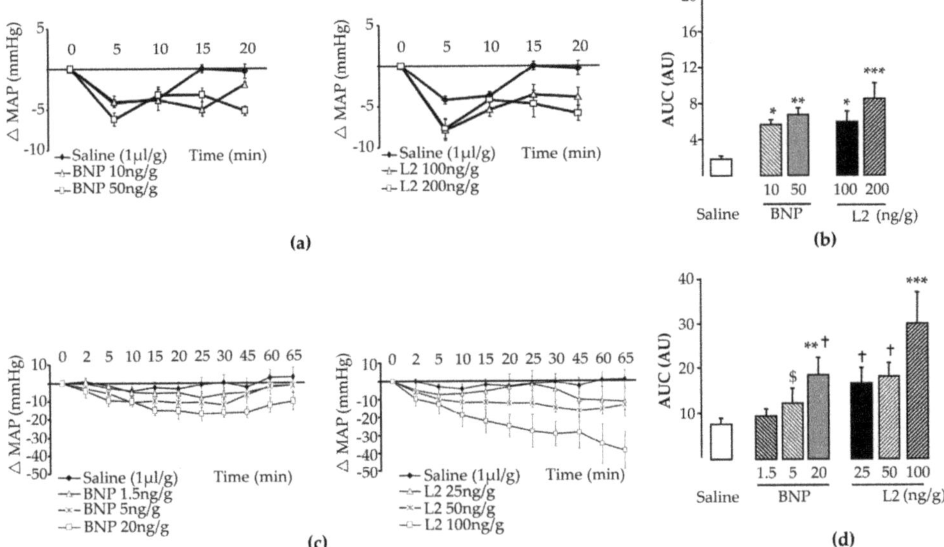

Figure 1. Effects of Lebetin 2 (L2) and B-type natriuretic peptide (BNP) on blood pressure. (**a**) Dose-dependent hypotensive effects of BNP (10 or 50 ng/g) and L2 (100 or 200 ng/g) in rats; (**b**) AUCs in rats, absolute decrease in MAP × time; (**c**) dose-dependent hypotensive effects of BNP (1.5, 5 or 20 ng/g) and L2 (25, 50 or 100 ng/g) in mice; (**d**) AUCs in mice, absolute decrease in MAP × time. Data are mean ± SEM. For the number of animals, see Table 1. *, $p < 0.05$, **, $p < 0.01$, ***, $p < 0.001$ vs. saline (control) group, \$, $p < 0.05$ vs. BNP (20 ng/g), †, $p < 0.05$ vs. L2 (100 ng/g). ΔMAP, variation in mean arterial blood pressure.

Table 1. Effects of B-type natriuretic peptide (BNP) and Lebetin 2 (L2) on blood pressure, heart rate, and post-ischemic areas at risk.

Group		Hemodynamic Study					IR Study		
		MAP (mmHg)		HR (bpm)			AR/LV (%)		
	n	Before	After	Before	After	n	R, 120 min	n	R, 14 days
Rats									
Control (saline 1 µL/g)	7	108 ± 5	96 ± 2	321 ± 6	310 ± 7	5	65.3 ± 3.0	10	82.4 ± 1.8
BNP (10 ng/g)	6	105 ± 7	92 ± 4	323 ± 6	315 ± 6				
BNP (50 ng/g)	6	104 ± 6	59 ± 6 *	335 ± 5	326 ± 6	5	66.6 ± 3.2	6	85.6 ± 1.3
L2 (100 ng/g)	7	105 ± 5	63 ± 3 *	326 ± 6	317 ± 6	6	66.4 ± 3.5	7	84.8 ± 1.1
L2 (200 ng/g)	7	106 ± 5	34 ± 7 *	330 ± 7	320 ± 4				
Mice							R, 120 min		
Control (saline 1 µL/g)	11	72 ± 3	70 ± 4	447 ± 27	422 ± 16	6	25.8 ± 2.5		
BNP (1.5 ng/g)	10	70 ± 3	62 ± 2 *	436 ± 14	467 ± 21				
BNP (5 ng/g)	5	67 ± 2	57 ± 3 *	432 ± 32	415 ± 36				
BNP (20 ng/g)	7	71 ± 6	54 ± 5 *	469 ± 35	473 ± 32	5	31.7 ± 1.2		
L2 (25 ng/g)	5	75 ± 4	64 ± 3	518 ± 9	414 ± 21	5	31.0 ± 2.6		
L2 (50 ng/g)	4	79 ± 4	63 ± 3 *	497 ± 42	412 ± 19				
L2 (100 ng/g)	5	69 ± 1	36 ± 8 *	511 ± 19	454 ± 7				

Data are mean ± SEM. For the hemodynamic study, blood pressure [mean arterial pressure (MAP)] and heart rate (HR) were recorded before and after treatment at maximal effect following BNP or L2 injection in each protocol. For the acute ischemia-reperfusion (IR) study, the area at risk (AR) was determined after 30 min of ischemia followed by 120 min of reperfusion in rats and mice. For the prolonged IR study, AR was determined after 35 min of ischemia followed by a 14–day reperfusion period in rats. bpm, beats per minute; R, reperfusion. *, $p < 0.05$ vs. corresponding value before treatment.

2.2. L2 Decreases LV Infarct Size Following IR Injury

After IR, the area at risk (AR) did not differ statistically among experimental groups (Table 1). After 120 min of reperfusion, infarct size (IS) in controls averaged 24.9% ± 1.8% in rats (Figure 2a) and 41.3% ± 3.5% in mice (Figure 2b). At 100 ng/g in rats or 25 ng/g in mice, L2 significantly reduced IS/AR to 15.3% ± 1.2% (−39%, $p < 0.001$) and 21.3% ± 2.8% (−48%, $p < 0.01$) respectively (Figure 2a,b). The BNP IS-limiting effect was about 60% in rats (50 ng/g, 9.9% ± 1.9%, $p < 0.001$; Figure 2a) and 41% in mice (20 ng/g, 24.3% ± 4.7%, $p < 0.01$; Figure 2b).

At day 14 post-reperfusion, the L2 (100 ng/g) cardioprotective effect was maintained in rats. IS fell to 3.7% ± 0.7% [−80%, vs. control non-treated animals (17.7% ± 1.4%), $p < 0.001$; Figure 3a]. This effect was more marked than that obtained with BNP at 50 ng/g (7.2% ± 1.0%, $p < 0.01$; Figure 3a).

2.3. L2 Exerts Post-Ischemic Anti-Fibrotic Effect

At day 14 post-reperfusion, interstitial fibrosis, corresponding to collagen accumulation relative to the total LV area, averaged 8.1% ± 0.9% in control animals (Figure 3b). Treatments with L2 (100 ng/g) and BNP (50 ng/g) decreased fibrosis by 52% (3.9% ± 0.5%, $p < 0.001$) and 34% (5.3% ± 0.7%, $p < 0.01$) respectively, compared to the control group (Figure 3b,c).

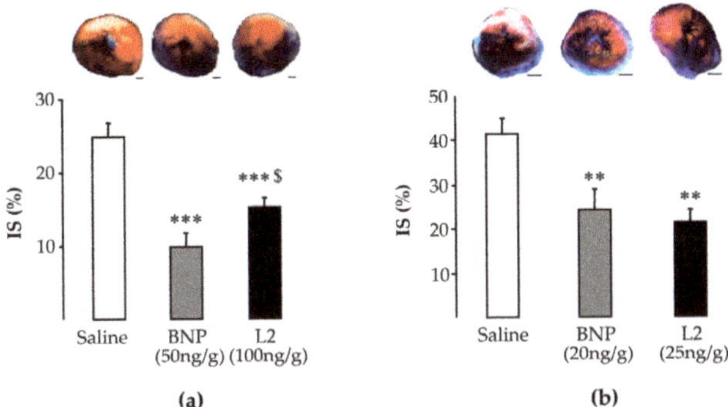

(a)   **(b)**

Figure 2. Effects of Lebetin 2 (L2) and B-type natriuretic peptide (BNP) on acute infarct size. Myocardial infarction was induced by 30-min ischemia followed by 120-min reperfusion. (**a**) Rats received saline (1 µL/g), BNP (50 ng/g) or L2 (100 ng/g) intraperitoneally, 5 min before reperfusion. (**b**) Mice received saline (1 µL/g), BNP (20 ng/g) or L2 (25 ng/g) intravenously, 5 min before reperfusion. Infarct size (IS) was measured relative to area at risk (IS/AR%) after 2,3,5-Triphenyltetrazolium chloride (TTC) staining. Images above histograms represent heart sections corresponding to experimental groups. Areas stained with Evans blue dye correspond to the tissue that had not undergone ischemia-reperfusion (IR). Red and white areas were obtained after TTC staining and correspond, respectively, to tissue that remained viable after IR and to the infarcted zone. Data are mean ± SEM. For the numbers of animals, see Table 1. **, $p < 0.01$, ***, $p < 0.001$ vs. saline (control) group, \$, $p < 0.05$ vs. BNP corresponding group. Scale bars = 1 mm.

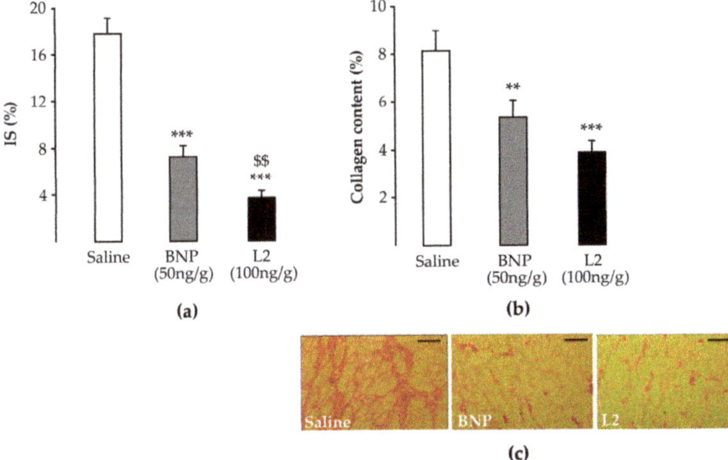

(c)

Figure 3. Effects of Lebetin 2 (L2) and B-type natriuretic peptide (BNP) on infarct size and fibrosis. Myocardial infarction was induced in rats by 35 min of ischemia followed by a 14-day reperfusion period. Saline (1 µL/g), BNP (50 ng/g) or L2 (100 ng/g) was administered intraperitoneally, 5 min before reperfusion. (**a**) Infarct size (IS) determined after Sirius red staining was expressed as a percentage ((endocardial + epicardial circumference of the infarcted tissue)/(endocardial+epicardial circumference of the LV) × 100); (**b**) density of collagen I and III used as a marker of fibrosis, was calculated after Sirius red staining as the area occupied by collagen/the surface of the image. On each section, several fields were photographed (15–20 images/rat) and the mean collagen density was calculated; (**c**) representative microphotographs of collagen content in Sirius red-stained myocardial slides from control and treated animals. Data are mean ± SEM. For the numbers of animals, see Table 1. **, $p < 0.01$, ***, $p < 0.001$ vs. saline (control) corresponding group, \$\$, $p < 0.01$ vs. BNP corresponding group. Scale bars = 50 µm, ×40.

2.4. L2 Modulates the Post-Ischemic Inflammatory Response

In control (untreated) rats, the number of CD45-positive cells was significantly enhanced in the infarcted LV area (353 ± 21 cells/mm^2) compared to the viable LV area (133 ± 11 cells/mm^2, $p < 0.001$; Figure 4a). Both L2 (100 ng/g) and BNP (50 ng/g) treatments decreased leukocyte infiltration in the infarcted area by 51% (173 ± 21 cells/mm^2, $p < 0.001$) and 34% (234 ± 33 cells/mm^2, $p < 0.001$), respectively, compared to controls (Figure 4a,b).

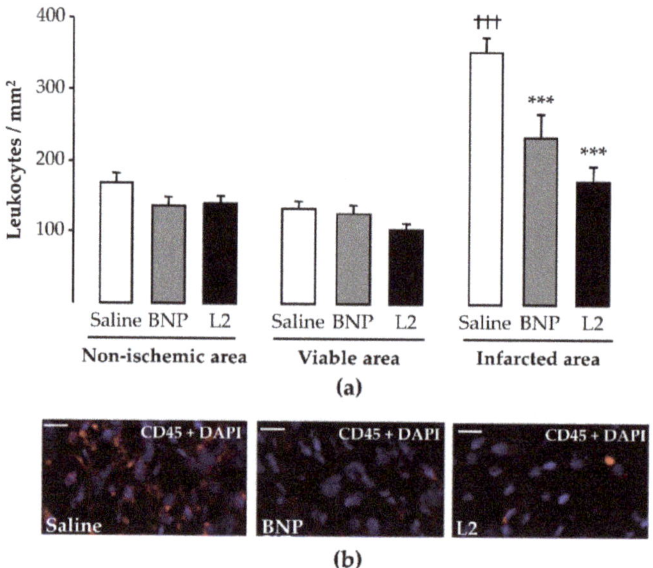

Figure 4. Effects of Lebetin 2 (L2) and B-type natriuretic peptide (BNP) on post-ischemic leukocyte infiltration. Myocardial infarction was induced by 35 min of ischemia followed by a 2-day reperfusion period. (**a**) Infiltrated leukocytes were assessed by CD45 immunolabeling in the infarcted area and its border zone (viable area), as well as in the septum (non-ischemic area) of rats treated with saline (1 μL/g), BNP (50 ng/g) or L2 (100 ng/g) intraperitoneally, 5 min before reperfusion. On each section, several fields were photographed (12–30 images/rat, $n = 5$–6) and the mean number of leukocytes per field was calculated; (**b**) representative microphotographs showing variable density of the leukocytes (CD45 in red and nuclei in blue) in the infarcted area from control and treated animals. Data are mean ± SEM. ***, $p < 0.001$ vs. saline (control) corresponding group, †††, $p < 0.001$ vs. corresponding other LV areas. Scale bars = 50 μm. ×20. DAPI, 4′, 6′-diamidino-2-phenylindole.

As with leukocytes, the number of CD68-positive cells was significantly increased in the infarcted LV area (357 ± 18 cells/mm^2) compared to the viable LV area (117 ± 17 cells/mm^2, $p < 0.001$; Figure 5a). Both L2 (284 ± 27 cells/mm^2, $p < 0.05$) and BNP (185 ± 26 cells/mm^2, $p < 0.001$) decreased the number of macrophages in the infarcted LV area by 20% and 48% respectively, compared to the control group (Figure 5a,b).

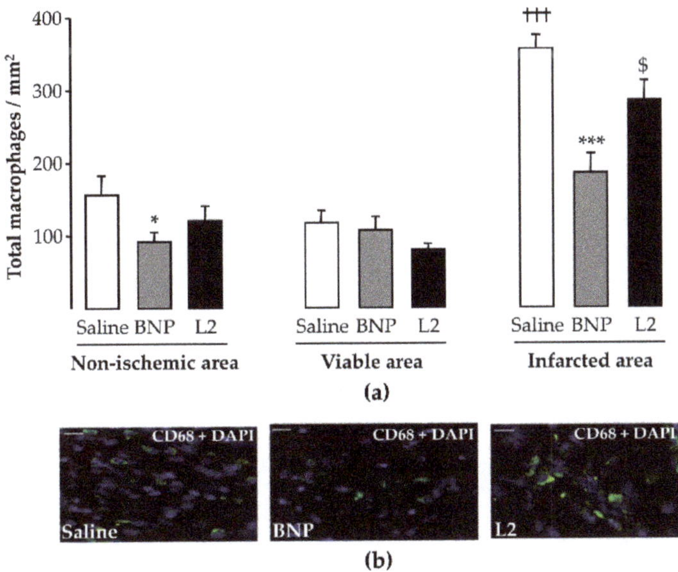

Figure 5. Effects of Lebetin 2 (L2) and B-type natriuretic peptide (BNP) on post-ischemic macrophage recruitment. Myocardial infarction was induced by 35 min of ischemia followed by a 2-day reperfusion period. (**a**) Infiltrated macrophages were assessed by CD68 immunolabeling in the infarcted area and its border zone (viable area), as well as in the septum (non-ischemic area) of rats treated with saline (1 μL/g), BNP (50 ng/g) or L2 (100 ng/g) intraperitoneally, 5 min before reperfusion. On each section, several fields were photographed (12–30 images/rat, $n = 5$–6) and the mean number of macrophages per field was calculated; (**b**) representative microphotographs showing the variable density of macrophages (CD68 in green and nuclei in blue) in the infarcted area from control and treated animals. Data are mean ± SEM. *, $p < 0.05$, ***, $p < 0.001$ vs. saline (control) corresponding group, \$, $p < 0.05$ vs. BNP corresponding group, †††, $p < 0.001$ vs. corresponding other LV areas. Scale bars = 50 μm. ×20. DAPI, 4′, 6′-diamidino-2-phenylindole.

To detect total versus alternatively activated macrophages, M2 macrophage polarization was determined by counting CD68 and MRC-1 double-labeled cells. For both M1 (236 ± 16 cells/mm^2) and M2 (120 ± 11 cells/mm^2) macrophages counted in control rats, a significant increase was observed in the infarcted LV area compared to the viable LV area ($p < 0.001$; Figure 6a,c). L2 (23 ± 4 cells/mm^2) and BNP (28 ± 5 cells/mm^2) decreased M1 infiltration in all LV areas with a sustained reduction in the LV infarcted area (−90% and −80% respectively, $p < 0.001$ vs. control group; Figure 6c). However, only L2 (242 ± 25 cells/mm^2, $p < 0.001$), but not BNP (139 ± 25 cells/mm^2), increased the number of M2-like macrophages in the infarcted area compared to control animals (Figure 6a,b).

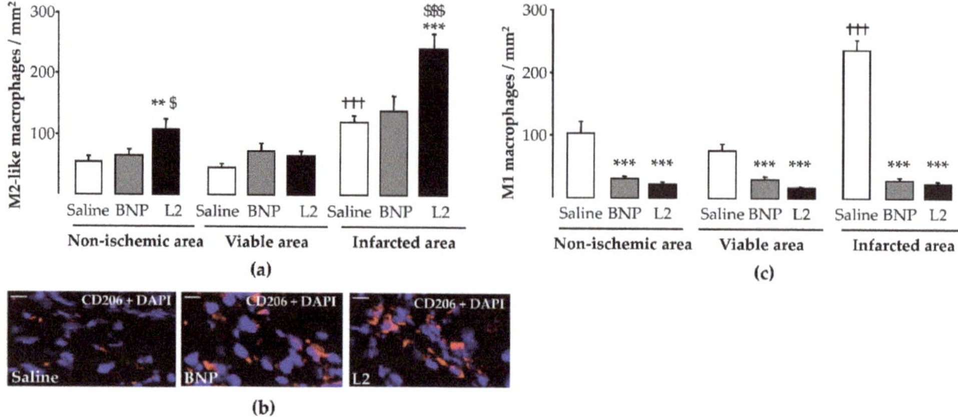

Figure 6. Effects of Lebetin 2 (L2) and B-type natriuretic peptide (BNP) on macrophage polarization and on recruitment of M2-like macrophages. Myocardial infarction was induced by 35 min of ischemia followed by a 2-day reperfusion period. (**a**) Infiltrated M2-like macrophages were assessed by double immunolabeling of CD68 and CD206/MRC-1 in the infarcted area and its border zone (viable area), as well as in the septum (non-ischemic area) of rats treated with saline (1 µL/g), BNP (50 ng/g) or L2 (100 ng/g) intraperitoneally, 5 min before reperfusion. On each section, several fields were photographed (12–30 images/rat, $n = 5$–6) and the mean number of M2-like macrophages per field was calculated; (**b**) representative microphotographs showing the variable density of M2-like macrophages (CD206/MRC-1 in red and nuclei in blue) in the infarcted area from control and treated animals; (**c**) proinflammatory M1 macrophages were identified as CD68$^+$ or CD206$^-$ cells in different LV areas of control and treated animals. Data are mean ± SEM. **, $p < 0.01$, ***, $p < 0.001$ vs. saline (control) corresponding group, \$, $p < 0.05$, \$\$\$, $p < 0.001$ vs. BNP corresponding group, †††, $p < 0.001$ vs. corresponding other LV areas. Scale bars = 50 µm. ×20. DAPI, 4′, 6′-diamidino-2-phenylindole.

2.5. L2 Enhances Endothelial Cell and Cardiomyocyte Densities upon MI

L2 treatment induced an increase in endothelial cell density after IR in all LV areas, with an increase of 64% in the infarcted area (254 ± 26 cells/mm^2, $p < 0.001$), while BNP (160 ± 10 cells/mm^2) failed to induce any change in endothelial cell density in this area compared to non-treated rats (155 ± 6 cells/mm^2) (Figure 7a,b). However, cardiomyocyte density was significantly increased by both L2 and BNP in all LV areas, with an increase of 75% (568 ± 23 cells/mm^2, $p < 0.01$) and 113% (691 ± 98 cells/mm^2, $p < 0.001$), respectively, in the infarcted area, compared to control group (325 ± 21 cells/mm^2) (Figure 7c,d).

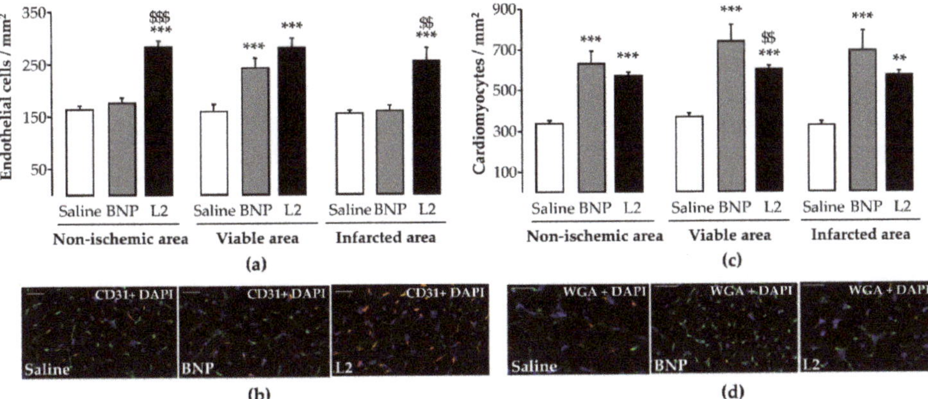

Figure 7. Effects of Lebetin 2 (L2) and B-type natriuretic peptide (BNP) on post-ischemic endothelial cell and cardiomyocyte densities. Myocardial infarction was induced by 35 min of ischemia followed by a 2-day reperfusion period. (**a**) Endothelial cell density was assessed by CD31 immunolabeling in the infarcted area and its border zone (viable area), as well as in the septum (non-ischemic area) of rats treated with saline (1 µL/g), BNP (50 ng/g) or L2 (100 ng/g) intraperitoneally, 5 min before reperfusion. On each section, several fields were photographed (12–30 images/rat, $n = 5$–6) and the mean number of endothelial cells per field was calculated; (**b**) representative microphotographs showing the variable density of endothelial cells (CD31 in red and nuclei in blue) in the infracted area from control and treated animals; (**c**) cardiomyocyte cell density was assessed with wheat germ agglutinin (WGA) immunolabeling in the infarcted area and its border zone (viable area), as well as in the septum (non-ischemic area) of rats treated with saline (1 µL/g), BNP (50 ng/g) or L2 (100 ng/g) intraperitoneally, 5 min before reperfusion. On each section, several fields were photographed (12–30 images/rat, $n = 5$–6) and the mean number of cardiomyocytes per field was calculated; (**d**) representative microphotographs showing the variable density of the cardiomyocytes (WGA in green and nuclei in blue) in the infarcted area from control and treated animals. **, $p < 0.01$, ***, $p < 0.001$ vs. saline (control) corresponding group, \$\$, $p < 0.01$, \$\$\$, $p < 0.001$ vs. BNP corresponding group. Scale bars = 50 µm. ×40. DAPI, 4′, 6′-diamidino-2-phenylindole.

3. Discussion

Previous (pre-)clinical studies have confirmed the beneficial effects of NPs on myocardial repair [5,11], but the underlying mechanisms of action are still poorly understood. Anti-inflammatory pathways, in particular, have not been fully investigated, and so far, no report has addressed the role of these peptides in regulating infiltrating leukocytes and macrophage subtypes after MI. We show here that L2 has the ability to repair heart damage, as evidenced by attenuation of both necrosis and IS, reduction of fibrosis, and promotion of endothelial-cell regeneration. L2 also significantly reduced leukocyte recruitment, while increasing M2-macrophage recruitment in infarcted myocardium. While other NPs required sustained administration to achieve cardioprotection [21–23], we show here that a single injection of L2 prior to reperfusion efficiently improved acute IR injury, with prolonged effects on days 2 and 14 post-reperfusion, highlighting the efficacy of L2 in IR, in vivo, and confirming our previous ex vivo data [18]. Our study provides novel insights into mechanisms of NPs in myocardial repair.

NPs exert vasodilating, natriuretic, and diuretic actions [24] as well as inhibitory effects on sympathetic nervous and renin-angiotensin-aldosterone systems [25], leading to a decrease in pre- and after-load. Thus, in our work, L2 exerts acute beneficial effects by maintaining the blood supply to cardiac tissue. Blood pressure-lowering effect is unlikely to account for cardioprotection in our study, since L2 and BNP were used at doses inducing only a mild, transient hypotensive effect; with no effect on HR. This is consistent with previous studies dissociating cardioprotective and blood pressure effects of vasoactive compounds in this model [26,27]. The L2 IS-limiting effect was still observed on day 14 post-reperfusion, and interestingly, was more pronounced than after acute IR, which suggests that L2, beside its effects on necrosis, likely induces other regulatory actions, the effects of which could not have been observed under acute conditions. Significantly, this was further confirmed by the reduction of post-ischemic fibrosis upon L2 treatment, as evidenced by the inhibition of collagen synthesis following prolonged reperfusion [28]. This anti-fibrotic effect may also be related to the inhibition of cardiac fibroblast growth [29] or the activation of matrix metalloproteinases [30], as reported for BNP, which accelerated infarct healing.

After MI, many inflammatory cells infiltrate the infarcted heart to convert necrotic tissue into scar tissue. However, enhanced inflammatory cell infiltration into the myocardium may exacerbate cardiac injury and worsen post-MI remodeling [29,30]. Macrophages are essential mediators of the inflammatory response, with an important role in the initiation and resolution of inflammation. It has been supposed that the healing of infarcted myocardium occurs through macrophage transition, a conversion from a pro-inflammatory (M1) phenotype to an anti-inflammatory (M2) phenotype [29]. This suggests that the modulation of the inflammatory response may improve healing and alleviate LV remodeling of the injured heart [30]. On day 2 post-reperfusion, L2 and BNP significantly reduced both leukocyte and macrophage infiltration into the infarcted area, where inflammatory cells are prominent [31,32]. Previous work has reported an association between the cardioprotective action of BNP and its inhibitory effect on proinflammatory cytokines and related mechanisms [11,13]. However, caution has been raised about transgenic mouse experiments by Kawakami et al. [12] and Izumi et al. [9], where the BNP/NPR-A axis was shown to stimulate polymorphonuclear infiltration during ischemic injury [9,12]. In the first case, a possible explanation could be that the excessive expression of BNP may be harmful, since NP-induced cardioprotection is effective at low-dose administration [33]. For the second study [9], the cardioprotection observed in NPR-A deficient mice, which was linked to a decrease of infiltrating neutrophils, may be due to compensatory mechanisms following loss of dominant receptor function [34]. Particularly, NPR-A downregulation was positively associated with increased expression of NPR-B protein in rat hearts [34], which alternatively could ensure cardioprotection [35].

Inflammatory-cell modulation by NPs has not been demonstrated yet. Therefore, our finding that the M2 macrophage population increased following L2 treatment was quite surprising. This was particularly interesting since BNP, although effective in reducing the M1 inflammatory subset, failed to induce M2 polarization. Expansion of the M2-macrophage population probably explains the increased level of total macrophages following L2 treatment (Figure 5). Altogether, our data explain the greater efficacy of L2, since M2 enrichment was associated with cardiac regeneration after MI [36]. Furthermore, many studies have reported the essential role of M2 macrophages in repair of MI and attenuation of post-MI remodeling [37,38].

IR triggers endothelial and coronary microvascular inflammation, leading to the impairment of NO-mediated, endothelium-dependent vasodilation [1]. NPs mitigate MI-induced endothelial dysfunction and subsequent release of NO [23], thereby increasing the coronary flow and blood supply to injured myocardial cells. However, despite the anti-apoptotic action of BNP [39], controversies still exist regarding the effect of the NP/cGMP signaling system on endothelial cell regeneration after vascular damage, with claims ranging from enhancement to protection [28,40]. In the present study, only the venom peptide, but not BNP, enhanced the endothelial cell number in the infarcted area, while both molecules similarly increased the number of cardiomyocytes at this site. This effect of L2 could be mediated by M2-like macrophage signaling, since both promote tissue repair and healing, including capillary network formation and proliferation capacity of endothelial cells [41]. This hypothesis is strengthened by the fact that in our work, BNP, which does not promote M2-like macrophage polarization, does not increase endothelial cell density either. This suggests that BNP in that context was not able to preserve endothelial cell viability or to induce their proliferation in injured myocardium. These results accord with previous work showing that only L2, and not BNP, efficiently increased the coronary flow [18], suggesting opposite pharmacologies of the two peptides on endothelial function.

It would therefore appear that unlike BNP, L2 enhances the integrity or viability of endothelial cells. Mechanistically, these discrepancies could be attributed to differences in the affinities of L2 and BNP to natriuretic receptors. The low affinity of BNP for NP receptor type C (NPR-C) [42], which triggers PI3K/Akt/eNOS signaling [21–23], may provide an explanation for why BNP does not have any effect on endothelial cell function, post-MI. More mechanistic investigations are required to further explore the potential affinity of L2 to NPR-C. Such affinity has been reported for other snake venom NPs [16].

Although our results provide novel insights into mechanisms of myocardial repair triggered by NPs, they leave some unanswered questions. First, our study was primarily designed to determine whether prolonged post-MI effects of L2 and BNP are associated with the modulation of inflammatory cell response. However, we did not explore how NPs reduce leukocyte recruitment and induce the switch to M2 macrophages. Previous studies [43] reported complex mechanisms involved in leukocyte recruitment and M1-to-M2 macrophage transformation, but how NPs trigger these events specifically is unclear and awaits further investigation. Decreased leukocyte and inflammatory macrophage infiltration could be achieved through BNP-mediated inhibition of interleukin (IL)-1, IL-6 and tumor necrosis factor alpha (TNFα) secretion [44], which, in turn, could inhibit chemokine signaling via macrophage-chemoattractant protein-1 (MCP-1) and macrophage inflammatory protein 1 (MIP-1), known for their role in promoting proinflammatory macrophage infiltration [45]. On the other hand, L2 treatment may increase M2 macrophages by increasing IL-10 release, which mediates the switch from proinflammatory macrophage infiltration to alternatively activated anti-inflammatory macrophages [36], as previously reported for NPs [46]. Second, L2 may improve the survival and recruitment of M2 macrophages in the infarcted heart; however, it is still unclear whether and how the increased number of M2 macrophages actually contributes to myocardial repair.

4. Conclusions

The present study shows that L2 potently alleviates acute, prolonged heart damage by attenuating both post-ischemic necrosis and fibrosis. L2 also promotes endothelial-cell regeneration in infarcted myocardium and significantly reduces inflammatory cell recruitment, while enhancing M2-macrophage post-MI. These findings provide novel insights into mechanisms of NPs in myocardial repair, since many of these effects have not been observed with BNP treatment. Despite a single systemic administration of L2 at the time of reperfusion, the L2 acute, protective effects persisted up to 14 days post-MI. This can be explained by the activation of the NPR/cGMP pathway, which is known to induce acute effects such as vasorelaxation/vasodilation, but also chronic effects such as anti-fibrotic action during cardiac injury [47]. Moreover, as inflammation triggers interstitial fibrosis [48,49], it is likely that L2 provides prolonged actions (inhibition of fibrosis) as a consequence of its inhibitory effect on inflammation.

Given that L2 limits leukocyte trafficking, inhibits non-resolving inflammatory cells, and enhances pro-resolving cells, these findings are clinically relevant, as the repair response can be subjected to pharmacological interventions directed toward regressing inflammation and subsequent post-MI fibrosis. Modulation of the inflammatory response by NP therapy will be of great significance in enhancing the treatment of HF, especially in patients with intense and prolonged inflammatory responses.

5. Materials and Methods

5.1. Drugs Used in the Study

All experiments were performed with synthetic L2 (isoform α, 38 amino acids) purchased from Genosphère Biotechnologies (Paris, France), the production of which is based on native L2 (4 kDa), from the venom of *Macrovipera lebetina*. Prior to use of synthetic L2, we confirmed that both synthetic and native L2 exerted similar effects on the cardiac NPR/cGMP signaling and cardiodynamics of isolated rat hearts [18]. BNP (human recombinant peptide, 32 amino acids) was purchased from

Sigma–Aldrich (Saint Quentin Fallavier, France). L2 and BNP share 54% structural homology based on their amino acid sequences (Figure 8).

```
L2 (38 aa)   :  GDNKPPKKGPPNGCFGHKIDRIGSHSGLGCNKVDDNKG
BNP (32 aa): ----NSKMAHSSSCFGQKIDRIGAVSRLGCDGLRLF--
```

Figure 8. Homology between the amino acid sequences of Lebetin 2 (L2, 38 amino acids) and B-type natriuretic peptide (BNP, 32 amino acids). Identical amino acid positions are shaded in black while conserved or similar residues are in gray. L2 and BNP display 54% similarity. Alignment was performed using GeneDoc, version 2.7.000.

5.2. Animals

Male Wistar rats (280–300 g) and male C57BL6 mice (20–30 g) were purchased from Janvier Labs (Le Genest-Saint-Isle, France). They were housed in climate-controlled conditions and had unrestricted access to standard rat chow and drinking water. Animal experiments were performed in accordance with the Guide for the Care and Use of Laboratory Animals of the National Institutes of Health (NIH Pub. No.85–23, Revised 1996), the Council of the European Communities (86/609/EEC; November 24th 1986), the French National Legislation (Ethical Approval No.76–114/ 01181.01) and also by the Biomedical Ethics Committee of the Pasteur Institute in Tunis (Ethical Approval No. 2015/08/I/LR11IPT08/V0; 9 November 2015).

5.3. In Vivo Dose-Response Effect on Blood Pressure

Anesthetized (60 mg/kg sodium pentobarbital, by intraperitoneal route) animals were placed on a thermally controlled heating pad (37 ± 1 °C). A catheter was inserted into the right carotid artery for blood pressure and HR recording (Iox2 software, Emka Technologies, Paris, France). Effective doses of L2 and BNP were chosen for their ability to induce mild hypotension, such as is usually observed with other vasocative compounds in IR studies [26,27], to minimize the effect of blood pressure decrease on IR outcomes. The L2 dose range was chosen based on a previous study showing that the equivalent active dose of L2 was higher than that of BNP [18]. Based thereon, blood pressure changes triggered by BNP or L2 were measured in rats treated with increasing doses of BNP (10 or 50 ng/g, $n = 6$), L2 (100 or 200 ng/g, $n = 7$) or saline (NaCl 9‰, $n = 7$). Blood pressure response was also assessed in mice treated with BNP (1.5, 5 or 20 ng/g, $n = 5$–10), L2 (25, 50 or 100 ng/g, $n = 4$–5) or saline (NaCl 9‰, $n = 11$). Drugs were injected as a single intravenous bolus (1 µL/g body weight).

The blood pressure was recorded, and the MAP was calculated at intervals after injection. AUCs (MAP absolute variations × time) were calculated in each animal according to the trapezoidal rule and averaged within each experimental group.

5.4. In Vivo Murine Models of Myocardial Ischemia-Reperfusion

5.4.1. Surgical Preparation

For acute IR studies, animals were anesthetized with sodium pentobarbital (60 mg/kg, by intraperitoneal route). For prolonged IR studies, rats were anesthetized intraperitoneally with a mix of ketamine (50 mg/kg) and xylazine (3.6 mg/kg). They were ventilated and a thoracotomy was performed. Body temperature was monitored with a rectal probe connected to a digital thermometer, and maintained at 37 °C using a heating pad. An electrocardiogram (ECG) was recorded throughout experiments (Iox2 software, Emka Technologies, Paris, France). The left main coronary artery was occluded close to its origin using reversible ligation, as previously described [31,50]. At the end of ischemia, the coronary blood flow was restored and reperfusion achieved by loosening the suture. The lungs were then re-inflated by increasing positive end expiratory pressure, and the chest was closed during the reperfusion period.

For acute IR studies, animals were kept on the heating pad throughout the experiment and hearts were immediately excised at the end of reperfusion to assess IS. Animals submitted to prolonged reperfusion, were returned to their cages (2–3 rats/cage) after recover on a heating pad for later assessment of post-ischemic inflammatory response, IS, and fibrosis.

5.4.2. Experimental Protocols

Animals underwent either acute or prolonged reperfusion (Figure 9). In the first set of experiments, in an attempt to validate acute beneficial effects of L2 that have been previously demonstrated only ex vivo [18], we subjected either rats (n = 5–7) or mice (n = 5–6) to acute IR injury induced by 30 min of ischemia followed by 120 min of reperfusion (Figure 9) and assessed IS. In a second set of experiments, once the acute effect was demonstrated, we subjected rats to coronary ischemia for 35 min followed by either 2 days of reperfusion to assess cellular inflammatory response, endothelial cells, and cardiac myocytes (n = 5–6), or 14 days of reperfusion, to evaluate IS and fibrosis (n = 7–10) (Figure 9).

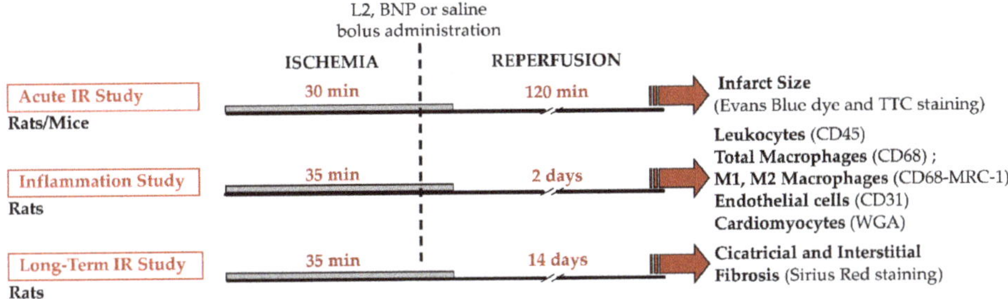

Figure 9. Experimental design. Three protocols of myocardial ischemia-reperfusion (IR) were applied. For acute infarct size study (upper panel), rats and mice underwent 30 min ischemia (full solid line) followed by 120 min coronary reperfusion (thin solid line). In two other additional series of experiments, rats underwent 35 min of ischemia followed by either 2 days (middle panel) or 14 days (lower panel) of reperfusion, to study respectively post-ischemic inflammation and fibrosis. Rats and mice received either B-type natriuretic peptide (BNP, 50 ng/g and 20 ng/g respectively), Lebetin 2 (L2, 100 ng/g and 25 ng/g respectively) or saline (NaCl 9‰, controls), as a single systemic bolus (1 μL/g body weight), 5 min before starting reperfusion.

Rats and mice received either BNP (50 ng/g and 20 ng/g; respectively), L2 (100 ng/g and 25 ng/g; respectively) or saline (control group; NaCl 9‰). Drugs were given as a single systemic bolus (1 μL/g body weight), 5 min before reperfusion. The intravenous route was selected for drug administration in acute experiments in mice, whereas intraperitoneal injection was performed in rats.

5.5. Histological Analysis

5.5.1. Acute Infarct Size

Acute IS was determined after 120 min of reperfusion, as described previously [31,50]. Briefly the chest was reopened at the end of reperfusion and the coronary artery was reoccluded. IS, expressed as a percentage of AR, was quantified on transverse LV slices after staining with Evans Blue and buffered 2,3,5-triphenyltetrazolium chloride (TTC) solutions. AR and IS were quantified using a computerized planimetric technique (ImageJ software, NIH, Bethesda, Maryland, MD, USA) by an observer blinded to treatment groups [31,50].

5.5.2. Cardiac Morphometry: Infarct Size and Fibrosis

At day 14 post-reperfusion, rats were euthanized with a lethal dose of sodium methohexital. Hearts were rapidly excised and arrested in ice-cold saturated potassium chloride buffer, and LV conserved in Bouin's fixative solution, as described previously [51]. After fixation, the LV was cut into several slices perpendicular to the apex-to-base axis. LV sections were dehydrated and embedded in paraffin. Then, 10-μm histological slices were stained with Sirius red for IS and fibrosis assessment.

For IS, LV sections were photographed under a light microscope (×1.25, Carl Zeiss, GmbH, Jena, Germany). Endocardial and epicardial circumferences of infarcted tissue and of the LV were determined and IS was calculated as (endocardial + epicardial circumference of the infarcted tissue)/(endocardial+epicardial circumference of the LV) × 100 using Image Pro Plus 6.3 software (Media Cybernetics, Rockville, Maryland, MD, USA) [51].

For fibrosis, cardiac collagen density was measured. Sections were photographed (×40, Carl Zeiss, GmbH, Jena, Germany), analyzed (Image Pro Plus 6.3 software, Media Cybernetics, Rockville, Maryland MD, USA) and collagen density was calculated as the surface occupied by collagen/the surface of the image [51]. On each section, several fields were photographed (15–20 images/rat) and mean collagen density was calculated.

5.6. Immunohistochemistry

5.6.1. Post-Ischemic Inflammation

At day 2 post-reperfusion, rats were euthanized with a lethal dose of sodium methohexital to assess leukocyte and macrophage infiltration, these being maximal at approximately 48 h reperfusion time [31,32]. Hearts were rapidly excised and arrested in ice-cold saturated potassium chloride buffer, frozen in liquid nitrogen, and stored at −80 °C pending study of post-ischemic inflammatory response. Cryosections (10 μm) were post-fixed in acetone and stained according to standard protocols [52]. Leukocyte infiltration was detected using rat anti-mouse CD45 antibody (1:200; Santa Cruz, Dallas, TX, USA) reacting against most leukocytes. For the detection of total vs. M2-like macrophages, mouse anti-rat CD68 (1:200; Serotec, Oxford, UK) was used in combination with rabbit anti-mouse CD206/MRC-1 (1:300; Santa Cruz, Dallas, TX, USA). Proinflammatory M1 were identified as CD68$^+$ and CD206$^-$ cells. Primary antibodies were revealed using FITC-conjugated donkey anti-mouse IgG (1:300; Jackson Immunoresearch Laboratories, Inc., West Grove, PA, USA) or Cy3-conjugated donkey anti-rabbit IgG (1:300; Jackson Immunoresearch Laboratories, Inc., West Grove, PA, USA). Nuclei were stained with Vectashield mounting-medium with DAPI (Vector Laboratories, Inc., Burlingame, CA, USA) in all rat heart cryosections. Micrographs were captured using a fluorescence microscope equipped with an Apotome (×20, AxioImager Z1, Carl Zeiss GmbH, Jena, Germany) and analyzed using Image Pro Plus 6.3 software (Media Cybernetics, Rockville, MD, USA) by an operator blinded to treatment groups. Inflammatory cells were quantified as cell number per mm^2 in various LV sections [infarcted area and its border zone (viable area) as well as in the septum (non-ischemic area)]. Several fields per section were photographed and the mean inflammatory cell number was calculated.

5.6.2. Endothelial Cell and Cardiomyocyte Densities

To assess cardiac endothelial cell and myocyte densities at day 2 post-reperfusion, LV cryosections (10 μm) fixed in acetone were stained according to standard protocols [53] using rat anti-mouse CD31 (PECAM-1, 1:100; BD Biosciences-Pharmingen, San Diego, CA, USA) for the detection of blood vessel endothelial cells and wheat germ agglutinin (1:100; WGA-A488, Invitrogen, Courtaboeuf, France) for the detection of cardiomyocytes. Secondary reagents included Streptavidin–Cy3 (1:1500; Jackson Immunoresearch Laboratories, Inc., West Grove, PA, USA). Nuclei were stained with Vectashield mounting-medium with DAPI (Vector Laboratories, Inc., Burlingame, CA, USA) in all rat heart cryosections. Micrographs were captured using a fluorescence microscope (×40, AxioImager Z1, Carl Zeiss GmbH, Jena, Germany) and analyzed by an operator blinded to treatment groups using Image Pro Plus 6.3 software (Media Cybernetics, Rockville, MD, USA). Endothelial cell and cardiomyocyte densities were quantified as the number of CD31 labeled cells per mm^2 and transversally sectioned cardiomyocytes per mm^2, respectively in various LV sections [infarcted area and its border zone (viable area), as well as in the septum (non-ischemic area)]. Several fields per section were photographed and the mean endothelial cell or cardiomyocyte number was calculated.

5.7. Statistical Analysis

All results are presented as means ± S.E.M. AUCs (MAP absolute variations × time), IS, and collagen density were compared by one-way analysis of variance (ANOVA). Two-way (Treatment × LV area) repeated measure ANOVA was performed to analyze inflammation, endothelial cells, and cardiomyocytes. Differences in parameters obtained at each time point were evaluated using a Student's *t* test or Dunnett's post-hoc test in case of significance, using JMP software (JMP, SAS Institute Inc., Cary, NC, USA). Values of $p < 0.05$ were considered statistically significant.

Author Contributions: Conceived and designed the experiments: E.M., E.B. and P.M.; performed the experiments: B.T., E.M., A.D. and S.G.; analyzed the data: B.T., A.D., E.M., E.B. and P.M.; wrote the paper: E.M.; Edited the paper: E.M., D.G.-R., N.M. and V.R.

Funding: This study was supported by the Institut Pasteur de Tunis, Inserm and the Ministère de l'Enseignement Supérieur et de la Recherche Scientifique de Tunisie. Bochra Tourki was supported by a grant-in-aid from the Institut Pasteur International Network.

Conflicts of Interest: The authors declare that they have no conflict of interest.

Abbreviations

AR	Area at Risk
AUC	Area under curve
BNP	B-type natriuretic peptide
cGMP	Cyclic guanosine monophosphate
DNP	*Dendroaspis* natriuretic peptide
HF	Heart failure
HR	Heart rate
IR	Ischemia-reperfusion
IS	Infarct size
L2	Lebetin 2
LV	Left ventricle
MAP	Mean arterial blood pressure
MI	Myocardial infarction
NO	Nitric oxide
NP	Natriuretic peptide
NPR	Natriuretic peptide receptor

References

1. Richard, V.; Kaeffer, N.; Tron, C.; Thuillez, C. Ischemic preconditioning protects against coronary endothelial dysfunction induced by ischemia and reperfusion. *Circulation* **1994**, *89*, 1254–1261. [CrossRef] [PubMed]
2. Kramer, D.G.; Trikalinos, T.A.; Kent, D.M.; Antonopoulos, G.V.; Konstam, M.A.; Udelson, J.E. Quantitative evaluation of drug or device effects on ventricular remodeling as predictors of therapeutic effects on mortality in patients with heart failure and reduced ejection fraction: A meta-analytic approach. *J. Am. Coll. Cardiol.* **2010**, *56*, 392–406. [CrossRef] [PubMed]
3. Nahrendorf, M.; Swirski, F.K. Monocyte and macrophage heterogeneity in the heart. *Circ. Res.* **2013**, *112*, 1624–1633. [CrossRef]
4. Andreadou, I.; Iliodromitis, E.K.; Koufaki, M.; Kremastinos, D.T. Pharmacological pre- and post- conditioning agents: Reperfusion-injury of the heart revisited. *Mini Rev. Med. Chem.* **2008**, *8*, 952–959. [CrossRef]
5. Colucci, W.S.; Elkayam, U.; Horton, D.P.; Abraham, W.T.; Bourge, R.C.; Johnson, A.D.; Wagoner, L.E.; Givertz, M.M.; Liang, C.S.; Neibaur, M.; et al. Intravenous nesiritide, a natriuretic peptide, in the treatment of decompensated congestive heart failure. Nesiritide Study Group. *N. Engl. J. Med.* **2000**, *343*, 246–253. [CrossRef] [PubMed]

6.   McMurray, J.J.; Packer, M.; Desai, A.S.; Gong, J.; Lefkowitz, M.P.; Rizkala, A.R.; Rouleau, J.L.; Shi, V.C.; Solomon, S.D.; Swedberg, K.; et al. Angiotensin-neprilysin inhibition versus enalapril in heart failure. *N. Engl. J. Med.* **2014**, *371*, 993–1004. [CrossRef] [PubMed]

7.   Publication Committee for the VMAC Investigators. Intravenous nesiritide vs nitroglycerin for treatment of decompensated congestive heart failure: A randomized controlled trial. *JAMA* **2002**, *287*, 1531–1540.

8.   Yancy, C.W.; Krum, H.; Massie, B.M.; Silver, M.A.; Stevenson, L.W.; Cheng, M.; Kim, S.S.; Evans, R.; Investigators, F.I. Safety and efficacy of outpatient nesiritide in patients with advanced heart failure: Results of the Second Follow-Up Serial Infusions of Nesiritide (FUSION II) trial. *Circ. Heart Fail.* **2008**, *1*, 9–16. [CrossRef] [PubMed]

9.   Izumi, T.; Saito, Y.; Kishimoto, I.; Harada, M.; Kuwahara, K.; Hamanaka, I.; Takahashi, N.; Kawakami, R.; Li, Y.; Takemura, G.; et al. Blockade of the natriuretic peptide receptor guanylyl cyclase-A inhibits NF-kappaB activation and alleviates myocardial ischemia/reperfusion injury. *J. Clin. Investig.* **2001**, *108*, 203–213. [CrossRef]

10.   Tamura, N.; Ogawa, Y.; Chusho, H.; Nakamura, K.; Nakao, K.; Suda, M.; Kasahara, M.; Hashimoto, R.; Katsuura, G.; Mukoyama, M.; et al. Cardiac fibrosis in mice lacking brain natriuretic peptide. *Proc. Natl. Acad. Sci. USA* **2000**, *97*, 4239–4244. [CrossRef]

11.   George, I.; Morrow, B.; Xu, K.; Yi, G.H.; Holmes, J.; Wu, E.X.; Li, Z.; Protter, A.A.; Oz, M.C.; Wang, J. Prolonged effects of B-type natriuretic peptide infusion on cardiac remodeling after sustained myocardial injury. *Am. J. Physiol. Heart Circ. Physiol.* **2009**, *297*, H708–H717. [CrossRef] [PubMed]

12.   Kawakami, R.; Saito, Y.; Kishimoto, I.; Harada, M.; Kuwahara, K.; Takahashi, N.; Nakagawa, Y.; Nakanishi, M.; Tanimoto, K.; Usami, S.; et al. Overexpression of brain natriuretic peptide facilitates neutrophil infiltration and cardiac matrix metalloproteinase-9 expression after acute myocardial infarction. *Circulation* **2004**, *110*, 3306–3312. [CrossRef] [PubMed]

13.   Hu, G.; Huang, X.; Zhang, K.; Jiang, H.; Hu, X. Anti-inflammatory effect of B-type natriuretic peptide postconditioning during myocardial ischemia-reperfusion: Involvement of PI3K/Akt signaling pathway. *Inflammation* **2014**, *37*, 1669–1674. [CrossRef] [PubMed]

14.   Huang, Y.; Ng, X.W.; Lim, S.G.; Chen, H.H.; Burnett, J.C., Jr.; Boey, Y.C.; Venkatraman, S.S. In vivo Evaluation of Cenderitide-Eluting Stent (CES) II. *Ann. Biomed. Eng.* **2016**, *44*, 432–441. [CrossRef] [PubMed]

15.   Chen, H.H.; Lainchbury, J.G.; Burnett, J.C., Jr. Natriuretic peptide receptors and neutral endopeptidase in mediating the renal actions of a new therapeutic synthetic natriuretic peptide *Dendroaspis* natriuretic peptide. *J. Am. Coll. Cardiol.* **2002**, *40*, 1186–1191. [CrossRef]

16.   Johns, D.G.; Ao, Z.; Heidrich, B.J.; Hunsberger, G.E.; Graham, T.; Payne, L.; Elshourbagy, N.; Lu, Q.; Aiyar, N.; Douglas, S.A. *Dendroaspis* natriuretic peptide binds to the natriuretic peptide clearance receptor. *Biochem. Biophys. Res. Commun.* **2007**, *358*, 145–149. [CrossRef]

17.   St Pierre, L.; Flight, S.; Masci, P.P.; Hanchard, K.J.; Lewis, R.J.; Alewood, P.F.; de Jersey, J.; Lavin, M.F. Cloning and characterisation of natriuretic peptides from the venom glands of Australian elapids. *Biochimie* **2006**, *88*, 1923–1931. [CrossRef]

18.   Tourki, B.; Mateo, P.; Morand, J.; Elayeb, M.; Godin-Ribuot, D.; Marrakchi, N.; Belaidi, E.; Messadi, E. Lebetin 2, a Snake Venom-Derived Natriuretic Peptide, Attenuates Acute Myocardial Ischemic Injury through the Modulation of Mitochondrial Permeability Transition Pore at the Time of Reperfusion. *PLoS ONE* **2016**, *11*, e0162632. [CrossRef]

19.   Barbouche, R.; Marrakchi, N.; Mansuelle, P.; Krifi, M.; Fenouillet, E.; Rochat, H.; el Ayeb, M. Novel anti-platelet aggregation polypeptides from Vipera lebetina venom: Isolation and characterization. *FEBS Lett.* **1996**, *392*, 6–10. [CrossRef]

20.   Vink, S.; Jin, A.H.; Poth, K.J.; Head, G.A.; Alewood, P.F. Natriuretic peptide drug leads from snake venom. *Toxicon* **2012**, *59*, 434–445. [CrossRef]

21.   Breivik, L.; Jensen, A.; Guvag, S.; Aarnes, E.K.; Aspevik, A.; Helgeland, E.; Hovland, S.; Brattelid, T.; Jonassen, A.K. B-type natriuretic peptide expression and cardioprotection is regulated by Akt dependent signaling at early reperfusion. *Peptides* **2015**, *66*, 43–50. [CrossRef] [PubMed]

22.   Burley, D.S.; Baxter, G.F. B-type natriuretic peptide at early reperfusion limits infarct size in the rat isolated heart. *Basic Res. Cardiol.* **2007**, *102*, 529–541. [CrossRef]

23.   Ren, B.; Shen, Y.; Shao, H.; Qian, J.; Wu, H.; Jing, H. Brain natriuretic peptide limits myocardial infarct size dependent of nitric oxide synthase in rats. *Clin. Chim. Acta* **2007**, *377*, 83–87. [CrossRef]

24. Costello-Boerrigter, L.C.; Boerrigter, G.; Cataliotti, A.; Harty, G.J.; Burnett, J.C., Jr. Renal and anti-aldosterone actions of vasopressin-2 receptor antagonism and B-type natriuretic peptide in experimental heart failure. *Circ. Heart Fail.* **2010**, *3*, 412–419. [CrossRef] [PubMed]

25. Nakanishi, M.; Saito, Y.; Kishimoto, I.; Harada, M.; Kuwahara, K.; Takahashi, N.; Kawakami, R.; Nakagawa, Y.; Tanimoto, K.; Yasuno, S.; et al. Role of natriuretic peptide receptor guanylyl cyclase-A in myocardial infarction evaluated using genetically engineered mice. *Hypertension* **2005**, *46*, 441–447. [CrossRef] [PubMed]

26. Messadi-Laribi, E.; Griol-Charhbili, V.; Pizard, A.; Vincent, M.P.; Heudes, D.; Meneton, P.; Alhenc-Gelas, F.; Richer, C. Tissue kallikrein is involved in the cardioprotective effect of AT1-receptor blockade in acute myocardial ischemia. *J. Pharmacol. Exp. Ther.* **2007**, *323*, 210–216. [CrossRef] [PubMed]

27. Messadi, E.; Aloui, Z.; Belaidi, E.; Vincent, M.P.; Couture-Lepetit, E.; Waeckel, L.; Decorps, J.; Bouby, N.; Gasmi, A.; Karoui, H.; et al. Cardioprotective effect of VEGF and venom VEGF-like protein in acute myocardial ischemia in mice: Effect on mitochondrial function. *J. Cardiovasc. Pharmacol.* **2014**, *63*, 274–281. [CrossRef]

28. Moilanen, A.M.; Rysa, J.; Mustonen, E.; Serpi, R.; Aro, J.; Tokola, H.; Leskinen, H.; Manninen, A.; Levijoki, J.; Vuolteenaho, O.; et al. Intramyocardial BNP gene delivery improves cardiac function through distinct context-dependent mechanisms. *Circ. Heart Fail.* **2011**, *4*, 483–495. [CrossRef]

29. Cao, L.; Gardner, D.G. Natriuretic peptides inhibit DNA synthesis in cardiac fibroblasts. *Hypertension* **1995**, *25*, 227–234. [CrossRef] [PubMed]

30. Tsuruda, T.; Boerrigter, G.; Huntley, B.K.; Noser, J.A.; Cataliotti, A.; Costello-Boerrigter, L.C.; Chen, H.H.; Burnett, J.C., Jr. Brain natriuretic Peptide is produced in cardiac fibroblasts and induces matrix metalloproteinases. *Circ. Res.* **2002**, *91*, 1127–1134. [CrossRef] [PubMed]

31. Messadi, E.; Vincent, M.P.; Griol-Charhbili, V.; Mandet, C.; Colucci, J.; Krege, J.H.; Bruneval, P.; Bouby, N.; Smithies, O.; Alhenc-Gelas, F.; et al. Genetically determined angiotensin converting enzyme level and myocardial tolerance to ischemia. *Faseb J.* **2010**, *24*, 4691–4700. [CrossRef] [PubMed]

32. Chang, C.; Ji, Q.; Wu, B.; Yu, K.; Zeng, Q.; Xin, S.; Liu, J.; Zhou, Y. Chemerin15-Ameliorated Cardiac Ischemia-Reperfusion Injury Is Associated with the Induction of Alternatively Activated Macrophages. *Mediat. Inflamm.* **2015**, *2015*, 563951–563959. [CrossRef] [PubMed]

33. Padilla, F.; Garcia-Dorado, D.; Agullo, L.; Barrabes, J.A.; Inserte, J.; Escalona, N.; Meyer, M.; Mirabet, M.; Pina, P.; Soler-Soler, J. Intravenous administration of the natriuretic peptide urodilatin at low doses during coronary reperfusion limits infarct size in anesthetized pigs. *Cardiovasc. Res.* **2001**, *51*, 592–600. [CrossRef]

34. Manivasagam, S.; Subramanian, V.; Tumala, A.; Vellaichamy, E. Differential expression and regulation of anti-hypertrophic genes Npr1 and Npr2 during beta-adrenergic receptor activation-induced hypertrophic growth in rats. *Mol. Cell. Endocrinol.* **2016**, *433*, 117–129. [CrossRef] [PubMed]

35. Burley, D.S.; Cox, C.D.; Zhang, J.; Wann, K.T.; Baxter, G.F. Natriuretic peptides modulate ATP-sensitive K(+) channels in rat ventricular cardiomyocytes. *Basic Res. Cardiol.* **2014**, *109*, 402. [CrossRef] [PubMed]

36. Dayan, V.; Yannarelli, G.; Billia, F.; Filomeno, P.; Wang, X.H.; Davies, J.E.; Keating, A. Mesenchymal stromal cells mediate a switch to alternatively activated monocytes/macrophages after acute myocardial infarction. *Basic Res. Cardiol.* **2011**, *106*, 1299–1310. [CrossRef] [PubMed]

37. Shiraishi, M.; Shintani, Y.; Shintani, Y.; Ishida, H.; Saba, R.; Yamaguchi, A.; Adachi, H.; Yashiro, K.; Suzuki, K. Alternatively activated macrophages determine repair of the infarcted adult murine heart. *J. Clin. Investig.* **2016**, *126*, 2151–2166. [CrossRef] [PubMed]

38. Tsujita, K.; Kaikita, K.; Hayasaki, T.; Honda, T.; Kobayashi, H.; Sakashita, N.; Suzuki, H.; Kodama, T.; Ogawa, H.; Takeya, M. Targeted deletion of class A macrophage scavenger receptor increases the risk of cardiac rupture after experimental myocardial infarction. *Circulation* **2007**, *115*, 1904–1911. [CrossRef]

39. Talha, S.; Bouitbir, J.; Charles, A.L.; Zoll, J.; Goette-Di Marco, P.; Meziani, F.; Piquard, F.; Geny, B. Pretreatment with brain natriuretic peptide reduces skeletal muscle mitochondrial dysfunction and oxidative stress after ischemia-reperfusion. *J. Appl. Physiol.* **2013**, *114*, 172–179. [CrossRef]

40. Kook, H.; Itoh, H.; Choi, B.S.; Sawada, N.; Doi, K.; Hwang, T.J.; Kim, K.K.; Arai, H.; Baik, Y.H.; Nakao, K. Physiological concentration of atrial natriuretic peptide induces endothelial regeneration in vitro. *Am. J. Physiol. Heart Circ. Physiol.* **2003**, *284*, H1388–H1397. [CrossRef]

41. Barbay, V.; Houssari, M.; Mekki, M.; Banquet, S.; Edwards-Levy, F.; Henry, J.P.; Dumesnil, A.; Adriouch, S.; Thuillez, C.; Richard, V.; et al. Role of M2-like macrophage recruitment during angiogenic growth factor therapy. *Angiogenesis* **2015**, *18*, 191–200. [CrossRef] [PubMed]

42. Suga, S.; Nakao, K.; Hosoda, K.; Mukoyama, M.; Ogawa, Y.; Shirakami, G.; Arai, H.; Saito, Y.; Kambayashi, Y.; Inouye, K.; et al. Receptor selectivity of natriuretic peptide family, atrial natriuretic peptide, brain natriuretic peptide, and C-type natriuretic peptide. *Endocrinology* **1992**, *130*, 229–239. [CrossRef] [PubMed]

43. Mantovani, A.; Sica, A.; Sozzani, S.; Allavena, P.; Vecchi, A.; Locati, M. The chemokine system in diverse forms of macrophage activation and polarization. *Trends Immunol.* **2004**, *25*, 677–686. [CrossRef] [PubMed]

44. Mann, D.L. Stress-activated cytokines and the heart: From adaptation to maladaptation. *Annu. Rev. Physiol.* **2003**, *65*, 81–101. [CrossRef] [PubMed]

45. Ndisang, J.F.; Jadhav, A. Hemin therapy improves kidney function in male streptozotocin-induced diabetic rats: Role of the heme oxygenase/atrial natriuretic peptide/adiponectin axis. *Endocrinology* **2014**, *155*, 215–229. [CrossRef]

46. Chiurchiu, V.; Izzi, V.; D'Aquilio, F.; Carotenuto, F.; Di Nardo, P.; Baldini, P.M. Brain Natriuretic Peptide (BNP) regulates the production of inflammatory mediators in human THP-1 macrophages. *Regul. Pept.* **2008**, *148*, 26–32. [CrossRef] [PubMed]

47. Xue, M.; Li, T.; Wang, Y.; Chang, Y.; Cheng, Y.; Lu, Y.; Liu, X.; Xu, L.; Li, X.; Yu, X.; et al. Empagliflozin prevents cardiomyopathy via sGC-cGMP-PKG pathway in type 2 diabetes mice. *Clin. Sci.* **2019**, *133*, 1705–1720. [CrossRef]

48. Sontia, B.; Montezano, A.C.; Paravicini, T.; Tabet, F.; Touyz, R.M. Downregulation of renal TRPM7 and increased inflammation and fibrosis in aldosterone-infused mice: Effects of magnesium. *Hypertension* **2008**, *51*, 915–921. [CrossRef]

49. Van Linthout, S.; Riad, A.; Dhayat, N.; Spillmann, F.; Du, J.; Dhayat, S.; Westermann, D.; Hilfiker-Kleiner, D.; Noutsias, M.; Laufs, U.; et al. Anti-inflammatory effects of atorvastatin improve left ventricular function in experimental diabetic cardiomyopathy. *Diabetologia* **2007**, *50*, 1977–1986. [CrossRef]

50. Wang, Y.; Li, X.; Wang, X.; Lau, W.; Wang, Y.; Xing, Y.; Zhang, X.; Ma, X.; Gao, F. Ginsenoside Rd attenuates myocardial ischemia/reperfusion injury via Akt/GSK-3beta signaling and inhibition of the mitochondria-dependent apoptotic pathway. *PLoS ONE* **2013**, *8*, e70956.

51. Mulder, P.; Devaux, B.; Richard, V.; Henry, J.P.; Wimart, M.C.; Thibout, E.; Mace, B.; Thuillez, C. Early versus delayed angiotensin-converting enzyme inhibition in experimental chronic heart failure. Effects on survival, hemodynamics, and cardiovascular remodeling. *Circulation* **1997**, *95*, 1314–1319. [CrossRef] [PubMed]

52. Besnier, M.; Galaup, A.; Nicol, L.; Henry, J.P.; Coquerel, D.; Gueret, A.; Mulder, P.; Brakenhielm, E.; Thuillez, C.; Germain, S.; et al. Enhanced angiogenesis and increased cardiac perfusion after myocardial infarction in protein tyrosine phosphatase 1B-deficient mice. *FASEB J.* **2014**, *28*, 3351–3361. [CrossRef] [PubMed]

53. Contard, F.; Glukhova, M.; Sabri, A.; Marotte, F.; Sartore, S.; Narcisse, G.; Schatz, C.; Guez, D.; Rappaport, L.; Samuel, J.L. Comparative effects of indapamide and hydrochlorothiazide on cardiac hypertrophy and vascular smooth-muscle phenotype in the stroke-prone, spontaneously hypertensive rat. *J. Cardiovasc. Pharmacol.* **1993**, *22*, S29–S34. [CrossRef] [PubMed]

Article

A Novel Bradykinin-Related Peptide, RVA-Thr^6-BK, from the Skin Secretion of the Hejiang Frog; *Ordorrana hejiangensis*: Effects of Mammalian Isolated Smooth Muscle

Yue Wu [1,†], Daning Shi [1,2,†], Xiaoling Chen [1,*], Lei Wang [1,*], Yuan Ying [1], Chengbang Ma [1], Xinping Xi [1], Mei Zhou [1], Tianbao Chen [1] and Chris Shaw [1]

[1]   Natural Drug Discovery Group, School of Pharmacy, Queen's University Belfast, Belfast BT9 7BL, Northern Ireland, UK
[2]   School of Government, Peking University, No 114, The Leo KoGuan Building, Beijing 100871, China
*   Correspondence: x.chen@qub.ac.uk (X.C.); l.wang@qub.ac.uk (L.W.); Tel.: +44-28-9097-2200 (X.C.); Fax: +44-28-9024-7794 (L.W.)
†   These authors contributed equally to this work.

Received: 26 May 2019; Accepted: 25 June 2019; Published: 28 June 2019

Abstract: A novel naturally-occurring bradykinin-related peptide (BRP) with an N-terminal extension, named RVA-Thr^6-Bradykinin (RVA-Thr^6-BK), was here isolated and identified from the cutaneous secretion of *Odorrana hejiangensis (O. hejiangensis)*. Thereafter, in order to evaluate the difference in myotropic actions, a leucine site-substitution variant from *Amolops wuyiensis* skin secretion, RVA-Leu^1, Thr^6-BK, was chemically synthesized. Myotropic studies indicated that single-site arginine (R) replacement by leucine (L) at position-4 from the N-terminus, altered the action of RVA-Thr^6-BK from an agonist to an antagonist of BK actions on rat ileum smooth muscle. Additionally, both BK N-terminal extended derivatives (RVA-Thr^6-BK and RVA-Leu^1, Thr^6-BK) exerted identical myotropic actions to BK, such as increasing the frequency of contraction, contracting and relaxing the rat uterus, bladder and artery preparations, respectively.

Keywords: bradykinin-related peptide (BRP); RVA-Thr^6-BK; site-substitution variant; agonist; antagonist; myotropic actions

Key Contribution: The full-length kininogen cDNA encoding skin kinins from *O. hejiangensis* was first cloned and characterized. Position-4 from the N-terminus was found as the core site for converting RVA-Thr^6-BK from an agonist to an antagonist of BK receptors on rat ileum smooth muscle.

1. Introduction

Bradykinin (BK) is ubiquitous in mammals and its generation from precursor kininogens by plasma/tissue kallikreins in the human kallikrein-kinin system (KKS), has been well-researched [1–3]. Currently, BK has been proved to be a critical participant associated with human physiological and pathological processes, such as adjusting the functions of the cardiovascular, renal and nervous systems, smooth muscle contraction, glucose metabolism, cell proliferation, inflammation and the production of pain [3,4]. In contrast to mammalian KKS, there is no compelling evidence of the existence of KKS among anura [5–7]. Surprisingly, following the initial discovery mammalian BK from *Rana temporaria* in 1965, numerous identical BK and its structural analogues with potent smooth muscle stimulant have been found in anuran amphibian venom apparatus-skin [7–12]. Moreover, an extraordinary structural diversity was observed in anuran skin bradykinin-related peptides (BRPs). In addition to single/multiple site-substitution that frequently occurred at Arg^1 and Ser^6, there is some deficiency of a certain amino-acid residue, such as the deletion of arginine at N-/C- terminus (des-Arg[9] and des-Arg[1]) [9,13–15]. On the other hand, hydroxylation of the proline residue often appeared among anuran BRPs, typical example like Hyp[3]-BK successively isolated from three species *Rana temporaria*, *Phyllomedusa azurea* and *Hylarana guentheri* [9,14,16]. In comparison of original BK, its structural isoforms, in general, displayed N-terminal extension and/or C-terminal extension, as well as segment insertion [9,17]. These variables regularly happened in multiples, ultimately conferring BPRs structural diversity [9,12]. However, the reasons for their diverse expression patterns and physiological roles of venom BK and BRPs in non-mammalian vertebrate, of the many the anuran skin BRPs, are not clear yet.

A naturally-occurring leucine-substituted BK analogue, RVA-Leu[1], Thr[6]-BK, also known as amolopkinin W1, was identified in *Amolops wuyiensis* skin secretion and has been reported as an antagonist in inhibiting contractility of BK on rat isolated ileum [18]. In this study, a novel BK structural variant was identified from *O. hejiangensis*. It is the first BRP from the defensive skin secretion in this frog species. This peptide primary structure, RVA-Thr[6]-BK, share high sequence similarity to RVA-Leu[1], Thr[6]-BK initially found able to antagonize the relaxant effect observed following BK application to a rat arterial smooth muscle preparation in vitro [18]. Their corresponding chemically synthesized replicate was subjected to our myotropic activity evaluation on rat isolated vascular and non-vascular smooth muscles respectively.

2. Results

2.1. Molecular Cloning of Biosynthetic Precursor-Encoding cDNA and Structural Characterization of RVA-Thr[6]-BK from RP-HPLC Fractions of Skin Secretion

A full-length biosynthetic BRP-precursor cDNA was consistently cloned from the cDNA library constructed from the skin secretion of *O. hejiangensis* (Figure 1). Similarly, in comparison of nine known BRP cDNA transcripts with the same length (Figure 2), this skin-kininogen precursor had 61 amino-acid residues encoding a single 12-mer BK-like peptide following a C-terminal extension sequence VAPEIV. The typical signal peptide consisting of 22 amino-acid residues in the translated open-reading frame was identified through NCBI-BLAST, whereas the convertase processing site, lysine, is not the most common. In accordance with the presence of a segment with 3 amino-acid residues (RVA) at the N-terminal and a threonine residue at position 6 in the BK-like sequence, the novel BK-like was named RVA-Thr[6]-BK. The RVA-Thr[6]-BK biosynthetic precursor-encoding cDNA has been deposited in the NCBI Database (accession code; MK863068).

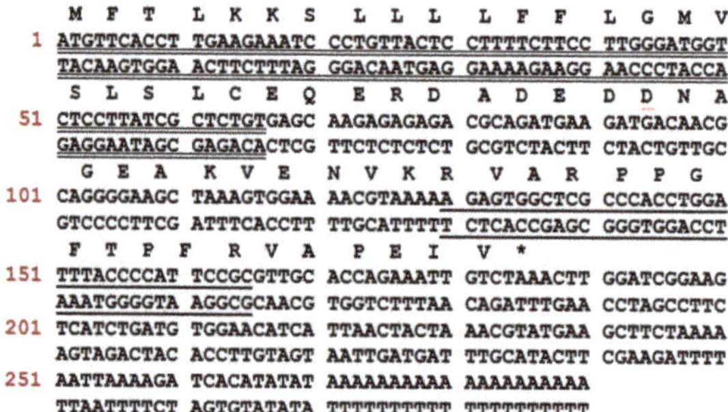

Figure 1. Nucleotide and translated open-reading frame amino-acid sequence of cDNA encoding the RVA-Thr[6]-BK precursor cloned from a skin secretion library of *O. hejiangensis*. The putative signal peptide and mature peptide are labelled by double-underlining and single-underlining, respectively. The asterisk indicates the stop codon.

RVA-T^6-BK	MFTLKKSLLLLFFLGMVSLSLCEQERDADEDDNACEAKVENVKRVARPPGFTPFRVAPEIV
RVAL-L^1, T^6, L^8-BK	MFTIKKSILLLFFLGTISLSLCEQERDANEDETEGEAEVKNVKRVALPPGFTPLRVAPEIV
RVA-L^1, T^6-BK	MFTSKKSILLLFFLGAISLSLCEEERDADEDDTEGEAKVENVKRVALPPGFTPFRVAPEIV
Amolopkinin-TR1	MFTLKKSLLVLFFLGAISVSLCEQERDADEEENGGEAKEESVKRAAVPPGFTPFRVAPEIV
RAP-L^1, T^6-BK	MFTLKKSLLLLFFLGAINLSLCEQERDADEEENEREAKVETVKRAPLPPGFTPFRVAPEIV
Amolopkinin-GN1	MFTMKKSLLLLFFLGAVSLSLCEQERDADEEETEGGAKVETVKRAPRPPGFTPFRVAPEIV
BK-MT	MFTSKKSLLLLFFLGAVSLSLCEQERDADEEETEGGAKVETVKRAPRPPGFTPFRVAPEIV
BK-TR1	MFTLKESLLLLFFLGAISLSLCEQERDADEEETEGGAKVETVKRAPVPPGFTPFRVAPEIV
Amolopkinin	MFTLKESLLLLFFLGAISLSLCEQERDADEEETEGGAKVETVKRAPVPPGFTPFRVAPEIV
Amolopkinin-LF1	MFTLKKSLLLLFFLGAISLSLCEQERDADEEETEGGAKVETVKRAPLPPGFTPFRVAPEIV
	*** * * * ***** **** **** * * *** ****** *******
	MFT K S L LFFLG SLCE ERDA E A VKR PPGFTP RVAPEIV

Figure 2. Amino-acid sequence alignment of RVA-Thr^6-BK precursor and other BRP precursors from the skin secretion of several frog species. These data were obtained from GenBank and their corresponding accession numbers from top to bottom are CDO58214, CAR82590, ADV36199, SBN54116, ADM34221, ADM34238, ABF74734, ADV36198 and ADM34255, respectively. Single- and double-underlining represent signal and mature peptide amino-acid sequences, respectively. Asterisks indicate consensus residues.

2.2. Isolation and Structural Characterization of RVA-Thr^6-BK from RP-HPLC Fractions of Skin Secretion

The average molecular mass of the putative mature peptide from the cloned precursor was calculated as 1400.64 Da. Thereafter, the presence of mature RVA-Thr^6-BK with 12 amino-acids, was further confirmed by RP-HPLC isolation from within the secretion (Figure 3) and tandem mass spectrometry (MS/MS) fragmentation sequencing (Table 1), with the retention time of 89 min.

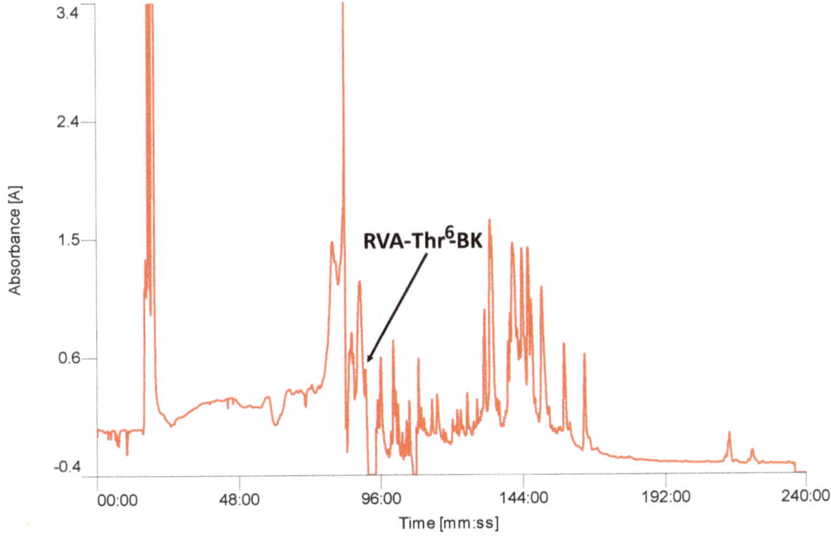

Figure 3. Region of RP-HPLC chromatogram of *O. hejiangensis* skin secretion indicating the retention time of RVA-Thr^6-BK. Absorbance wavelength was set at 214 nm.

Table 1. Calculated singly-charged and doubly-charged b-ions and y-ions produced by MS/MS fragmentation of RVA-Thr^6-BK identified in skin secretion. Observed ions following MS/MS fragmentation are indicated in red and blue typefaces.

#1	b(1+)	b(2+)	b(3+)	Seq.	y(1+)	y(2+)	y(3+)	#2
1	157.10840	79.05784	53.04098	R				12
2	256.17682	128.59205	86.06379	V	1244.68992	622.84860	415.56816	11
3	327.21394	164.11061	109.74283	A	1145.62150	573.31439	382.54535	10
4	483.31506	242.16117	161.77654	R	1074.58438	537.79583	358.86631	9
5	580.36783	290.68755	194.12746	P	918.48326	459.74527	306.83260	8
6	677.42060	339.21394	226.47838	P	821.43049	411.21888	274.48168	7
7	734.44207	367.72467	245.48554	G	724.37772	362.69250	242.13076	6
8	881.51049	441.25888	294.50835	F	667.35625	334.18176	223.12360	5
9	982.55817	491.78272	328.19091	T	520.28783	260.64755	174.10079	4
10	1079.61094	540.30911	360.54183	P	419.24015	210.12371	140.41823	3
11	1226.67936	613.84332	409.56464	F	322.18738	161.59733	108.06731	2
12				R	175.11896	88.06312	59.04450	1

2.3. Myotropic Activities of RVA-Thr^6-BK and RVA-Leu^1, Thr^6-BK

Myotropic activities of RVA-Thr^6-BK and RVA-Leu^1, Thr^6-BK were investigated in four different types of rat smooth muscle preparations (Table 2). These two peptides displayed BK-like contractile effects on isolated bladder preparations (rat urinary) (Figure 4A) and were found to increase the frequency of contraction of the uterus (Figure 4B). Moreover, at a higher concentration (10 µM), RVA-Thr^6-BK caused greater contraction of bladder preparations than its leucine-substituted isoform; however, it required higher threshold concentration of RVA-Thr^6-BK to produce a constriction response (Table 2 and Figure 4B). Interestingly, RVA-Thr^6-BK was capable of causing potent contraction of rat ileum preparations, whereas, there was no significant response from rat ileum smooth muscle with RVA-Leu^1, Thr^6-BK (Figure 4C). In contrast to the contractility of the non-vascular isolated smooth muscle of rat, both BK structural variants used here demonstrated relaxing action on artery preparations of the rat (Figure 4D).

Table 2. Myotropic activities of RVA-Thr^6-BK and RVA-Leu^1, Thr^6-BK in four different types of rat smooth muscle preparations. E max indicates the maximum response of isolated tissue inducing by BRPs at a peptide concentration of 10^{-5}M. TC and EC$_{50}$ represent threshold concentration and the concentration of BRPs giving the half-maximal response respectively. N means undetectable value under 10^{-5}M of BRPs.

Isolated Tissue	E Max	TC (nM)	EC50 (nM)
	RVA-Thr^6-BK/RVA-Leu^1, Thr^6-BK		
Bladder	0.56/0.38 (g)	100.00/10.00	7102.00/636.70
Uterus	7.00/7.00 (peaks/5 min)	10.00/100.00	817.20/3270.00
Ileum	0.43/N (g)	1.00/N	93.70/N
Tail artery	−0.13/−0.08 (g)	1.00/1	8.28/30.91

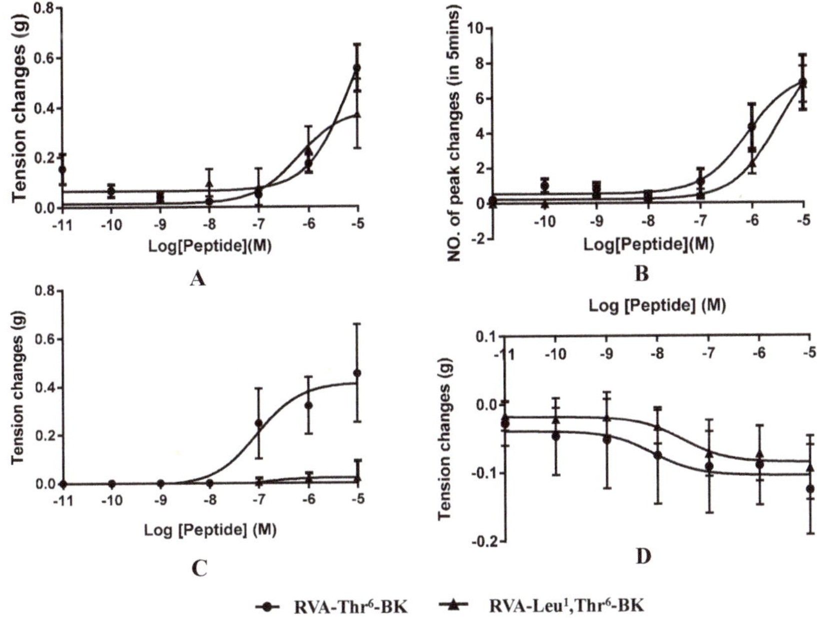

Figure 4. Dose–response curve of RVA-Thr^6-BK (●) and RVA-Leu^1, Thr^6-BK (▲) using (**A**) rat bladder smooth muscle preparations (N = 4, n = 8), (N refers to the number of animals being study, n refers to the number of tests conducted), (**B**) rat uterus smooth muscle preparations (N = 4, n = 7), (**C**) rat ileum smooth muscle preparations (N = 3, n = 4) and (**D**) rat tail artery smooth muscle preparations (N = 4, n = 8), respectively. Data indicate tension changes compared with non-peptide exposed controls.

2.4. Assessment of Cytotoxic Effect on Mammalian Erythrocytes

To evaluate the potential toxicity of RVA-Thr^6-BK and its analogue, RVA-Leu^1, Thr^6-BK, the healthy mammalian red blood cells were treated with these two peptides (Figure 5). The results indicated that both peptides showed no haemolytic activity in vitro.

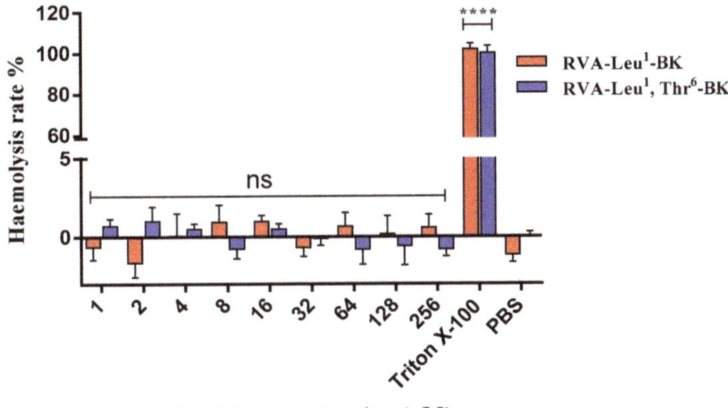

Figure 5. Haemolytic activities of RVA-Thr^6-BK (red) and RVA-Leu^1, Thr^6-BK (blue) using mammalian erythrocytes. **** $p < 0.0001$; ns; no significance.

3. Discussion

Some efforts have been made in anuran amphibian skin research and conservation over several decades [11,12,19,20]. As a consequence of these investigations, an increasing number of biologically-active peptides (frequently anti-microorganism peptides but also neuroactive peptides and protease inhibitor peptides) with a wide range of functions have been discovered [11,12,19–21]. Among these, it is revealed that defensive neuroactive peptides with myotropic activity are remarkably varied in anuran skin secretions, and these small molecule peptides can evoke contractions and relaxations on smooth muscles, BK and BRPs being a typical example [7–9,12]. Herein, a novel BK-like peptide and its corresponding precursor-encoding cDNA were identified. It was found that RVA-Thr^6-BK precursor has a highly conserved structure consisting of an N-terminal signal peptide domain followed by an acidic amino-acid-rich (spacer) domain and terminating in a single mature BK-like peptide coding domain (Figure 2). This result is similar to some of the BRP precursors in other frog species [7,14,18]. In contrast, there are only two types of BK peptides found in human (BK and Lys^0-BK). Conlon and Yano proved that there are no blood-derived kinins in the frog plasma after treatment with trypsin [11]. Another convincing piece of evidence of no detectable kininogens in toad plasma was that many amphibians express BK and BRPs by processing skin-kininogens/preprobradykinins in their elaborate skin kinin system, and such a producing site differs from that of human KKS [5]. Some identical BK derivatives found in anuran skin were also identified in other non-mammalian vertebrate species, such as Arg^0, Trp^5, Leu^8-BK and Thr^6, Leu^8-BK [21,22]. Therefore, it has been proposed that non-mammalian vertebrates develop similar BK and BRPs for their defensive systems, instead of having physiological roles as in mammals [7,11]. However, it is still a point of contention whether the amphibian skin-kininogens/preprobradykinins and plasma-kininogens of humans, are indeed expressed by homologous genes.

In this investigation, RVA-Thr^6-BK was shown to have the opposite effect of RVA-Leu^1, Thr^6-BK on isolated rat ileum smooth muscle in vitro, where RVA-Leu^1, Thr^6-BK had no direct myotropic effects on rat ileum smooth muscle, which is consistent with our previous study [18]. Moreover, another homologous analogue RAA-Leu^1, Thr^6-BK also found in *Amolops wuyiensis* skin secretion at the same time was also reported as an inhibitor of BK-induced contraction of isolated rat ileum smooth muscle [18], indicating that the Val^2 amino-acid replacement is less likely to result in the alteration of myotropic action among BRPs. The difference between of RVA-Thr^6-BK and RVA-Leu^1, Thr^6-BK solely involves arginine/leucine site-substitution at position 4 from the N-terminus, suggesting this is the core site for converting RVA-Thr^6-BK from an agonist to an antagonist of BK on rat ileum smooth muscle. Besides, BK-like vasodilator action, as well as contraction effects on bladder and uterus preparations of RVA-Leu^1, Thr^6-BK, are first described here. It was shown that the biological effects of BK are mediated by specific receptors, the B$_1$ and B$_2$ receptors which are classified as G-protein-coupled receptors having seven transmembrane domains crossing the lipid bilayer, three extracellular and intracellular loops, an exposed N-terminal domain and a cytoplasmic C-terminal domain [4,23–25]. Previous literature demonstrated that the conformation of BK was oriented randomly in aqueous environments, but nevertheless, a specific segment in the BK sequence from residues 2–5, PPGF, was shown to have a preference to form a β-turn conformation [26,27]. However, the unambiguous bioactive conformation for BK binding to its receptor remains inconclusive.

In the haemolytic test, the haemoglobin release in erythrocytes was not observed in both myotropic peptides. However, haemolytic activity in vitro is the initial test for evaluating toxic potential and this assay is not sufficient to completely exclude these peptides with the toxicity. Further investigation on cytotoxicity in more complex experimental models, such as cell lines [28] and insect models (e.g., *Galleria mellonella* larva) [29], need to be taken into account to ensure the development of safe for pharmaceutical use.

In summary, a novel BK analogue is described here which has not been reported previously and this is also the first BRP isolated from *O. hejiangensis*. Both arginine/leucine site-substitution in BK N-terminal extension analogues displayed similar myotropic properties to BK, except that Leu-4 site-substitution produced an antagonist of BK on the rat ileum. Our discovery and exploration of the myotropic activities of anuran skin BRPs provide unique opportunities to better understand the key role of venom BK and BRPs in non-mammalian vertebrates. Additionally, the structural diversity and BK-functional-related actions suggest that anuran amphibians BRPs are unique tools for studying structure-activity relationships as well as developing new or even more potent drugs with therapeutic potential.

4. Materials and Methods

4.1. Skin Secretion Acquisition

Specimens of *O. hejiangensis* (n = 8) of undetermined sex were obtained in China and were kept in a purpose-designed amphibian facility prior to secretion harvesting. The skin secretions were harvested using gentle transdermal electrical stimulation initially described by Tyler and co-workers in 1992 [30]. Briefly, following stimulation by platinum electrodes (6V, 4ms pulse-width, 50 Hz, 20s duration), the secretions from the dorsal skin of frogs were washed with deionised water, snap-frozen in liquid nitrogen and lyophilised. The lyophilised sample was stored at −20 °C before analysis. This study was performed under the UK Animal (Scientific Procedures) Act 1986, project license PPL 2694, issued by the Department of Health, Social Services and Public Safety, Northern Ireland. The procedures had been overseen by the IACUC of Queen's University Belfast, and approved on 1 March 2011.

4.2. "Shotgun" Cloning of cDNA Encoding RVA-Thr^6-BK Biosynthetic Precursor from Skin Secretion

Polyadenylated mRNA was extracted from 5 mg of lyophilised *O. hejiangensis* skin secretion using magnetic oligo-dT beads (Dynal Biotech, Bromborough, UK) and then subjected to 3′- and 5′-RACE procedures to obtain a full-length cDNA employing a 3′-/5′-SMART-RACE kit (Clontech, Mountain View, CA, USA) essentially as described previously [31]. Briefly, the 3′-RACE reactions utilized a Nested Universal Primer A (NUP) (provided with the kit) and a degenerate sense primer (S1: 5′-CCRVCNGGGTTYASSCCWTTY-3′) that was designed to a nucleotide sequence encoding the common internal sequence of anuran skin-derived BRPs, Pro-Pro-Gly-Phe-X-Pro-Phe, where X is either serine or threonine. Thereafter, PCR products were gel-purified, cloned using a pGEM-T vector system (Promega Corporation, Masison, WI, USA) and sequenced using an ABI 3100 automated capillary sequencer. Accordingly, the 5′-RACE reactions employed an antisense primer (S2: 5′-CATCAGATGACTGCCGATCCAA-3′), which was designed to a domain located within the 3′ non-translated region of the obtained 3′-RACE transcripts, and a Universal Primer A Mix (UPM) (supplied with the kit). Similarly, 5′-RACE PCR products were gel-purified, cloned using a pGEM-T vector system and finally sequenced.

4.3. Isolation and Structural Characterization of RVA-Thr^6-BK from Skin Secretion

The isolation and structural characterization of RVA-Thr^6-BK from the lyophilised *O. hejiangensis* skin secretion was carried out as previous study [32]. Briefly, a Cecil CE4200 Adept (Cambridge, UK) gradient reverse-phase High Performance Liquid Chromatography (RP-HPLC) system fitted with a column (Phenomenex C-5, 0.46 cm × 25 cm) was employed to isolate the skin peptides from 5 mg of lyophilise skin secretion. The fractions were collected automatically at 1 min intervals. Thereafter, each fraction was analyzed using MALDI-TOF MS (Perseptive Biosystems, Foster City, MA, USA) and an LCQ-Fleet electrospray ion-trap mass spectrometer (Thermo Fisher Scientific, San Jose, CA, USA) for MS/MS fragmentation sequencing. The primary structure of the novel peptide, RVA-Thr^6-BK, was thus established.

4.4. RVA-Thr^6-BK and Its Site-Substituted Analogue, RVA-Leu^1, Thr^6-BK: Synthesis and Purification

Here, the peptide, RVA-Leu^1, Thr^6-BK (RVALPPGFTPFR), was synthesized for investigating its myotropic activity directly in comparison with that of RVA-Thr^6-BK (RVARPPGFTPFR). Sufficient quantities to detect the bioactivities of each peptide were obtained using the Tribute peptide synthesizer

(Protein Technologies, Tucson, AZ, USA) along with standard Fmoc-chemistry according to our previous study [32].

4.5. Myotropic Activity Evaluation on Smooth Muscles

Following asphyxiation (by CO_2) and cervical dislocation, healthy female Wistar rats (250–350 g) were dissected from the mid-ventral line and these procedures were performed as required by UK Animal Research guidelines. After cutting the intact isolated organs into the desired sizes (10mm-length and 5mm-width parallel muscle bundles for bladder, 5 mm-width rings for uterine horn, 5 mm-length rings for ileum and 2 mm-width rings for tail artery), each strip hooked on tail-to-head was linked with fine silk ligatures (0.2 mm), respectively, and mounted into 2 mL organ baths with transducers. Meanwhile, to mimic the internal physiological environment, organ baths were filled with Kreb's solution at 37 °C flowing through the bath at 2 mL/min with constant bubbling (95% O_2, 5% CO_2). After 20 min of a resting period without drugs, tissues were incubated with drugs from 10^{-11} to 10^{-5} M and the tension changes were recorded via pressure transducers with a PowerLab System (AD Instruments Pty Ltd., Bella Vista, NSW, Australia). The retention periods for each peptide concentration were at least 5 min.

4.6. Haemolysis Assay

Haemoglobin release in the horse erythrocytes (2% solution) (TCS Biosciences Ltd., Buckingham, UK) were determined to evaluate the potential toxicity of RVA-Thr^6-BK and its site-substituted analogue, which was described previously [32]. The horse erythrocytes were treated with each peptide at the intended concentrations (i.e., from 1 to 256 μM) for 2 h at 37 °C. 1% Triton X-100 and phosphate-buffered saline (PBS) were served as the negative and positive controls, respectively. The supernatant of each tube was obtained by centrifugation at 1000× *g* for 5 min and the OD value of supernatant was analyzed at a wavelength of 550 nm using the plate reader.

4.7. Statistical Analysis

Data were analyzed to obtain the mean and standard error of responses by Student's test and dose–response curves were constructed using a best-fit algorithm through the data analysis package Prism 6 (GraphPad Software, La Jolla, CA, USA). EC50 values were calculated from the normalized curves.

Author Contributions: Conceived of and designed the experiments: L.W., M.Z. and T.C.; Performed the experiments: Y.W., Y.Y. and D.S.; Analyzed the data: D.S., X.C., X.X. and C.M.; Wrote the paper: Y.W. and X.C.; Edited the paper: C.S. and L.W.

Funding: This research received no external funding.

Conflicts of Interest: The authors declare no conflict of interest.

References

1. Mandle, R.J.; Colman, R.W.; Kaplan, A.P. Identification of prekallikrein and high-molecular-weight kininogen as a complex in human plasma. *Proc. Natl. Acad. Sci. USA* **1976**, *73*, 4179–4183. [CrossRef] [PubMed]
2. Bhoola, K.D.; Figueroa, C.D.; Worthy, K. Bioregulation of kinins: Kallikreins, kininogens, and kininases. *Pharmacol. Rev.* **1992**, *44*, 1–80. [PubMed]
3. Kaplan, A.P.; Joseph, K.; Silverberg, M. Pathways for bradykinin formation and inflammatory disease. *J. Allergy Clin. Immunol.* **2002**, *109*, 195–209. [CrossRef] [PubMed]
4. Regoli, D.; Barabe, J. Pharmacology of bradykinin and related kinins. *Pharmacol. Rev.* **1980**, *32*, 1–46. [PubMed]
5. Rabito, S.F.; Binia, A.; Segovia, R. Plasma kininogen content of toads, fowl and reptiles. *Comp. Biochem. Physiol. A.* **1972**, *41*, 281–284. [CrossRef]

6. Seki, T.; Miwa, I.; Nakajima, T.; Erdos, E. Plasma kallikrein-kinin system in nonmammalian blood: Evolutionary aspects. *Am. J. Physiol.* **1973**, *224*, 1425–1430. [CrossRef] [PubMed]

7. Xi, X.; Li, B.; Chen, T.; Kwok, H. A review on bradykinin-related peptides isolated from amphibian skin secretion. *Toxins* **2015**, *7*, 951–970. [CrossRef] [PubMed]

8. Anastasi, A.; Erspamer, V.; Bertaccini, G. Occurence of bradykinin in the skin of *Rana temporaria*. *Comp. Biochem. Physiol.* **1965**, *14*, 43–52. [CrossRef]

9. Conlon, J.M.; Aronsson, U. Multiple bradykinin-related peptides from the skin of the frog, rana temporaria. *Peptides* **1997**, *18*, 361–365. [CrossRef]

10. Conlon, J.M.; Jouenne, T.; Cosette, P.; Cosquer, D.; Vaudry, H.; Taylor, C.K.; Abel, P.W. Bradykinin-related peptides and tryptophyllins in the skin secretions of the most primitive extant frog. *Ascaphus truei. Gen. Comp. Endocrinol.* **2005**, *143*, 193–199. [CrossRef] [PubMed]

11. Conlon, J.M. Bradykinin-related peptides from frog skin. In *Handbook of Biologically Active Peptides*, 1st ed.; Kastin, J.A., Ed.; Academic Press: Amsterdam, The Netherlands, 2006; pp. 291–294.

12. Samgina, T.Y.; Gorshkov, V.; Vorontsov, Y.A.; Artemenko, K.; Zubarev, R.; Lebedev, A. Mass spectrometric study of bradykinin-related peptides (BRPs) from the skin secretion of russian ranid frogs. *Rapid Commun. Mass Spectrom.* **2011**, *25*, 933–940. [CrossRef] [PubMed]

13. Chen, T.; Shaw, C. Cloning of the (Thr^6)-phyllokinin precursor from *Phyllomedusa sauvagei* skin confirms a non-consensus tyrosine O-sulfation motif. *Peptides* **2003**, *24*, 1123–1130. [CrossRef] [PubMed]

14. Zhou, J.; Bjourson, A.J.; Coulter, D.J.; Chen, T.; Shaw, C.; O'rourke, M.; Hirst, D.G.; Zhang, Y.; Rao, P.; McClean, S. Bradykinin-related peptides, including a novel structural variant, (Val^1)-bradykinin, from the skin secretion of guenther's frog, *Hylarana guentheri* and their molecular precursors. *Peptides* **2007**, *28*, 781–789. [CrossRef] [PubMed]

15. Rates, B.; Silva, L.P.; Ireno, I.C.; Leite, F.S.F.; Borges, M.H.; Bloch, C., Jr.; De Lima, M.E.; Pimenta, A.M. Peptidomic dissection of the skin secretion of *Phasmahyla jandaia* (bokermann and sazima, 1978)(Anura, Hylidae, Phyllomedusinae). *Toxicon* **2011**, *57*, 35–52. [CrossRef] [PubMed]

16. Thompson, A.H.; Bjourson, A.J.; Shaw, C.; McClean, S. Bradykinin-related peptides from *Phyllomedusa hypochondrialis azurea*: Mass spectrometric structural characterisation and cloning of precursor cDNAs. *Rapid Commun. Mass Spectrom.* **2006**, *20*, 3780–3788. [CrossRef] [PubMed]

17. Samgina, T.Y.; Artemenko, K.A.; Gorshkov, V.A.; Ogourtsov, S.V.; Zubarev, R.A.; Lebedev, A.T. De novo sequencing of peptides secreted by the skin glands of the caucasian green frog *Rana ridibunda. Rapid Commun. Mass Spectrom.* **2008**, *22*, 3517–3525. [CrossRef] [PubMed]

18. Zhou, X.; Wang, L.; Zhou, M.; Chen, T.; Ding, A.; Rao, P.; Walker, B.; Shaw, C. Amolopkinins W1 and W2—novel bradykinin-related peptides (BRPs) from the skin of the chinese torrent frog, *Amolops wuyiensis*: Antagonists of bradykinin-induced smooth muscle contraction of the rat ileum. *Peptides* **2009**, *30*, 893–900. [CrossRef]

19. Bevins, C.L.; Zasloff, M. Peptides from frog skin. *Annu. Rev. Biochem.* **1990**, *59*, 395–414. [CrossRef]

20. König, E.; Bininda-Emonds, O.R.P.; Shaw, C. The diversity and evolution of anuran skin peptides. *Peptides* **2015**, *63*, 96–117. [CrossRef]

21. Jensen, J.; Conlon, J.M. Effects of trout bradykinin on the motility of the trout stomach and intestine: Evidence for a receptor distinct from mammalian B1 and B2 subtypes. *Br. J. Pharmacol.* **1997**, *121*, 526–530. [CrossRef]

22. Li, Z.; Secor, S.M.; Lance, V.A.; Masini, M.A.; Vallarino, M.; Conlon, J.M. Characterization of bradykinin-related peptides generated in the plasma of six sarcopterygian species (african lungfish, amphiuma, coachwhip, bullsnake, gila monster, and gray's monitor). *Gen. Comp. Endocrinol.* **1998**, *112*, 108–114. [CrossRef] [PubMed]

23. Farmer, S.G.; Burch, R.M. Biochemical and molecular pharmacology of kinin receptors. *Annu. Rev. Pharmacol. Toxicol.* **1992**, *32*, 511–536. [CrossRef] [PubMed]

24. Marceau, F.; Hess, J.F.; Bachvarov, D.R. The B1 receptors for kinins. *Pharmacol. Rev.* **1998**, *50*, 357–386. [PubMed]

25. Marceau, F.; Regoli, D. Bradykinin receptor ligands: Therapeutic perspectives. *Nat. Rev. Drug Discov.* **2004**, *3*, 845–852. [CrossRef] [PubMed]

26. Lintner, K.; Fermandjian, S.; Regoli, D. Conformational features of bradykinin: A circular dichroism study of some peptide fragments and structural analogues of bradykinin. *Biochimie* **1979**, *61*, 87–92. [CrossRef]

27. Bonechi, C.; Ristori, S.; Martini, G.; Martini, S.; Rossi, C. Study of bradykinin conformation in the presence of model membrane by nuclear magnetic resonance and molecular modelling. *Biochim. Biophys. Acta Biomembr.* **2009**, *1788*, 708–716. [CrossRef]
28. Parasuraman, S. Toxicological screening. *J. Pharmacol. Pharmacother.* **2011**, *2*, 74. [CrossRef]
29. Megaw, J.; Thompson, T.P.; Lafferty, R.A.; Gilmore, B.F. *Galleria mellonella* as a novel in vivo model for assessment of the toxicity of 1-alkyl-3-methylimidazolium chloride ionic liquids. *Chemosphere* **2015**, *139*, 197–201. [CrossRef]
30. Tyler, M.J.; Stone, D.; Bowie, J.H. A novel method for the release and collection of dermal, glandular secretions from the skin of frogs. *J. Pharmacol. Toxicol. Methods* **1992**, *28*, 199–200. [CrossRef]
31. Xiang, J.; Wang, H.; Ma, C.; Zhou, M.; Wu, Y.; Wang, L.; Guo, S.; Chen, T.; Shaw, C. *Ex vivo* smooth muscle pharmacological effects of a novel bradykinin-related peptide, and its analogue, from Chinese large odorous frog, *Odorrana livida* skin secretions. *Toxins* **2016**, *8*, 283. [CrossRef]
32. Zhang, L.; Chen, X.; Wu, Y.; Zhou, M.; Ma, C.; Xi, X.; Chen, T.; Walker, B.; Shaw, C.; Wang, L. A Bowman-Birk type chymotrypsin inhibitor peptide from the amphibian, *Hylarana erythraea*. *Sci. Rep.* **2018**, *8*, 5851. [CrossRef] [PubMed]

MDPI
St. Alban-Anlage 66
4052 Basel
Switzerland
Tel. +41 61 683 77 34
Fax +41 61 302 89 18
www.mdpi.com

Toxins Editorial Office
E-mail: toxins@mdpi.com
www.mdpi.com/journal/toxins